BODY MODERN

BODY MODERN

FRITZ KAHN, SCIENTIFIC ILLUSTRATION, AND THE HOMUNCULAR SUBJECT

MICHAEL SAPPOL

MINNESOTA

UNIVERSITY OF MINNESOTA PRESS
MINNEAPOLIS LONDON

Published by the University of Minnesota Press
111 Third Avenue South, Suite 290
Minneapolis, MN 55401-2520
http://www.upress.umn.edu

Printed in the United States of America on acid-free paper

23 22 21 20 19 18 17 10 9 8 7 6 5 4 3 2 1

Library of Congress Cataloging-in-Publication Data
Names: Sappol, Michael, author.
Title: Body modern : Fritz Kahn, scientific illustration, and the homuncular subject / Michael Sappol.
Description: Minneapolis : University of Minnesota Press, 2017. | Includes bibliographical references and index.
Identifiers: LCCN 2016036909 (print) | ISBN 978-1-5179-0020-5 (hc) | ISBN 978-1-5179-0021-2 (pb)
Subjects: LCSH: Kahn, Fritz, 1888–1968. | Scientific illustration—History—20th century. | Visual communication in science—History—20th century. | Science in popular culture—History—20th century. | Modernism (Aesthetics). | Human beings.
Classification: LCC Q143.K24 S27 2017 (print) | DDC 741.6092—dc23
LC record available at https://lccn.loc.gov/2016036909

FOR EVA

who sustained me with love and tenderness
during the making of this book

Ever since people have had the idea that modern times
needed "modern styles," all the old rules of thumb
have stopped helping me.

—JOSEPH ROTH, "ARCHITECTURE,"
MÜNCHNER ILLUSTRIERTE PRESSE, 1929

The picture is a model of reality.

—LUDWIG WITTGENSTEIN,
TRACTATUS LOGICO-PHILOSOPHICUS, 1921

The fact that the world becomes picture . . .
is what distinguishes the essence of the modern age.

—MARTIN HEIDEGGER,
"THE AGE OF THE WORLD PICTURE," 1938

CONTENTS

Preface xi

Introduction
Fritz Kahn, Modernity, and the Invention of Conceptual Scientific Illustration 1

1. Reading Kahn and the Homunculus 11

2. "Much Better Than Words": *Pictured Knowledge* and the Rhetoric of Visuality 29

3. Ocularcentric! Conceptual Illustration at Work in the "Great Loop" 49

4. Variety Show: The Studio of Kahn and Its Visual Devices 71

5. Kahn's Take Away: Conceptual Scientific Illustration's Iconophilic Diaspora 133

6. "To Picture the Body": Kahn's Images in the Postmodern Afterlife 163

Epilogue
Toward a Theory of the Homunculus 185

Acknowledgments 193

Fritz Kahn: A Chronology 197

Notes 199

Index 239

PREFACE

This book is about the origins, development, uses, and effects of a new kind of picture: the conceptual scientific illustration. Unlike the descriptive anatomical and natural historical illustrations of the eighteenth and nineteenth centuries—images that show how things *look* (albeit always in some idealized or stylized way)—the conceptual scientific illustration visually explains how things *work* using concepts, metaphor, and allusion.

It emerged first in the United States in what now seems an enchanted moment: the early twentieth century, a time of heedless technological enthusiasm. Assembled out of the proliferating visual practices and tropes of newspaper and magazine graphics, the conceptual scientific illustration was instigated by the Progressive-era visual instruction reform movement and fueled by the growing belief that visual narrative—in books, magazines, pamphlets, posters, lantern shows, and motion pictures—was the most effective and modern way to reach workers, children, and the general public. Such was argued in the 1917 edition of a seminal publication, the encyclopedia/manifesto *Pictured Knowledge* and its article "The Body We Live In," which features drawings of little figures and machines doing work inside the cross-sectioned human body, rendered in bravura Sunday-magazine illustration style.

In 1919, those images crossed the Atlantic and appeared in a German-language Swiss publication called *Wunder in Uns.* One of the contributors, a German Jewish physician-journalist named Fritz Kahn, was quick to seize on the possibilities of conceptual illustration. In collaboration with a stable of commercial artists, he went way beyond "The Body We Live In" and built a career out of the new genre. Kahn presented thousands of pictures. In popular books and articles, his artists infused conceptual scientific illustration with a rich assortment of modern and modernist styles: figurative realism, Dada, Art Deco, photomontagery, Bauhaus functionalism. By the late 1930s, conceptual scientific illustrations, in works authored or influenced by Kahn, were appearing in publications and exhibitions in Europe, the Soviet Union, the Middle East, North and South America, even China. In his own land, however, things were complicated: initially the Nazis denounced Kahn and banned his books, but after a few years, they appropriated his illustrations for their own health publications, adding material on racial hygiene and suppressing any evidence of Jewish authorship. By the early 1950s, conceptual scientific illustration was incorporating nearly every style of aesthetic modernism, even abstract expressionism, and was cropping up in all sorts of cultural spaces—pharmaceutical advertisements, animated cartoons, military instructional manuals, exhibition halls, sexology guides, and children's books.

Body Modern is neither a biography of Fritz Kahn nor a comprehensive study of his work. Rather, what readers will find here is a mix of historical narration, analysis of images, and essayistic meditation on the dialogical relation between embodied life, subjectivity, and pictures. In other words, this book is about how images cast their spell, how technologies shape the mode and content of images, and how people use images and images use people.

In the twentieth century, there was a politics of text versus image: debates over the social meaning, moral effects, desirability, instrumental uses, and control of images. That politics, I argue, was all about how to get modern. Fritz Kahn was an impresario of the modern. He exploited his readers' ardent desire to become modern, their worries over how to be modern, and their pleasure in seeing themselves as moderns. He did so with great effect, in part because his career was itself a performance of the modern—a wallow in the sheer pleasure of his own proliferating images of modernity.

The picture practices of *Pictured Knowledge* and Kahn, then, present us with an opportunity to rethink modernity and its images. Scholarly arguments about when modernity begins—its essential technologies and characteristics, what's right and wrong with it—are now threadbare. Kahn and his readers defined the modern in opposition to the primeval and recent historical past; they figured it as a condition that dissolves the historicity of present-day experience; and yet they had a deeply historicist understanding of themselves as moderns, inhabitants of the modern age. We, of course, know that modernity was never outside history. And so we need to treat modernity as a historical artifact, to consider *their* modernity, a powerfully seductive (and contested and unstable) set of beliefs, performances, and practices in a particular place and time. The modern was a protean, shape-shifting identity formation made out of ever-changing experiences with its own prolific imaginary.

Kahn catered to a public that eagerly devoured modernist and modernizing productions. His trademark was the medical and scientific text, richly illustrated with pictures that used every trick in the book to visually explain the modern body and physical world. His readers avidly sought materials that could contribute to their own identification with modernity, looked for props that might help them to perform their modernness to onlookers, in every location of social and cultural life, and to themselves before the mirror. These performances of the modern had viral effects on both performer and audience.

Being modern meant living in a world of proliferating images, image technologies, and styles. There was a pictorial turn. Modernizers in motion pictures, advertising, popular publication, and pedagogy performed themselves as modern by taking up the cause of the picture, introducing pictures and more pictures into their field, making pictures that were modern and more modern still. There was a kind of pictorial one-upmanship: as image followed on image, the picture had to be ever more dynamic, visually arresting, and more semantically, maybe more aesthetically, modern. But the modernizers never had an entirely free hand, even in the realm of pictorial imagination. Their image productions were material, required resources, pushed against all sorts of constraints, had costs and benefits, and had unanticipated consequences.

In this milieu, images took on a privileged role in the conception and production of subjectivity. The image became a technology of the self, a device for modeling the subject in formation. Kahn was especially famous for featuring pictures of the body as a factory, with workers tending switchboards and controls, working assembly lines, and doing manual labor. His 1926 four-color poster "Der Mensch als Industriepalast" (Man as industrial palace) became an emblematic image of humanity in the industrial age—a microcosmos. The body was industrial and technological, operated by little homuncular figures. In turn, images like "Der Mensch als Industriepalast" functioned as a higher-level homunculus, served Kahn's readers as a representation of the self. That self was, of course, never solely derived from Kahn's homuncular figures or anyone else's. The reading public moved in a swarm of images. Some homuncular figures resonated more strongly than others, but the haphazard consequences of contact with images en masse were felt cumulatively and collectively.

MY ROAD TO KAHN

I first stumbled on a folded and yellowed copy of "Der Mensch als Industriepalast" in a volume of *Das Leben des Menschen* around the year 2001 while doing research for *Dream Anatomy,* a National Library of Medicine exhibition. *Dream Anatomy* focused on imaginative, evocative representations. It displayed anatomical dissection, the dissected body, and the body's interior. In the exhibition text, I argued that "anatomy is us": it supplies us with an authoritative scientific representation of the body as well as a mirror image that signifies monstrosity, pathology, death, and desire. Anatomy is our inner truth; it maps our inner geography. The anatomist intensely focuses on the particular characteristics of the specimen using the timeless principles, methods, and categories of anatomical science. At the same time, anatomical illustrations have been produced for hundreds of years and are intensely and utterly historical, always visibly marked by their time and place. In focusing on the innards of the body, anatomists always carry with them the particulars of the world they live in: etiquette, theatrical presentations and forms, hierarchical social relations, conventions of gender performance, death practices, architectural styles, fashions in dress, ideals of beauty and ugliness, and popular entertainments such as freak shows.

This matters because over the course of several centuries, an anatomical conception of self came to be almost universal. Before the nineteenth century, only a few people thought of themselves as anatomical beings. Now we all do. We carry with us a conceptual effigy, a homunculus made up of skeleton, organs, nerves, and blood vessels, each bone and organ precisely named and delineated. Even if we may not know exactly where the spleen sits in relation to the liver, that anatomical *vade mecum* refers back to centuries of dissection, illustrated books, and now X-rays, MRIs, and sonograms, which describe us as we truly are (or think we are).

When I curated *Dream Anatomy,* I didn't know much about Kahn or his work. In one way, his illustrations seemed to fit perfectly the theme of the exhibition. Yet in other ways, they were a complete departure. His images of the interior of the body were too entertaining, his pleasure in showing the functions of the body too emancipated from any requirement to show its structures. I loved his pictures and picked "Der Mensch als Industriepalast" and three or four others for inclusion in the show, but I did so with a guilty conscience. I sensed that Kahn was over the edge—outside the category that the exhibition defined.

Having now spent a good part of a decade studying Kahn and his image practice, that's where I locate him now. He was steeped in the discourse of medical and scientific self-making and improvement, which in the nineteenth century was known in English-speaking lands as "popular anatomy," a genre of medical and scientific showmanship. However, by the 1920s, that discourse was shopworn; anatomy no longer had much mystique. In the public eye, it had been devalued by too much entrepreneurship and too much sensationalism—and in any case it was no longer cutting-edge science. The anatomical illustration was increasingly linked to disreputable popular anatomy museum and fairground exhibitions and obsolete home health guides and schoolbooks.

Kahn found a way to take advantage of that discontent; he invented a new way of doing things. Like his anatomical forebears, he was in the business of making effigies of self for readers to identify with—effigies that simultaneously bore the imprimatur of scientific authority and entertained with views of bodies and body parts that were beautiful, orderly, monstrous, funny, and/or morbid. But Kahn did not limit himself to pictures of cadavers, specimens, and models. He presented bodies made of physics and chemistry, machine engines and electricity, cities and factories, automobiles and telephones, X-rays and Victrolas; he showed bodies that contained all of the ingenious methods, processes, and forces of modern science and technology. That branded him as a modern and made him a useful intermediary between medical experts and the lay public, a credible guide for readers who were already working hard to modernize.

The cultural logic of all that did not only belong to Kahn and his readers. It also belonged to my parents and grandparents—and, by inheritance, to me. Kahn was a diasporic Jew. He grew up in Germany and the United States; he lived more than half of his life outside Germany. Like every member of the Jewish diaspora, he faced in two directions. In his private life and in Jewish publications, he was active in the Jewish community, a Zionist, and a frequent writer on Jewish topics. He thought of himself as the issue of an ancient race and a faithful upholder of its ancient traditions. Yet in his medical and scientific writings, he was a tireless exponent of everything modern. He wrote as a universal man on the latest science for a universal public in a modern language (German), using the most modern means of educating that public: the profusely illustrated article or book, full of images that carried the look and zeitgeist of modernity.

My history connects to his. My grandparents came to Brooklyn from Eastern Europe only a few years before the Great War (1914–18). They were Jews, with a strong sense of their Jewishness. They mostly kept the faith, but they wanted to be modern. Their children (my parents, aunts, and uncles) retained and enjoyed their parents' Yiddish, but they mostly spoke New York–accented English. They selectively upheld tradition, more ambivalently and less fervently than their parents, and they desperately, passionately embraced modernity. For my mother and father, that took the form of a love of modern art and a belief in communism. My parents rejected the religious/ethnic past and looked forward to the secular scientific universalist future, which they incorporated into their Jewish identity. They took pride in the idea that emancipated Jewry vitally contributed to the progressive development of civilization. Their privileging of the modern was transmitted to me as I wandered through the byways and battlefields of modernity into the postmodern present. So I read Kahn the way I do because I come to it already sensitized to the performance of modernity: I was born into it.

The intertwined prime directives in Kahn are: Get modern. Be modern. Explain the world and self in pictures. This book examines the history of visuality in modernity and Fritz Kahn's role in that history. It also echoes Kahn's method in its use of images—mainly Kahn's images—to illustrate, comment on, and enliven the text. It is both a study of iconophilia and an iconophilic performance. In the end, I hope that it contributes to a fuller picture of how our culture acquired its compulsive visuality. Certainly it serves as a vehicle for me to indulge in my own compulsion to show pictures. We feel uneasy about our desire to binge on images, yet the imperative to depict stands as common sense. When students commence to write or work in other media, we tell them: "Don't explain. Show." From there, the demand "show me" takes on a second exhibitionistic sense: "Show *me.*"

Now to begin.

INTRODUCTION

FRITZ KAHN, MODERNITY, AND THE INVENTION OF CONCEPTUAL SCIENTIFIC ILLUSTRATION

Not as the cadaver of the older anatomy,
not the skeleton and the specimen preserved in alcohol,
but instead man as he lives, acts, and thinks . . .
this is the subject of modern biology!

—FRITZ KAHN, *MAN IN STRUCTURE AND FUNCTION*, 1939

"DER MENSCH ALS INDUSTRIEPALAST" AS CONCEPTUAL ILLUSTRATION

There is a kind of illustration that is utterly familiar, but rarely brought to account. Nowadays it goes by the name "infographics," "data visualization," "visual display," or "visual explanation"—terms that have overlapping and inconsistent definitions and usages.[1] Among an older generation of professional illustrators, such illustrations were sometimes called "conceptual," which is the term we'll provisionally adopt here, along with "visual explanation." But none of these terms does justice to the rich history and variety of explanatory illustrations devoted to scientific, technical, and medical (and also historical, geographical, and cultural) topics.

Conceptual illustrations combine images, text, data, diagram. Their mission is to describe the scientific or technical workings of the material world and our material bodies. Designed mainly but not exclusively

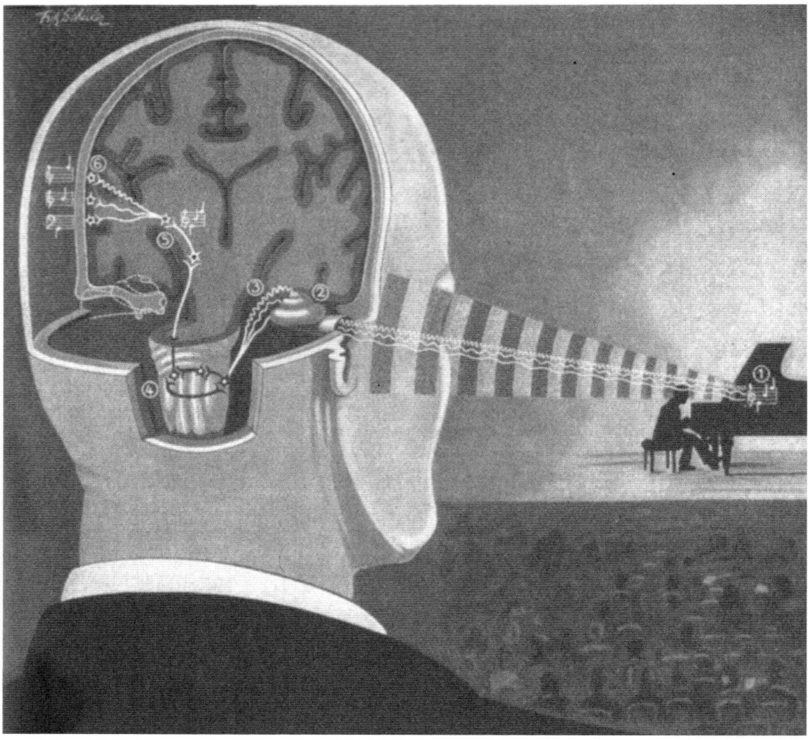

FIGURE I.1. *A conceptual illustration that visually explains the physics and physiology of what goes on in your ears and brain when you hear a piano concert. "Sound Perception," Fritz Kahn,* Das Leben des Menschen *(Stuttgart: Kosmos/Franckh'sche Verlagshandlung, 1929), 4:197. Artist: Fritz Schüler. National Library of Medicine. Copyright Kosmos Verlag, Stuttgart.*

for general readers, they appear in newspapers, magazines, textbooks, instructional videos, advertisements, children's books, and television programs, as well as on classroom walls and websites. Mostly they don't aim to show how things *look* but rather explain how things *work*. They model, celebrate, or narrate processes; they dramatize data and statistical correlations, disease progression, natural cycles, and events; or they just allude to or evoke the operations of living things, chemical reactions, and technical devices.

Such illustrations use various strategies and styles. They can be sober or whimsical; they can be executed with varying levels of scientific and aesthetic sophistication (Plate 5). Yet casual readers rarely think of them as art and almost never as rhetoric. Historians, scholars of visual culture, and sociologists of science have not attended very much to their ideological provenance or role in the circulation of scientific knowledge or aesthetic principles. In everyday practice, the artfulness of such illustrations, as well as their status as cultural artifact, is superseded by their claim to show us what is true and real. Their history and workings, seemingly obvious, tend to disappear from view. Their stratagems camouflage and sometimes subvert their rhetorical goals.[2]

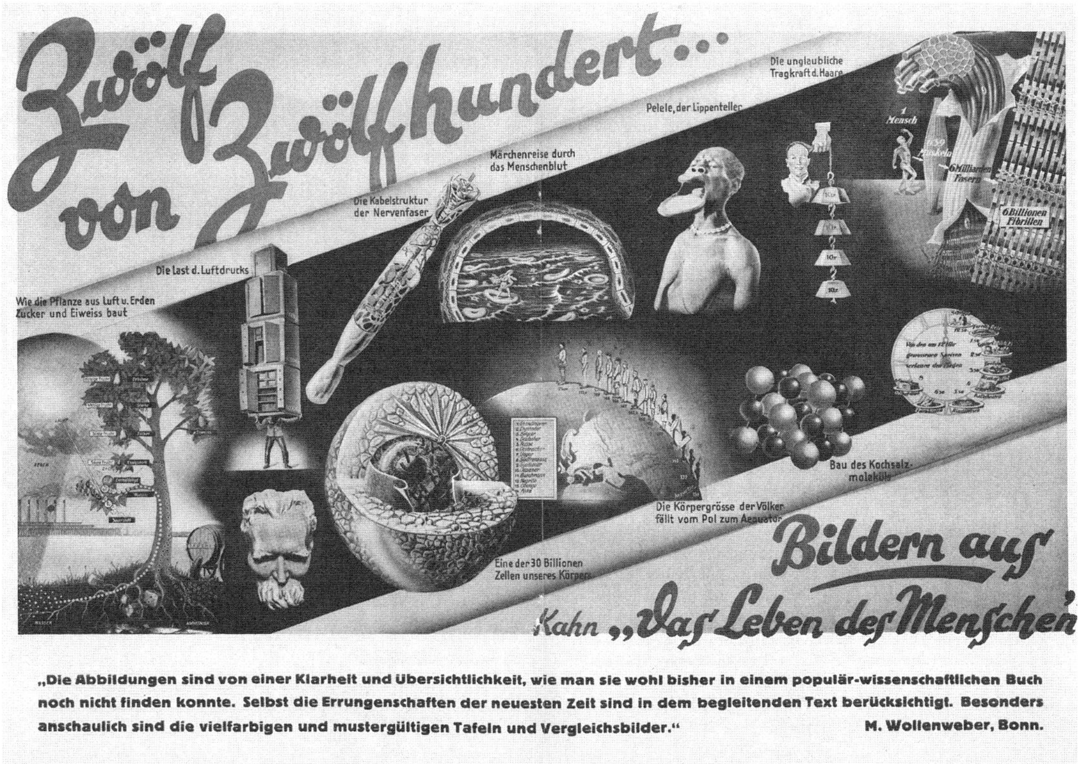

FIGURE I.2 *"Twelve of twelve hundred . . . illustrations from* 'Das Leben des Menschen'" *(ca. 1931). Two-color promotional insert. Artist: Roman Rechn. Leo Baeck Institute, New York. Copyright Kosmos Verlag, Stuttgart.*

It seems as though conceptual illustration has been around forever, but it hasn't. It had a long gestation in sixteenth-century treatises and compendia, eighteenth-century encyclopedias, and nineteenth-century illustrated weeklies. But it was truly born in the early decades of the twentieth century, and its first great exponent was Fritz Kahn (1888–1968), a German Jewish physician. Kahn's career as a popular science author blossomed in the 1920s with illustrated articles in *Kosmos* (Germany's most popular science magazine), *Berliner Illustrierte Zeitung* (Germany's most popular weekly newspaper), *Uhu* (an American-style general interest mass-circulation monthly magazine), and especially the five-volume *Das Leben des Menschen* (The life of man).[3] Published between 1922 and 1931, *Das Leben des Menschen* sold over 70,000 copies and featured some 1,200 illustrations, many of them commissioned by Kahn, others lifted from a jumble of sources.[4] Kahn's later publications were even more image laden. Over the course of a long career, in collaboration with a cadre of never adequately acknowledged commercial artists, Kahn developed entirely new genres and tropes of conceptual scientific illustration. Eclectic pictoriality was his working method.[5] Kahn's artists depicted subjects that were familiar or exotic or both, and they worked in whatever style suited their talents and temperament—surrealism, Art Deco, neoclassicism, Jugendstil, expressionism, photomontagery, Bauhaus functionalism, *Neue Sachlichkeit* (New Objectivity), commercial realism.

Kahn's first publications appeared in the 1910s and continued on past his death in 1968.[6] He dealt with a variety of scientific subjects, but he was best known for illustrated books and articles on the human body, biology, and sexology. After the Nazis expelled him from Germany in 1933, he went to Palestine, then to France, and, after the fall of France, to the United States. His works were translated into many languages, and he attained a modicum of fame. In the 1970s, he lapsed into obscurity, but a small revival is now in progress. In the last decade, Kahn has been discussed in several monographs and treated as a main subject in a handful of articles. His illustrations are scattered all over the Internet. There are a couple of monographic coffee-table books and a dissertation. More studies are on the way.[7]

His most celebrated work is "Der Mensch als Industriepalast" (Man as industrial palace) (Plate 1). A collaboration with artist Fritz Schüler (who was never credited), the poster shows a human figure in profile, schematically rendered as an industrial complex with conveyor belts, pulleys, wiring, switchboards, pipes, mixing machines, storage areas, managers, office workers, and factory operatives.[8] Today it appears on dozens of websites, posters, T-shirts, and coffee mugs, and it graces the covers of several books, including a recent English translation of Deleuze and Guattari's *Anti-Oedipus.*[9] It was also selected for inclusion in *Modernism: Designing a New World, 1914–1939,* a 2006 exhibition at the Victoria and Albert Museum in London.[10]

"Der Mensch als Industriepalast" was easy to overlook amid the pile of modernist objects on view at the V&A. But it could have easily been deployed as a keystone, a crucial piece of evidence for the exhibition's central thesis. Twentieth-century modernism, the show asserted, had grandiose utopian and universalist aims. It was international and internationalist, but carried forth by a jumble of national and local movements, all of them responding to the exigencies of national and local entanglements. What linked them together was a common agenda: modernist designers wanted to apply an Occam's razor of functionalist aesthetics to everything. The modernists aimed to reject the past and to remake everyday life—to modernize every object, every subject. Knives and forks, toilet bowls, large buildings, the alphabet, railroad trains, medical instruments, clothespins, the classroom, the city, the nation—nothing was too large or too small to be functionalized and aestheticized.[11]

Not even the self. And this transformation of humanity brought things full circle. The incessantly dreaming, desiring, conniving human subject, with its body and body parts, full of pulsing blood and guts and convoluted particularities, was, in the final instance, modernism's target. Modernizers aimed to reconstitute everything, but also everybody. In the modern age, the self had to undergo a redesign—or maybe just a design.[12] And like any design, this required something to show, something to sell: a persuasive representation, renderings. The body became both the subject and object of industrial "image growth," the telling phrase that enjoyed currency among early twentieth-century American pedagogical modernizers in a signature publication, *Pictured Knowledge* (1916), the illustrated encyclopedia that inspired "Der Mensch als Industriepalast" and transformed Kahn's career.[13]

MODERNITY, THE MODERN, MODERNISM

Kahn did not theorize modernity or the body. Addressing a mass readership, his publications dwell on the science and technical properties of embodied life, using pictures laced with headlines, labels, and captions—image/texts (W. J. T. Mitchell's useful formulation) that situate modernity in the body and the body in modernity.[14] Scenes of the modern, references to contemporary fashion, science, devices, and experiences recur insistently, repetitively, throughout. The modern was his "bolero." His foreword to the 1939 *Der Mensch Gesund und Krank* (Man in health and sickness) revels in the modernness of "modern science":

> The human biology of 1940 is not the anatomy of 1880 . . . is not the biology of 1900 . . . ; the human biology of 1940 is the study of . . . an entirely new, completely unimagined world which has been disclosed to us during the past few decades by the discoveries of modern science, and with the recognition of which there begins a new era in the study of man and the shaping and organization of human life.[15]

Kahn's modernity infused both the outer world and body interior. It was millennial, dynamic, transformative, breathless:

> Still beating hearts of dead creatures, bacilli in the bodies of bacilli, suckling males—this is the biology of 1940. Man as the centre of the world—not as the cadaver of the older anatomy, not the skeleton and the specimen preserved in alcohol, but instead man as he lives, acts and thinks, man at the work-bench and at the family table, man at the wheel of his car and as an athlete in training: you, just as you live and are, as you work and suffer, your perfections and inadequacies—this is the subject of modern biology![16]

The modern rejects what went before; it needs something old and dead ("the cadaver of the older anatomy") to define itself against. In modernizing discourse, this is first of all tradition (Kahn wrote his medical dissertation on folk beliefs about pregnancy). But the modern also opposes beliefs and practices of more recent vintage, the obsolete or failed modernization projects of prior generations and even the present.[17] "The spirit of modernity," early twentieth-century cultural critic Horace B. Samuel argued, "enjoys . . . only a temporary reign, and is speedily killed by its inevitable successor," a new spirit of modernity.[18] (Paradoxically, this unremittingly oedipal opposition to the imperfect present could lead to a revaluation of the distant past: modernist primitivism.) In other words, no matter what it chooses to oppose, modernity is a purification ideal; it purges past from present and present from future.

According to sociologist-philosopher Bruno Latour, a great deal of "sorting out, cleaning up and dividing up is required to obtain the impression of a modernization":

> The adjective "modern" designates a new regime, an acceleration, a rupture, a revolution in time. When the word "modern," "modernization," or "modernity" appears, we are defining, by contrast, an archaic and stable past. . . . The word is always being thrown into the middle of a fight, in a quarrel where there are winners and losers, Ancients and Moderns.[19]

But more often than not, the quarrel takes place among modernizers. In the early twentieth century, there were plenty of ways to be modern; there was no shortage of modern aesthetic movements, political programs, and business practices, no shortage of modern fashions, songs, appliances, and argot. There was disagreement, one-upmanship, disdain, and contradiction. But all of those things in some way presented a critique of past and present, in word or deed or image or object, and all of them contributed to an intoxicated vision of the transformed present and utopian future—the millennial dream time that would come after the destruction of the ancien régime and the end of history.

This "deflationary" poststructuralist interpretation of modernity has become necessary because modernity is a banana peel upon which even gifted scholars repeatedly slip. If "modernization and modernity" are "explanation and *explanandum,*" as media historian Lisa Gitelman cogently argues, then they "chas[e] each other" round and round, going nowhere, and end up explaining nothing.[20] Modernity was not a deus ex machina that descended on the nineteenth and twentieth centuries. Instead it was built by people, with resources and skills, in objects and performances, all of which asserted, and provided source material for, a strangely variable metaidentity. Like all identities, the modern was made and remade, dialogically, through the ongoing appraisals, punishments, and rewards of family members, peers, customers, voters, teachers, employers, and so on.[21]

The modern was a particularly unstable and proliferative identity because by definition it had to be the latest thing, a critique of what went before.[22] It was always braided with efforts to constitute generational identities and stage generational conflicts. But the modern was not just a badge or a costume. In performance, it had cognitive effects that ran deep. If the modern structured thought and action, it also worked to structure thinkers and actors. It defined what it means to be a person in the world and defined particular kinds of person and personhood, as well as the particular groups and institutions that such persons developed within and interacted with.

This book focuses on the rise of conceptual scientific illustration as a visual rhetoric of industrial modernity. While the aesthetic movements of previous decades and centuries favored bio- and anthropomorphic elements, Kahn and his collaborators favored the machine, the factory, the industrial product, the schematic, the abstraction, the statistic, and, in the hardest case, undecorated, precision-ruled geometry. The human body—with its messy organic structures and details, an anatomical rococo of textures and forms—was representationally remade into an assembly line, internal combustion engine, electromagnetic device, chemical plant, diagram, graph, or statistical table, and thus transformed into the locus of dynamic physical and industrial forces. Fittingly, these new images of the industrial body were made with industrial technologies (halftone photomechanical reproduction, photolithography, photomicrography, motion pictures, electric light, glass, plastic)—and with a modernizing aesthetic that in some way rejected inherited aesthetic conventions, such as anatomical naturalism. That modernizing aesthetic was, often enough, some version of modernism. But there was another possibility. Emancipated from the obligation to adhere to aesthetic conventions, images could have a pragmatic antiaesthetic that tacitly facilitated mass production for mass effect and profit. Either way, messy new political and cultural opportunities arose. There was an exuberant riptide that pleasurably or nightmarishly exceeded every utilitarian and functionalist calculus.

The modern, then, was not a coherent set of doctrines: modernizers easily appropriated anything useable that came their way. They ransacked mathematics, physics, chemistry, engineering, and geology, but also biology. There was a "biomorphic" modernism that prized the simplicity and functionality of evolved natural design and the dynamic vitalism of organisms, and that drew on organic forms for design elements and motifs (such as streamlining).[23] Kahn and his artists, while celebrated for their industrial body images, also cooked up surreal biolandscapes and bioarchitectures. Readers were untroubled by publications that featured technological materialism or organic vitalism, or by images that combined the two.

In the last half-century or so, modernism and modernity have become favorite topics for scholars, and the literature is vast. We have studies on modernity and death, interiority, boredom, space and time, fatigue, manners, toilet training, sexuality, education, spiritualism, media, nostalgia, alcoholism, urbanism, nationalism, literary realism, incarceration, optics, militarization, and, of course, art and technology. Scholars ask: what makes modernity different from what went before? When did it begin or end? What technological, economic, institutional, intellectual, cultural, political innovations and events inaugurated it, defined it, fostered it, thwarted it? Have we always or never been modern? Here I will try to sidestep such questions and instead treat modernity as a wobbly historical artifact—an artifact of artifacts (think of the plural title of Horace B. Samuel's 1914 essay collection *Modernities*).[24]

Though they were hardly the first generation to identify themselves as moderns,[25] in the 1920s and 1930s, people around the globe experienced the "new world of machines, radios and fast motor cars" as a condition of life radically unlike any that preceded them—a "machine age" of permanent technological revolution and breakthrough transformations wrought by science and reason, an era of unstoppable discovery and invention progressing into a future of wonders.[26] Uno Åhrén, a visionary young Swedish architect, declared that the "industrial spirit, the natural sciences, the new psychology, the new philosophy, all of these are at the same time engendering, and themselves the outcome of, a new world image, an alteration in our attitude to the world and to life."[27] This historicist belief pervaded both popular and elite discourse (and often enough bridged the gap between them).

People not only believed in modernity, they also lived in it—new cityscapes, manufactures, styles, media, images, and new political regimes, work regimes, domestic regimes, and visual regimes. They were caught up in, and took pleasure in, what futurist ideologue Filippo Marinetti called "the magnificent dynamism of contemporary life"—a new urban culture of time and space (to use historian Stephen Kern's formulation) that was fast and accelerating, powerful and proliferative, wonderful and terrible, perilous and inebriating, like a roller-coaster ride.[28] Joseph Roth, an acute observer of life in interwar Berlin, writing on a "triangular railway junction," celebrated the new order:

> It's the heart of a world whose life is belt drive and clockwork, piston rhythm and siren scream. It's the heart of a world, which spins on its axis a thousand times faster than the alternation of day and night . . . ; whose continuous and never-ending rotation looks like madness and is the product of mathematical calculation; whose dizzying velocity makes backward-looking sentimentalists fear the ruthless extermination of inner forces and healing balance but actually produces life-creating warmth and the benediction of movement.[29]

Roth's breathless commentary, like Marinetti's manifestoes and proclamations, was part of a larger cultural enterprise. Across the social spectrum, in cities all over the globe, people performed modernity.

Captivated by the idea of the modern—the new, the transformative—people sought to cast off old ways, to shed local identities and traditions, to emancipate themselves and define themselves against what came before. They did so by studying science, purchasing electromechanical appliances, driving cars, wearing fashions, dancing, listening to the radio, seeking out new medical treatments, going to the movies, and doing a thousand other things that branded them as moderns.[30] Siegfried Kracauer, a noted Weimar cultural critic, wryly observed, "Only the most modern is good enough for our masses."[31] There was a sense that the new order was defined by proliferating, bustling collective activity. Modernity was the cumulative result of relentless human and technological commotion in populous cities and nations, where masses of people lived and died.

Kracauer also remarked that "the movements of these masses . . . are today of an ambivalent nature."[32] The masses (and not only the masses) were anxious about how to be modern and how to cope with modern life. "The experience of this century mocks human laws," declared Joseph Roth, only half joking. "Man, surrounded by machines, is compelled to become a machine himself."[33] Modernity could be dangerously unstable, fast changing, and uncertain in some domains, and crushingly tedious, routine, mechanical, and rule bound in others. The polarity between pleasure and anxiety was especially intense in Weimar Germany, where people had lived through the catastrophic experience of industrial warfare, epidemic disease, boom–bust capitalism, political violence, and the destruction of the old order and the establishment of a shaky new republic. Not everyone wanted to be modern, and not everyone agreed on how to be modern. But in every case, you had to situate yourself in relation to modernity.[34]

Fueling the fire, people sought out productions that critiqued the old and the passé—productions that presented a utopian, emancipatory, intoxicated vision of the future. People thirsted for the modern. This included a cultural elite of artists, scientists, teachers, architects, doctors, lawyers, and other professionals, but also factory, shop, and department store workers, as well as students at university and *Gymnasium* (secondary school)—and a new group, office workers, which Kracauer dubbed "the salaried masses." The "masses," first a political and sociological construct and then an identity term, became the subject of much commentary. Without much precision, it came to denote a peculiarly modern group of people who were living a peculiarly atomized collective existence, mass life.

The masses were largely composed of people who had recently arrived in the city or who were only a generation or two away from the cultural periphery. They were a churning European diaspora—many diasporas—of Jews, Poles, Italians, and other ethnic groups, war veterans, young men and women, from the countryside or from formerly prosperous families ruined by the economic crisis.[35] Displaced by war, the collapse of empires, the redrawing of national boundaries, revolution, and economic depression—and seduced by the siren call of city life, mass media, and industrial consumer products—the masses led diasporic lives, took on diasporic identities. They continually sought to modernize themselves, to do things that worked to dissolve ethnic, class, and cultural particularities, or to carry those particularities forth into a continuously emergent and ever-changing hybrid mass culture.[36] Doing all that, they might easily happen to encounter the works of Fritz Kahn.

1

READING KAHN AND THE HOMUNCULUS

> One can only correctly comprehend something from the outside if one knows it on the inside; that is true for machines just as it is for living things.
>
> —WALTER BENJAMIN, "A VISIT TO THE BRASS WORKS," RADIO BERLIN BROADCAST, JULY 1930

AN IMPRESARIO OF THE MODERN

Modernizing aspirations—and the means of playing them out—didn't materialize out of thin air. There were entrepreneurs who saw the existing and potential audience for commodities and performances that signified the modern, and who made careers for themselves as impresarios of the modern. They offered modernizing advice, entertainment, experiences, and programs, in modernizing words and images. In their personas and productions, they crafted a lexicon and poetics of the modern.

Here it is important to make a distinction: modernists were modernizers, but modernizers were not necessarily modernists. Modernism has a pedigree in art and art history. Early twentieth-century modernists were self-conscious performers who created modern art and joined movements that opposed academic and/or commercial art, and that in some way challenged bourgeois propriety. They opposed the already constituted era of modernity as insufficiently or oppressively modern, and they offered distinctively critical modernist aesthetics

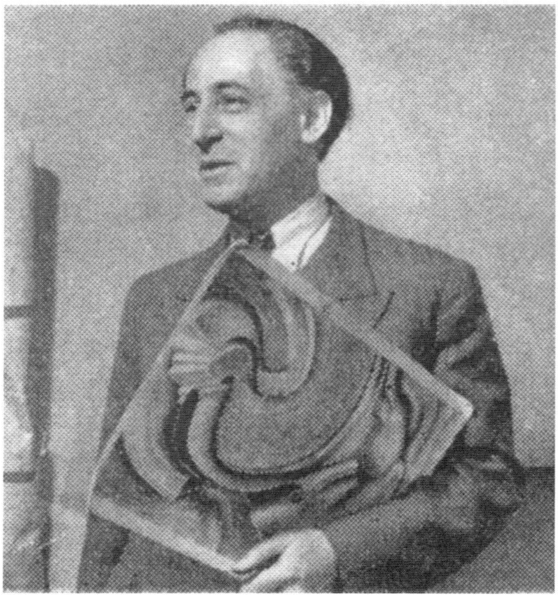

FIGURE 1.1. *Portrait of Fritz Kahn,* Life, *19 April 1943, 23. Photograph by James L. Hussey.*

(of various sorts) that were connected to larger cultural/political projects. Even if they denounced each other and competed for audiences and authority, their common purpose was to induce modern ways of doing, seeing, thinking, and being in every domain and register according to the particularities of their aesthetic and political programs. For modernists, the new era in its apotheosis should be in some fashion a *Gesamtkunstwerk,* a total integrated design—or more than that, a universal condition.[1] The poster "Der Mensch als Industriepalast" is modernist. The five-volume *Das Leben des Menschen,* with its hodgepodge of illustrations in a variety of styles, is not. It modernizes but is only modernist in parts.

Fritz Kahn was an impresario of the modern. He sought to make a name for himself as a disseminator of "the magic of modern science"—and as a modernizer of the image practices of popular medical and scientific publication.[2] More than any other science writer of his time, Kahn was an exponent of the profusely and imaginatively illustrated text, with pictures that made reference to industrial ways of seeing, doing, and living.

Kahn's images asked his readers to collude in their own modernization, to think of themselves as scientific and technological artifacts, with mind/bodies that were a site of production and exchange, full of manufacturing processes. His method, influenced by advertisements and mass-market newspapers and magazines, was to bombard the reader with pictures that evoked the experience of living in modernity, that pleasurably modeled and encouraged a kind of modernizing self-surveillance. Readers could see themselves (individually and collectively) and their cities, workplaces, homes, and cars in the images. To view and adopt the image was an easy self-affirming and self-improving way to be modern.

Like many of his readers, Kahn had a diasporic history and compound identity. To the larger public, he was a secular author of popular books on science, medicine, and health. But he was also a Zionist, participated in a Jewish Masonic lodge, and wrote books and articles on Jewish culture and history for a Jewish readership.[3] Growing up, his family was in motion, going from Germany (Halle in Anhalt-Saxony) to America (Hoboken, New Jersey, then New York City), then back to Germany, staying in Hamburg, Halle (again) and Bonn before settling down in Berlin. As an adult, his itinerary became even more complicated. After the rise of the Nazis, he was forced to flee from his home in

Berlin, settling first in Palestine, then France. After the Nazi conquest of France, he again had to flee. With the aid of American secret agent Varian Fry and a letter of reference from Albert Einstein, he managed to escape to the United States via Lisbon, settling in the New York metropolitan area, where he continued his work. He returned to Europe in the 1950s and spent the remainder of his years mostly in Denmark and Switzerland. It was a life in binaries and contradictions: traditional and modern, Jewish and German, national and cosmopolitan.

KAHN'S READERSHIP

That's a rough accounting of the identity of Kahn the author. But what about his readers? We know quite a bit about the broad demographic and cultural trends of interwar Germany: the swelling growth of cities, increased female participation in the industrial workforce, the enthusiasm for illustrated newspapers and magazines, movies, jazz, automobiles, radio, and airplanes. But we know very little about Kahn's readership, except that it was substantial. His publicity touted his large sales figures; his articles were printed in leading newspapers and magazines. He had a long-standing association with Franckh'sche Verlagshandlung (known as Franckh), a well-established imprint that specialized in science for the people in the Humboldtian tradition. Franckh published a large list of books and pamphlets, as well as *Kosmos*, Germany's most widely read magazine of popular science, with a circulation of more than 100,000, and many articles by Kahn. The publisher consolidated its market share through a network of clubs for readers, the Gesellschaft der Naturfreunde (Friends of Nature). Members received pamphlets on scientific subjects and a discount on books and subscriptions to *Kosmos*. They were also encouraged to go on nature walks and hikes, where they could observe such things as animals, plants, rock formations, and meteorological conditions. They were urged to hold meetings to talk about and celebrate what they read and observed.

How did such meetings go? Kahn's scrapbook of book reviews and letters from readers includes the program of a "Fritz Kahn Evening" held by a "science studies" group in the town of Mittweida, Saxony, in a high school auditorium, on 14 July 1927, a date probably chosen to coincide with Bastille Day. The event was a reverential evening of secular worship. Attendees were treated to a piano recital of selections from Schubert, Beethoven, Bizet, Grieg, and Rachmaninoff, interspersed with readings from the works of Kahn—the foreword to *Das Leben des Menschen*, "The Beauty of the Microcosm," "Manmade Food and the World's True Mystery," "A Special Aspect of Osteology," "Hymn to the Teeth Collective: The Future of Our Teeth," "Praise to the Diaphragm," "A Supernatural Journey through Realms Beyond the Human Hand," and "Five Scenes from the Seeds of Human Development."[4] It was a program that linked high romantic aesthetics—spirit!—to emancipatory science and scientific medicine. Modernity was the zenith of a millennial struggle between enlightenment and ignorance, the apotheosis of a civilizing process that stretched from antiquity to the present. Seen as a series of triumphs, modernity was more of a continuum than a rupture. The achievements of science and culture were to be celebrated as mutually supportive contributions to progress. To that end, Kahn equipped the beginning of each chapter in *Das Leben des Menschen* with quotations from Plato, Goethe, Emerson, Leonardo da Vinci, and other great men; he laced his science lessons with textual and visual references to art, music, philosophy, and literature.[5]

We don't know how many people belonged to the Mittweida group or who attended the Fritz Kahn Evening, but meetings of the Naturfreunde were probably common enough. In 1924, there were 84,000 members. The Fritz Kahn Evening was a special social occasion—we can imagine that both men and women attended—but Kahn's overall readership was probably disproportionately male.[6] In 1931, Walter Hofmann made a study of the German publishing market, based in part on an analysis of the borrowing patterns of Leipzig library patrons. He found that "men prefer educational literature," while women readers preferred fiction, "more than twice as much as educational literature."[7] Kahn's publications focused on the hard science of the body—topics that were, according to contemporary gender conventions, mainly coded as masculine. Kahn gave readers a modernizing course of instruction, which connected health and self-improvement to the acquisition of scientific knowledge. Edgar Eichbaum, who as an eleven-year-old Jewish refugee was sent from Germany to Palestine (and who went on to become a surgeon in America), put his enthusiasm for Kahn in the same frame as his enthusiasm for a Jack London bildungsroman:

> At the time [ca. 1939] I was reading . . . *Martin Eden*, which is about a fellow with no means who educated himself and became a world-class novelist. He became my literary hero. But the next book I discovered in the library was . . . *Der Mensch [Gesund und Krank]* . . . by Fritz Kahn. Kahn became the most influential authority in my life. I studied this book like the Bible, became fascinated by all its contents, and decided to become a physician. I theorized that . . . nothing would provide a better opportunity to develop an insight into human nature than the practice of medicine.[8]

But Kahn also dealt with child development, obstetrics, sex, nutrition, and preventive medicine—topics coded as feminine. In the early decades of the twentieth century, there was a large female readership for illustrated books on family health, such as *Die Frau als Hausärztin* (The wife as family doctor), which sold millions of copies in successive

to the performance of modernity, and who read articles and books on healthy living *(Lebensreform)*—diet, sanitation, medicine (orthodox and alternative), physical fitness, eugenics, even nudism *(Nacktkultur)*. These were among the crowds that flocked to the health exhibitions that sprang up in Germany after the war—the 1926 GeSoLei exhibitions in Düsseldorf and elsewhere, and the Deutsches Hygiene-Museum in Dresden (which sponsored traveling exhibitions and slide lectures, and which furnished GeSoLei with a substantial portion of its exhibits in the period before the museum's 1930–31 opening)—as well as commercial, municipal, and state-sponsored exhibits and illustrated lectures.[13]

Those exhibits were full of eye-catching headlines, objects, and images. It was a moment when the sheer insistence on the image—images in multiples—in any idiom, modernist or not, signified and performed the modern. Walter Benjamin was impressed: "A new, standardized, and wordless system of symbols seems to be arising in very different areas of life—in traffic, art, statistics. Here, a pedagogical problem touches upon a wide-sweeping cultural one, which could be expressed with the slogan: For the symbol, against the word!"[14] The twentieth century, Otto Neurath proclaimed, was "the century of the eye," an era of "visual hustle," an age of hyperillustrated newspapers, magazines, books, and posters, set within the electrically lit, hypervisual city of billboards, signs, kiosks, display windows, newsstands, and every kind of interior.[15] Given that lush environment, in which images ostentatiously exercised power over individuals and the masses, Kahn's extravagant commitment to pictorialism attracted notice. Novelist and social critic Alfred Döblin, reviewing *Das Leben des Menschen*, praised the usefulness, "originality and vigor of its visual presentation" and linked it to a democratizing political agenda:

> There is no more comprehensive or striking way to demonstrate the parts of which Man consists than through an image. "Man is made," unfolds: two hands throw substances into a crucible and add liquid; one sees a carbon-based shape forming in the bottom; the homunculus is created in a retort. . . . Some of the images are dramatic. For example, the fight . . . between invading bacteria and [immune responder] migratory cells looks like a battle in the primeval world. I very much recommend Kahn's book. More than a popularization of biological knowledge, it is instead a very original and moving presentation (*Durchfühlung*) . . . of that knowledge. To bring this knowledge to life, bring people closer to science, and take science out of the hands of [just the] scientists, appears to me one of the most desirable things we can do in today's world.[16]

DER MENSCH'S HOMUNCULUS

"Der Mensch als Industriepalast," the bonus give-away life-size color poster that came with *Das Leben des Menschen,* showcased Kahn's approach. The goal, he explained in a supplementary pamphlet, was to visually represent "industrial processes" in the human body, forces "which have an invisible character" and cannot be "directly observed" in real life or anatomical illustration. Only by showing "technical processes" could the reader see "a total picture of the inner workings of the human body."[17] That vision of the body as a productive array of industrial technologies was vitally important to Kahn and his readers, incited and affirmed their common desire to be modern—modern, but reassuringly familiar. Readers could see themselves represented by proxy, a whole staff of generic little *Mensch*-kin proxies, each working at some job, in some sector of the body. There were homunculi tending machinery, stocking shelves, operating controls and instruments, cleaning the shop floor, assembling components, directing and overseeing production lines. The term *homunculus* (Latin for "little man") goes back to sixteenth-century alchemy. The homunculus, a tiny preformed human, inhabits or constitutes the sperm or egg; upon insemination, it grows to become a fully formed human.[18] Unlike the alchemical homunculus, Kahn's homunculi stay small and don't grow into anything. Playfully legible social types (managers, switchboard operators, office and factory workers), they stand for stability, not transformation. They aren't ghosts in the machine; they don't haunt anything. They just go nicely about their business, doing the many tasks involved in the operation of the body. They do, however, have a secret mission outside the frame of the illustration: to gently persuade the reader to defer to scientific authority and to identify with the figure of the machine. If readers really take it in, the homuncular elements contribute to a kind of technological self-regard (which is not quite the same thing as self-surveillance, because it's founded more on pleasure than incitement to worry).

Even so, the homunculus potentially complicates things. In the division inside the body between machine and human, which side of the dyad stands for spirit and which for matter? Is spirit the inanimate, unfeeling machine, which embodies scientific laws, and thus stands as the soaring accomplishment of reason and inspiration? Or is it the pulsing sensate human who creates machines but who is reduced to the status of a cog in the industrial manufacturing process? Neither exactly fits, but either way, the human is mechanized and the machine humanized. As the distinction between machine and organic life breaks down, the factory, with its constituent machines and social relations, comes to serve as an effigy of self, a potential unifying principle.

"Der Mensch als Industriepalast" was a career-making achievement. But in the 1920s and 1930s, Kahn and his artists developed many other devices to explain corporeal structure and function. Some of these, like "Der Mensch," referred to the factory and the technologies housed within it. Others referred to the railway station, the city street, the dinner table, the elementary school, the concert hall, the sporting event; they featured cars, elevators, skyscrapers, phonographs, X-rays, telephones, radios, movies, and other iconic industrial inventions. Readers were handed a lush new visual vocabulary, a shelf full of ways to imagine "You . . . as you live and are," with the emphasis on the present tense: "you" as a modern in modernity. Everyday lives and locations receive representation in Kahn's images of the human body. In turn, the images offer up models of how daily activities ought to look and feel. An iconographic/cognitive Möbius strip runs between viewer and image.

"WHAT GOES ON IN YOU WHEN YOU VIEW THIS IMAGE?"

That relationship is intensified when Kahn and his artists depict another characteristically modern activity: looking at pictures. Take, for example, a fold-out monochrome plate in *Das Leben des Menschen*, "The Oxygen Cycle," which comes close to theorizing the relationship between reader and printed image.[19] The self, a physiological diagram in the shape of a man (entirely automated, no *Mensch*-kins at the controls), views a physiological wall chart. The subtitle, "What Goes On in You When You View This Image?," invites the reader—who is outside the frame of the picture but invoked by direct address—to see himself or herself as a textualized, regulated physiological entity. The reader is asked to adopt the image as an identity object, to identify with the schematic diagram of a man-machine seeing himself. The peculiar armlessness of the image makes every other function extraneous; it turns the human into a scope, a viewing device. You are a precision-engineered seeing mechanism. You see a precision-ruled, razor-sharp diagram of the mechanism when you look at yourself. And you use that mechanism to see yourself, in a precision-engineered, regulated way (those radiating dashed lines that go from the eye to the chart).

"The Oxygen Cycle" is a hall of mirrors inhabited by a singularity that has a face: the self. "Der Mensch als Industriepalast" takes a different tack. There the identity object is a faceless man-machine staffed by a workforce of nearly faceless homunculi. Differentiated only by function, they stand collectively, anonymously, for humanity in the mass. But the homunculus in Kahn is more than just charming little figures that work inside the body. Like "The Oxygen Cycle," many of the illustrations dispense with them entirely. Even when they are present, they don't really amount to much. Within the body, they have a status that mirrors the declining status of the individual within industrial mass society. Joseph Roth's comments on the effects of industrialization could easily have been a comment on "Der Mensch": "Everything

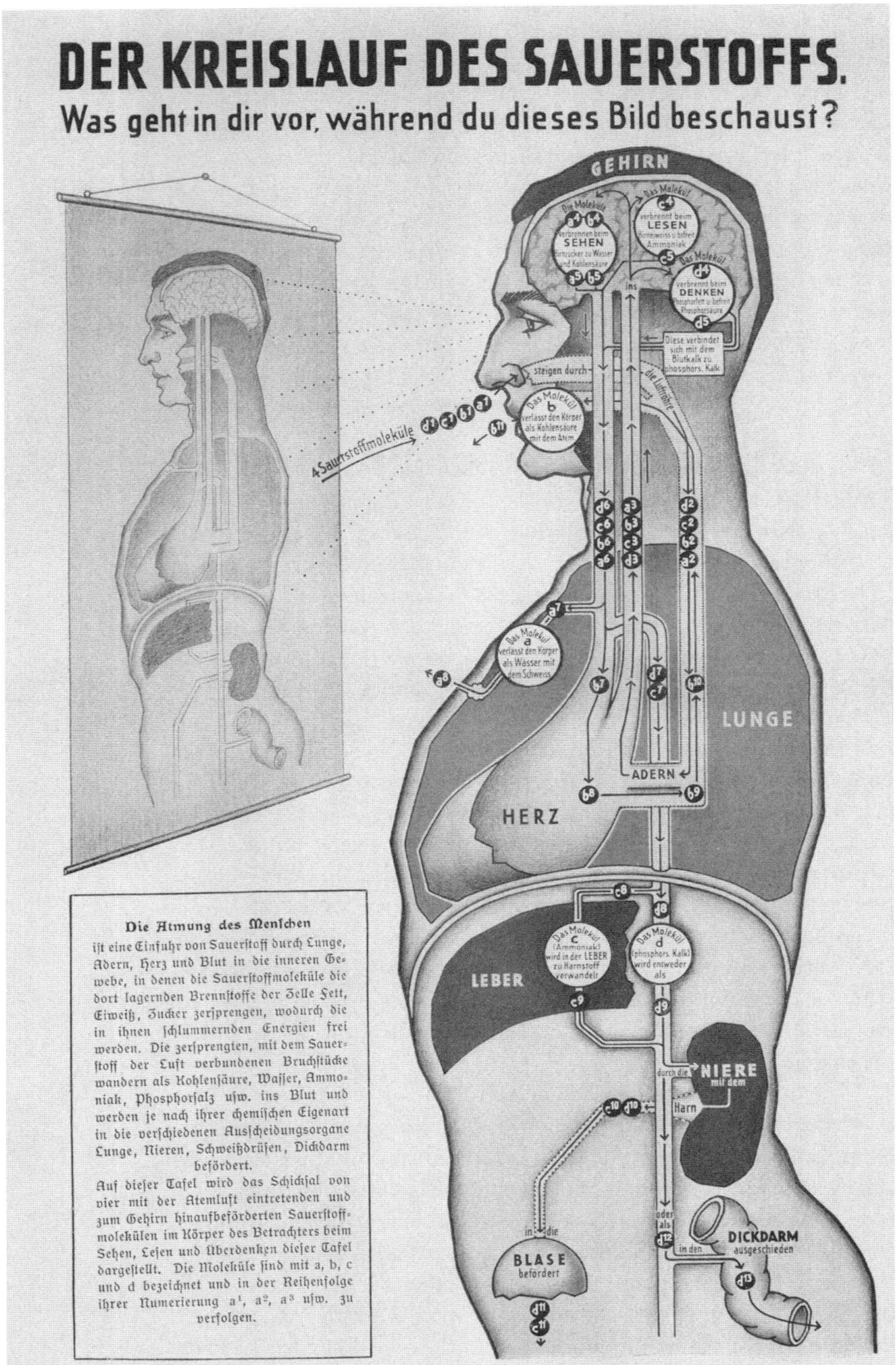

FIGURE 1.3 *"The Oxygen Cycle,"* Das Leben des Menschen *(1926), vol. 3, foldout, pl. 11. Artist uncredited (Fritz Schüler, with Otmar Trester?). National Library of Medicine. Copyright Kosmos Verlag, Stuttgart.*

human in this metal arena is small and feeble and lost, reduced to an insignificant supporting role in the grand enterprise. . . . A man in uniform wanders about . . . , a tiny human, . . . functioning only as a machine. His significance is no greater than that of a lever."[20] What is more significant than the "tiny human" is the illustration itself. Nearly every illustration in *Das Leben des Menschen* offers itself as a homunculus, a figure to be archived inside the reader. Homuncularity is a potential property of figurative presentation. Shrunken down to a figure on a page, almost any object can explain and conceptualize "what goes on in you." Figures of machines, houses, automobiles, animals—anything—can serve as a tiny identity object, an avatar that refers back to "you" and represents "you" as a universal human, as a modern.

That property, in which Kahn exploits so extravagantly elevates homuncularity into an industrial device, tacitly describes a principle. Kahn's illustrated books and articles are an archive of homuncular figures. Imaginatively distilled from social, economic, technological, and medical experiences, and shrunken down to the size of a picture on a page, the homunculi are returned to the reader as a scale model of his or her own living, breathing, life-size self. That feature is readily apparent in "Der Mensch als Industriepalast," Kahn's only full-size poster. The scale of the illustration is one to one: if you stand in front of it, it becomes your profile, an industrial shadow full of tiny humans going about their work.

To be clear, I'm not claiming that Kahn and his illustrators contributed in any way to critical theory or the philosophy of mind. Instead, I'm arguing that their tactic of showing the body in modernity by showing modernity in the body refigured subject formation as the work of new social relations, aesthetics, and technologies. If multiples of individuals in technological arrays made a mass society, then mass society in technological arrays reciprocally made the individual. The homunculus is a worker among (or inside) workers and a picture among (or inside) pictures. "Der Mensch als Industriepalast" models divided and distributed subjectivity. Curiously, this also models the process that produced "Der Mensch" and *Das Leben des Menschen*, which were collectively produced by author, artists, sources, publishers, printers, and crucially, if indirectly, by readers and critics. Along with the infinite regression of solo agents (which philosophers of mind term the "homuncular fallacy"), the reader gets an infinite extension of agent ensembles.[21] We are meta-*Mensch*-kins, part of a proliferating world of *Mensch*-kins. Like Kahn's illustrations, in our everyday lives and relations, and representations of same, we also appear as figures and do others the service of performing the role of homunculus.

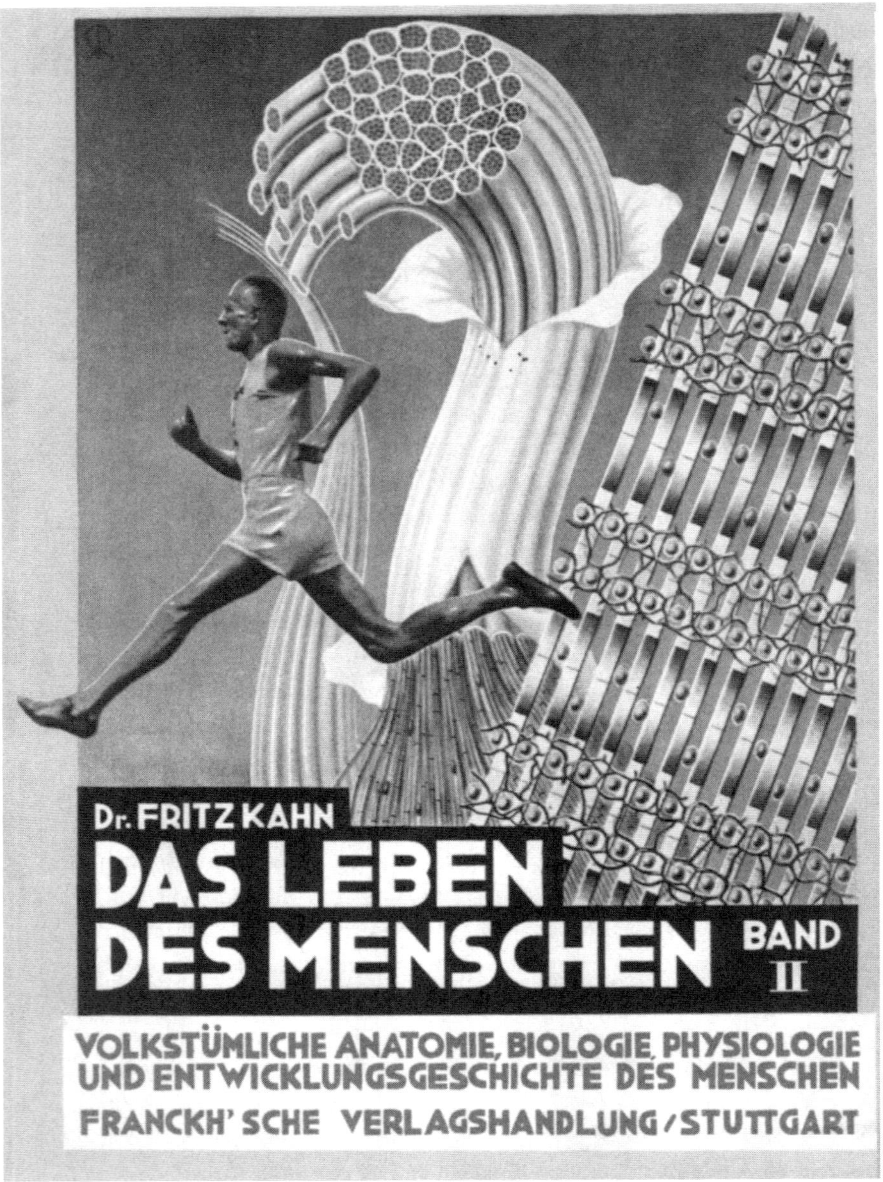

FIGURE 1.4. *Two-color dust jacket,* Das Leben des Menschen *(ca. 1931). Artist: Roman Rechn. Leo Baeck Institute, New York. Muscle fiber looks like industrial cable; nerves look like electrical wire and relays. The fluidity, dynamism, and beauty of the body in modernity—the athlete as homunculus—is rendered in a modernist style that combines photography and precision-ruled lines. Copyright Kosmos Verlag, Stuttgart.*

HOMUNCULAR APPLICATIONS: "DER MENSCH" AS UTOPIA

Kahn's pictures (and pictures within pictures) thus provided identity props for all concerned. They helped brand Kahn as a modernizing science writer, helped Franckh brand itself as a modernizing popular science publisher, and helped readers brand themselves as moderns.[22] If it worked, the popular scientific illustration, like the advertisements that increasingly appeared on walls and pages, would do more than just instruct readers. It would "install . . . itself into their bodies and souls"—a phrase used to describe how the masses were held captive by the manipulative visual tactics of modernity, in the 1930 novel *Blood and Celluloid*.[23]

In "Der Mensch als Industriepalast," managers in offices (the brain) monitor and direct production, a complex array of mixing vats, sprayers, pumps, chutes, storage areas, piping, pressure valves, furnaces, conveyor belts, pistons, pulleys, and workers. The poster is a kind of sequential art, an ordered progression of scenes in the body (which looks a bit like a pinball machine). In other illustrations, Kahn shows a sequence of figures, or he attempts a freeze-frame that dramatizes a dynamic process or action occurring in time and space—but not in "Der Mensch." There, the power and violence of industrial manufacturing is tamped down into a whirring, mechanical regularity. The poster declines to suggest the possibility of industrial accidents, speed-ups, strikes, explosions, or work stoppages. The incessant movement, speed, and pressure of manufacturing processes are represented, but in a measured way. Much of embodied industrial and organic life—acceleration, growth, disease, reproduction, contingency—go entirely unrepresented. There is no grease and grime, screech and clatter, smoke and stench, rupture and ruin—the disruptive, disfiguring, dystopic conditions that also afflicted industrial society in the interwar period.

Those omissions have a political valence. "The cell state," Kahn proclaimed, "is a republic under the hereditary hegemony of the mind's aristocracy. Its economic system is communism."[24] "Der Mensch als Industriepalast" transcends politics; it offers something for everyone, every position on the social and political spectrum. It could accommodate Hans Surén, the popular exponent of physical fitness and nudism who went on to support Hitler and the Nazis, and Alfred Döblin, the caustic left-wing social critic and popular novelist. Both celebrated *Das Leben des Menschen*.[25] In an era rife with harsh political rhetoric, sadistic beatings, and murders—an utter lack of social consensus—"Der Mensch" stands as an industrial/physiological utopia. The parts and systems of the body are in perfect harmony.

In some other illustrations, that utopia extends outward: Kahn puts the human body and body parts in global perspective, literally pairing globe and body. "Der Mensch als Industriepalast" doesn't do that; it doesn't feature any perspective at all. The life-size profile floats in the space of a featureless white background, outside of any troubling context. Legible, rational, beautiful—a work of modern science, technology, and aesthetics—"Der Mensch" stands outside of nature and history, perfect, a total design.

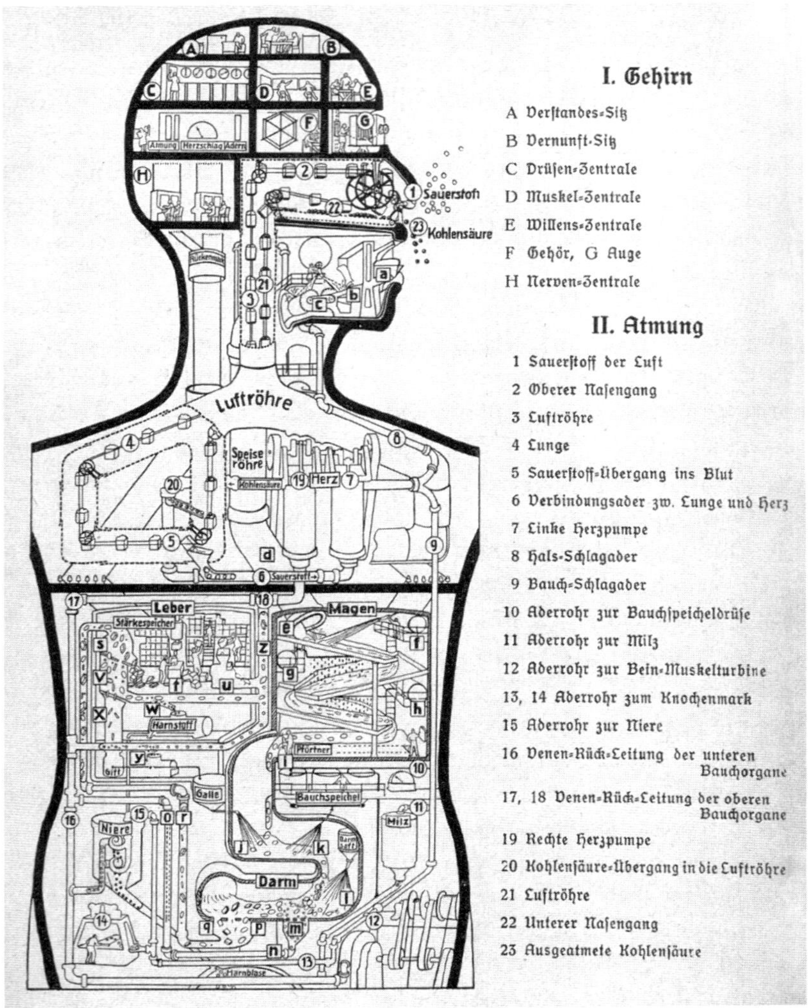

FIGURE 1.5. *Diagram of first edition poster, "A Technical Description of the Main Processes of Life," Supplement,* Das Leben des Menschen *(Stuttgart: Franckh'sche Verlagshandlung, 1926), 4:iv. National Library of Medicine. Copyright Kosmos Verlag, Stuttgart.*

WEIMAR AMERICANISM AND THE UNIVERSAL HUMAN

"Der Mensch als Industriepalast" aspires to timeless, placeless universality, a technoscientific universal, and in the parallel domain of aesthetics, a modernist universal. Even so, its universalism is stamped with its place of origin; it has Weimar written all over it. It was first published in 1926, in the middle of a brilliantly productive decade for *mitteleuropäische* functionalist aesthetics. In the interwar period, there was a concerted ideological effort to erase the ancien régime, to strip away fussy and grandiose architectural ornament, pave over cobblestone with asphalt, cover up overdecorated facades, straighten or overlay old irregularities. The new order of the day, historian Janet

Ward has shown, was schematicism, rectilinearity, smooth, undecorated surfaces, electric lighting, the polished luster of steel and glass—a shine (in German, *Glanz*) that glowed and sparkled. By these signs, it was affirmed that humanity had entered a new phase, with a new aesthetic, for the new republic, the new industrial order, and the newly citified masses. The past had to make way for mass production and spectacular productions for the masses. Most spectacular of all was the modern city.[26]

To be sure, the project to replace the discredited Wilhelmine cultural regime with a universal, refashioned built environment was controversial. Modernity took place in Paris, Chicago, New York, Buenos Aires, Stockholm, and Leningrad as much as any German city. There was nothing particularly German about the modernizers' critique of the past or the iconic inventions of modernity. Germans patriots could lay claim to the X-ray and the printing press, but not the railroad, typewriter, machine gun, telephone, radio, or motion pictures. Nazis and social conservatives took pride in Germany's accomplishments—the country was in many respects the most scientifically advanced in Europe—but they defined their patriotism mainly in opposition to modernity, which they negatively associated with transnational Jewry, transnational culture (jazz, "degenerate art," Hollywood), the department store (many of which were owned by Jews and featured products from all over the world), and internationalist Marxism. On the right, these targets comprised a morally corrupting, placeless modernity that served as the opposition for the Nazi politics of purification.[27]

A more benign program of purification came from the left: the internationalist machine style of design promoted by Walter Gropius and other exponents of the New Architecture *(neues Bauen)*. The Bauhaus fashioned its architecture out of basic geometrical forms and smooth or reflective surfaces made out of industrial glass, metal, and concrete.[28] "Der Mensch als Industriepalast" was under the influence, but not that rigorous. It had encyclopedic aspirations; it wanted to include everything (except the organs of generation), so it had to pack a lot in. It couldn't empty the body (as did the more focused "Oxygen Cycle"). Even so, its lines are sharply rendered. Detail is restrained; geometric regularity and schematic order govern. The body's surfaces are clean, smooth, and brightly colored. Blood (colored red and blue) is pumped in shiny, thick glass containers. Gas molecules are standardized round pills (color coded to show the conversion of oxygen to carbon dioxide); starch, fat, and protein are round and square modules, which break down into constituent modules.[29] The body parts are metal, glass, concrete, wire, and rubber, the favored materials of the new design.

"Form follows function" was Bauhaus's famous dictum, and "Der Mensch" has plenty of functions and functional devices: inputs and outputs, conveyor belts, sprays, chutes, pressure chambers, meters, gears, dials, switches. The rationalizing impulse is figured in the organization of the manufacturing process. At the front and top of the head (headquarters), the work of the body is directed by men in three offices: *Verstand* ("Understanding," the chief executive or manager), *Vernunft* ("Reason," three men arguing at a table with a cabinet of books in the room), and *Wille* ("Volition," three men around a table making decisions). Production is planned and laid out like a flowchart, neatly divided into sequentially numbered or lettered boxes, which is explained by keyed captions in the accompanying pamphlet.[30]

"Der Mensch" is mostly a chemical plant. Germany's leading role in the development and production of new chemicals and pharmaceuticals was a point of national pride—and something left and right could both agree on. New chemical inventions, in particular intensely brilliant inks and dyes, were celebrated as great accomplishments of the age. The focus on industrial chemistry, intense color coding (those brilliant inks), neat draftsmanship, omission of unnecessary detail, and polished surfaces all mark "Der Mensch als Industriepalast" as a particularly German version of the modern.

But not just German. In the 1920s, factory managers, efficiency experts, and business psychologists were reconceptualizing the office as a complex machine, just as Frederick Winslow Taylor and Henry Ford had earlier reconceptualized the factory. In Germany, this reorganization was dubbed the "American system," a set of practices that emphasized efficiency, time-motion study, assembly lines, and mass production (the principles of Fordism); it used flowcharts and schematics, pushpin graphs, and other devices, with inputs and outputs, to diagram, track, and reengineer the work process. Siegfried Kracauer, on a visit to the office of a "modern factory," described its colorful visual apparatus:

> The commercial director . . . explains the business. . . . "The . . . work process," he says, "is rationalized down to the last detail." He points to diagrams whose colorful networks of lines illustrate the whole operation. The plans hang in frames on the walls of his room. On the other wall there are two peculiar cases that look a bit like children's abacuses. Within them brightly colored balls, arranged on vertical cords, rise in close formation to varying heights. One glance at them and the director at once knows all about the firm's current situation. Every couple of days the little balls are repositioned by a statistics clerk. . . . A cupboard-like centerpiece studded with colored light-bulbs forms the principal ornament of the . . . office. In general, the sole purpose today of red, yellow and green tints is to organize an enterprise more rationally. From the flashing and dimming of the tiny bulbs, the manager can at all times deduce the state of work in the individual departments. In the . . . tour

> through the offices . . . , we gradually pace out the network of lines on the wall of the room. The marvelous thing is that the operation of the plan is set in motion by real people.[31]

The worker was also reconceptualized—not as a machine operant (too much agency there) but as a mechanical part of a larger entity, "society," which itself could be figured as a "working machine" (so said Georg Schlesinger, head of the Institute for Industrial Psychotechnics in Berlin).[32] This double reconceptualization (society as machine, worker as machine part) was not merely the attendant effect of the experience of living in modern times, and was not only derived from Taylorist designs imposed from above. It was also connected to feelings of affection for machines and mechanical contrivance, a delight in machine rhythm and machine power. Working people may have often hated their enslavement to machines, for good reason, but they also developed a kind of love for the machine. They knew machines intimately; they tinkered with them, took pride in their ability to run and fix them, and even identified with them. As Alfred Döblin noted, "The technical spirit goes through the streets, stirs things up and fashions things."[33]

This was not the "disenchantment" wrought by the rationalizing principles of modernity, as in Max Weber's scheme of things. The "technical spirit" had a humming dynamic energy, maybe even a jazz rhythm, infusing everything. It worked an enthusiastic physics of powers and transformations that animated the industrial world as well as the human.[34] Joseph Roth, the indispensable observer of life in Weimar Berlin, marked the mood: "What may have the appearance of a war against the elements"—the skyscraper as an icon of modern industrial technology—"is in fact union with the elements: man and nature becoming one":

> Exploitation of every dimension: exaltation, visible from outside, that also communicates itself to the spirit within. . . . Men striding out, purpose in their eyes, enterprise in their loose limbs; machine rattle and clatter of conveyances; shouted order of command; the even beat of mechanical perplexity, working toward a common end.[35]

Industrial modernity—the apotheosis of cold reason—unleashes hot primordial force. Humanity in the mass becomes one with the machine. Filled to the brim with manufactured objects, the new era is a joyful plenum that blithely ignores, or maybe reconciles, the quarrel between materialist reductionism and spiritual holism. For Germans, that technological revelation had a national identity: America.

Utopian thoughts such as these may seem to us poignantly delusionary when weighed against the fact of poverty, urban grime and decay, endemic disease, political strife, unemployment, economic instability, violence, and despair, the themes of so many Weimar-era plays, films, and novels, as well as what was to follow. But that imploding exploding political economic scene was precisely what fueled American dreams, giving them force and intensity.

"Der Mensch" thus stands as a four-color blueprint of the body on the American plan.[36] Weimar was keen to import all things American. Weimar modernizers—producers and consumers—were enchanted with jazz, Hollywood, the assembly line, the billboard, the phonograph, the skyscraper, and the Model T.[37] So even if "Der Mensch" wears a made-in-Germany heart on its sleeve (actually the figure in the poster has no arms or sleeves), that label is a sign of the Weimar-era infatuation with *Amerikanismus.* As we shall see in the next few chapters, Kahn and his artists copied America, just like everybody else. But their Americanism went way beyond anything the Americans did. It inaugurated a new form of visual explanation suited for life in the industrial age: *Amerika* the beautiful.

2

"MUCH BETTER THAN WORDS"

PICTURED KNOWLEDGE AND THE RHETORIC OF VISUALITY

WUNDER IN UNS

Fritz Kahn's breakthrough, the human body rendered as a cross-sectioned profile diagram of a factory, was in fact an American invention. So now the backstory: the origin of the conceptual scientific illustration, and how it came to Europe. And back of that, how the modernizing rhetoric of visuality preceded, and provided a rationale for, the visual rhetoric of modernity that Kahn propagated.

Conceptual scientific illustration made its European debut in 1921 in the pages of *Wunder in Uns* (The wonder in us), "a book on the human body for everyone." Edited by Hanns Günther (the pen name of prolific science journalist Walter de Haas) and published in Switzerland, the volume presented twenty-four illustrated essays on "recent developments" in medicine and "modern physiology." Five writers wrote the articles, all of them frequent contributors to *Kosmos* and allied publications; indeed, some of the articles originally appeared in *Kosmos*. Three of the essays were by Fritz Kahn.[1]

Central Europe was then still reeling from the mass destruction and disruptions of the Great War, as well as from the political and economic turmoil that followed. Yet even in troubled times, *Wunder in Uns* was still able to attract readers and quickly sold out its run of 10,000 copies. Two years later, it was reissued by a different Swiss publisher in a second edition of 10,000, with revised text and additional essays and plates. Eight of the plates from the first edition were redrawn, colorized, and

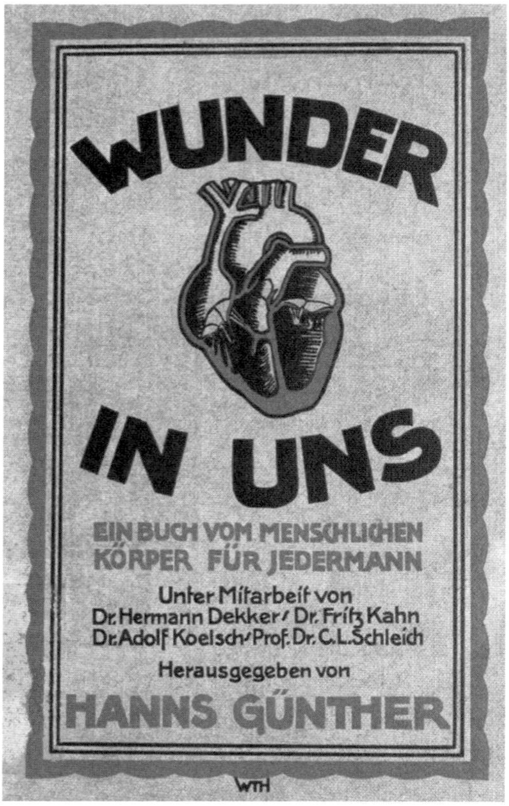

FIGURE 2.1. *Two-color book cover, Hanns Günther, ed.,* Wunder in Uns, *2nd ed. (Leipzig: Grethlein, 1923). Designer: Walter Thamm. National Library of Medicine.*

equipped with fancy translucent overlay pages imprinted with captions keyed to the details of the picture underneath.[2]

Wunder in Uns mostly conformed to the image–text practice already prevalent in Europe and North America. Between 1900 and 1920, German popular science publishing took a pictorial turn, but it was a slow pivot. In those years, Franckh and other publishers began to feature some graphic designs on their covers—an improvement over covers that only showed a banner on top of a table of contents, along with uninspiring, generic decorative elements.[3] Interior illustrations began to increase in number, but text still dominated; there were even still a few articles that were entirely unillustrated. In general, the visual rhetoric of popular science stuck to the straight and narrow, gesturing toward sober presentation: simple pictures corresponded, one to one, to the topics discussed in the text. Articles on astronomy, natural history, zoology, ethnography, chemistry, and other scientific topics mainly used photographs and drawings that did the work of literal depiction. Articles on the human body mainly used illustrations of bodily structures, both anatomical and microscopic.

However, a few of the plates in *Wunder in Uns,* the ones that were redone in the second edition, did something different: they used visual metaphors of industrial labor, technology, and architecture to explain how the body works. Oddly, none of them was connected to Kahn's three essays. Instead, the plates appeared in Günther's introduction ("In the Land of a Thousand Wonders") and in an essay titled "Man as Machine" (in the first edition written by Hermann Dekker, replaced in the second edition by an essay of the same title by Hans Hauri).[4]

Three of the plates depict industrial scenes. In one, a stylized profile cutaway of the head shows a brain made up of bundles of wires that connect the eyes, nose, mouth, ears, and lower body to tiny offices staffed by a controller, switchboard operator, announcer, and file clerks (Plate 2). In another, "Digestion," a stylized frontal cutaway of a torso, foods tumble off a stylized tongue down an esophageal chute into the stomach and intestines, which are depicted as a sweaty mine or furnace room tended by four manual laborers. In a third plate, "What Happens in Our Mouth When We Eat Bread," two men outside the mouth lift a large, jagged slab of stonelike bread onto the teeth, which will grind it into smaller pieces. Inside the mouth, a man stands ready to turn on faucets that supply saliva (a process that will turn the ground-up material into paste); a fourth man shovels a half-processed pile to move it under two saliva spigots. Taken together, the plates represent the organization of the human body as a class system in the industrial era: the clean, modern, electrified head office (the brain), staffed by men in suits, controls the mouth and body; in the mouth, men in aprons and work clothes feed furnaces, open and close valves, and push around material, supplying the muscle power needed for sweaty, industrial dirty work.

The visual metaphors are didactic and straightforward. The body is staffed by a skeleton crew, with five people in the head and five in the torso.[5] (In contrast, the body in "Der Mensch als Industriepalast," in its first edition, has a staff of twenty-seven; see Plate 1.) This is not yet the modernity of the assembly line (which in 1921 was still fairly recent, having been introduced at the Model T factory by Henry Ford in December 1913); the industrial production process is not clean, sophisticated, mechanical, or automated. Even so, the illustrations perform modernity as perceived in their moment, in form and genre as well as content. The plates have more than a slight resemblance to turn-of-the-century engineering and architectural display models, as well as vernacular brick factory architecture. The artwork has the look of Hearst and Pulitzer Sunday supplement magazine graphics.

The pictures are printed in bright colors on semigloss paper. The translucent paper overlaying the plates is an interactive device. Like deluxe wrapping (the packaging revolution for chocolates and other

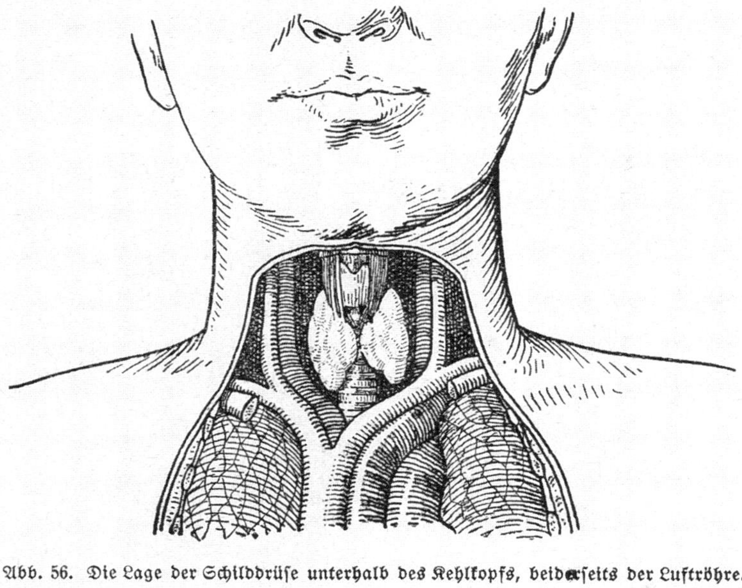

FIGURE 2.2. *Illustration in the* Kosmos *line-art house style before Kahn's innovations. "The Mystery of the Thyroid,"* Wunder in Uns *(2nd ed., 1923), 219. Artist uncredited. The same image appears in Kahn, "Die Schilddrüse,"* Kosmos Hand-weiser für Naturfreunde *(1918), 1:3. National Library of Medicine.*

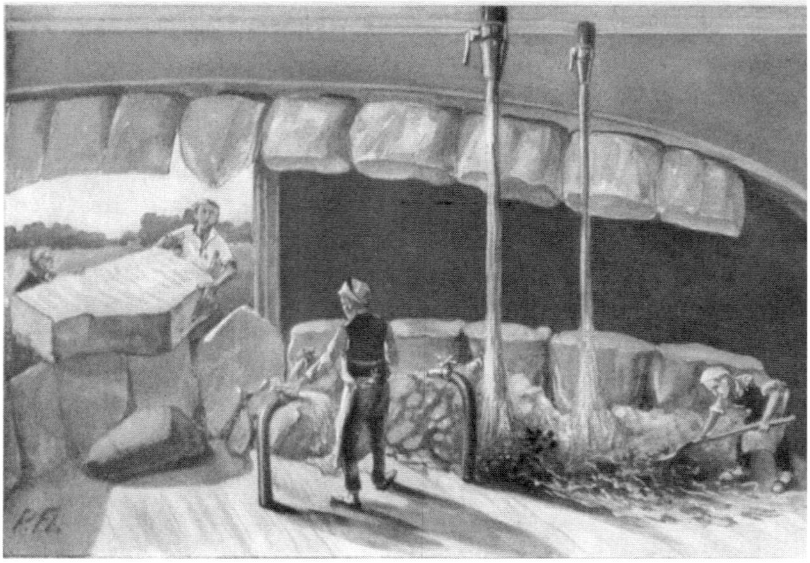

FIGURE 2.3. *"What Happens in Our Mouth When We Eat Bread" (based on* Picture Knowledge *[1917], 248 ["The Community Mill"], colorized and redrawn by "P.Fl" [Paul Flanerky]),* Wunder in Uns *(1923), pl. 1. National Library of Medicine.*

luxuries was then well underway), the paper veils the underlying image, tempting the reader to lift the page to see what is underneath. The appearance of visual metaphor is unusual, and industrial metaphor even more so.[6] Taken together, the features make *Wunder in Uns* an object of consumer desire, an appealing novelty.

PICTURED KNOWLEDGE AND THE MODERNIZING RHETORIC OF VISUALITY

The existence of *Wunder in Uns* demonstrates that Kahn did not invent the profile cutaway of the body as industrial complex. Neither did Günther. He got it from an article titled "The Bodies We Live In: What Strange Land Is This?," produced "under the direct supervision of Dr. Winfield Scott Hall" for the 1917 second edition of an American children's encyclopedia, *Pictured Knowledge.*[7] Did Kahn ever see the American publication? We don't know. But he certainly knew about it; Günther credited Hall and *Pictured Knowledge* on each plate. Whatever the case, it is certain that Kahn keenly attended to the images (and probably to the Dekker-Hauri "Man as Machine" articles that they accompany). The plates mark a dividing line in Kahn's career. It was only after *Wunder in Uns* that he began to present readers with metaphorical profile cutaways and began to make other kinds of conceptual illustration—visual explanation—his stock in trade.[8]

It's thus worthwhile to take a closer look at *Pictured Knowledge* and "The Bodies We Live In." The article begins by asking, "What Strange

Land Is This?" The answer: "A remarkable community of millions. . . . They have machines that run and repair themselves. Cameras that take pictures in color. Telephones that record sound. Policemen that never fail to do their duty. Citizens that love one another better than themselves and many other things you would hardly believe."[9] The human body, the article argues, naturally accomplishes what the utopian socialist community established at Oneida, New York, in 1848 aspired to. The

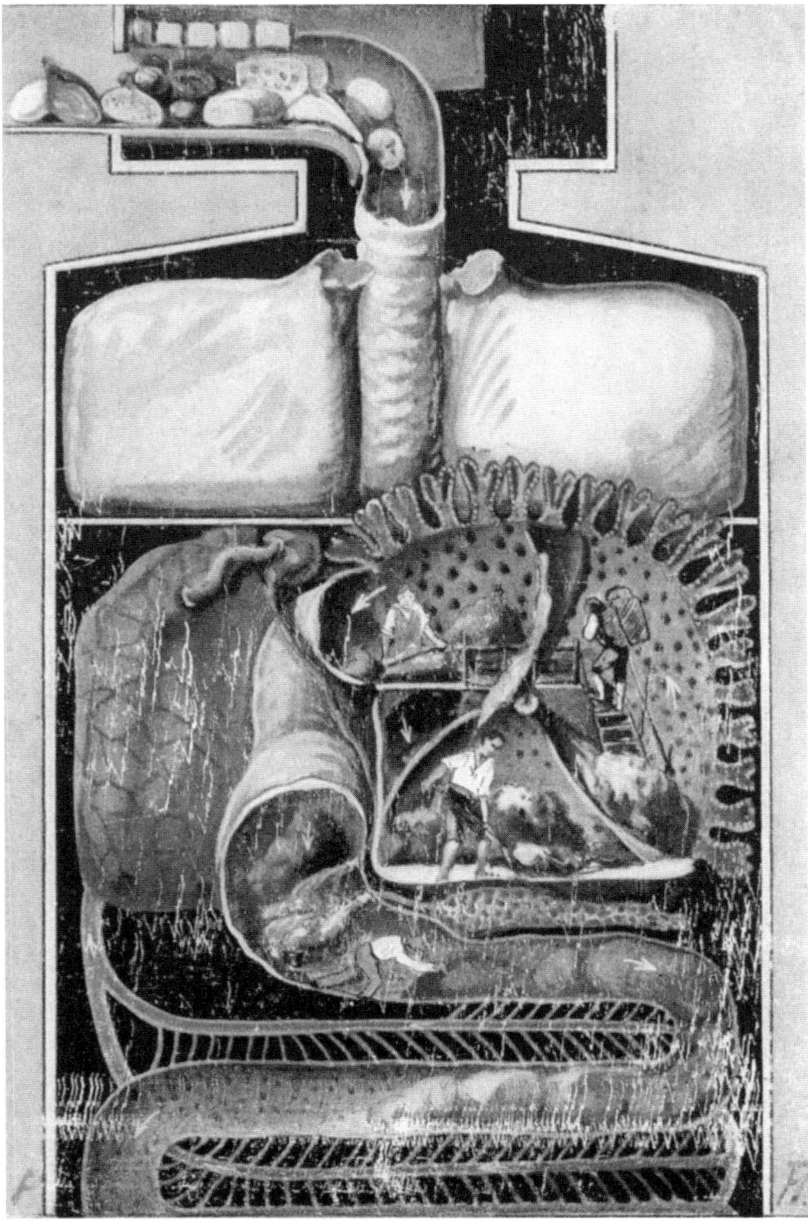

FIGURE 2.4. *Hard labor in the digestive tract. "Digestion" ("A Trip through the Food Factories," colorized and redrawn by Paul Flanerky),* Wunder in Uns *(1923), pl. 3. National Library of Medicine.*

scale of the body is simultaneously small and large. It is a harmonious "United States" of body parts, a town "where nobody is ever greedy" but also a mass society of "individuals which . . . are the millions of tiny cells." It has a "republican" government that employs "sharp-eyed" inspectors to oversee the body's many "factories," and it has "offices" staffed by "specialists" and "experts," "workers" and "consumers."[10]

The body is a benevolent industrial economy and a bureaucratic state—the rationalized, technologized, morally purified nation that

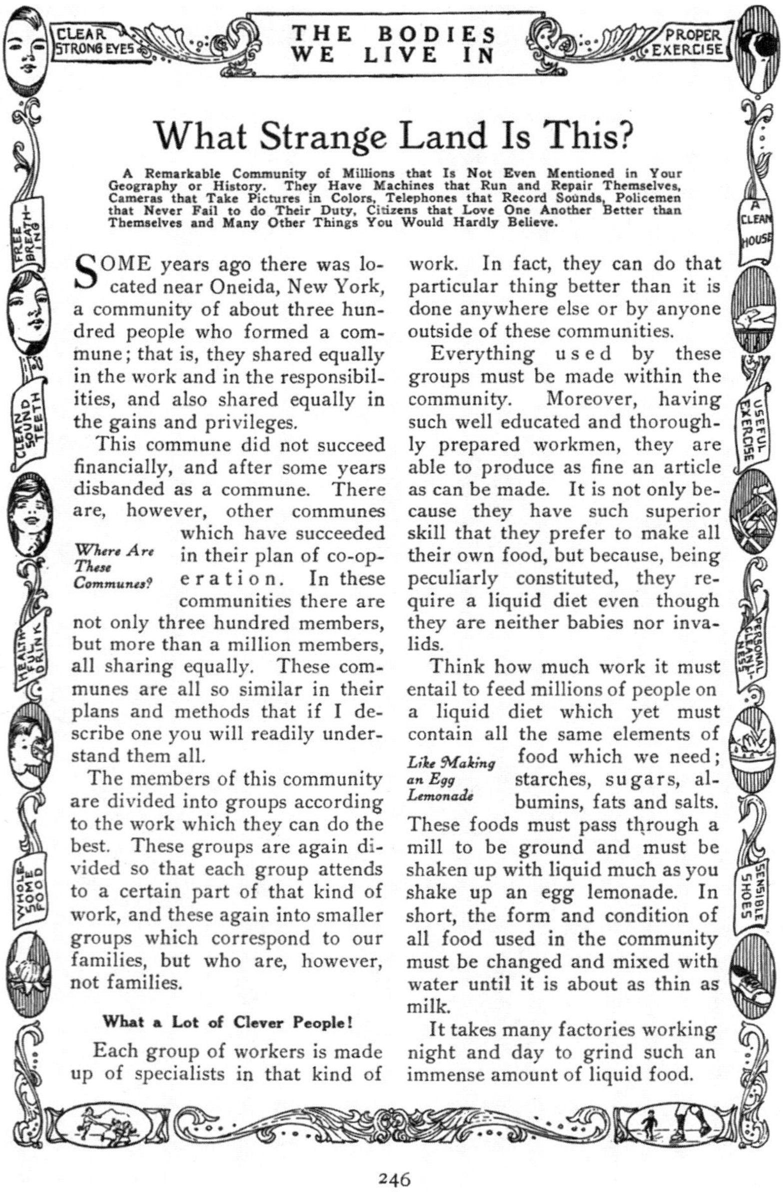

THE BODIES WE LIVE IN

What Strange Land Is This?

A Remarkable Community of Millions that Is Not Even Mentioned in Your Geography or History. They Have Machines that Run and Repair Themselves, Cameras that Take Pictures in Colors, Telephones that Record Sounds, Policemen that Never Fail to do Their Duty, Citizens that Love One Another Better than Themselves and Many Other Things You Would Hardly Believe.

SOME years ago there was located near Oneida, New York, a community of about three hundred people who formed a commune; that is, they shared equally in the work and in the responsibilities, and also shared equally in the gains and privileges.

This commune did not succeed financially, and after some years disbanded as a commune. There are, however, other communes which have succeeded in their plan of co-operation. In these communities there are not only three hundred members, but more than a million members, all sharing equally. These communes are all so similar in their plans and methods that if I describe one you will readily understand them all.

Where Are These Communes?

The members of this community are divided into groups according to the work which they can do the best. These groups are again divided so that each group attends to a certain part of that kind of work, and these again into smaller groups which correspond to our families, but who are, however, not families.

What a Lot of Clever People!

Each group of workers is made up of specialists in that kind of work. In fact, they can do that particular thing better than it is done anywhere else or by anyone outside of these communities.

Everything used by these groups must be made within the community. Moreover, having such well educated and thoroughly prepared workmen, they are able to produce as fine an article as can be made. It is not only because they have such superior skill that they prefer to make all their own food, but because, being peculiarly constituted, they require a liquid diet even though they are neither babies nor invalids.

Think how much work it must entail to feed millions of people on a liquid diet which yet must contain all the same elements of food which we need; starches, sugars, albumins, fats and salts. These foods must pass through a mill to be ground and must be shaken up with liquid much as you shake up an egg lemonade. In short, the form and condition of all food used in the community must be changed and mixed with water until it is about as thin as milk.

Like Making an Egg Lemonade

It takes many factories working night and day to grind such an immense amount of liquid food.

246

FIGURE 2.5. *"The Bodies We Live In: What Strange Land Is This?"* Pictured Knowledge, *2nd ed. (Chicago: Compton-Johnson, 1917), 246. National Library of Medicine.*

Progressive-era reformers were striving to bring about. It has kitchens and an "air station" (referencing the contemporary enthusiasm for pneumatic tubes), modern ventilation and sewage systems (both obsessions of Progressive reformers), along with "an old-fashioned mill" (for grinding and processing food intake). The capital of the body is the "Headquarters," outfitted with the most "remarkable system of telegraphy" ever devised, a "hall of records," and "laboratories" that analyze odors and tastes, and measure temperature and other vital signs.[11] (At the time, Progressives were agitating for the systematic collection of social and health statistics, as well as the establishment of state, county, and municipal laboratories that could perform chemical and microbiological tests to assure public health.)

All of this requires administration. The corporeal state is governed by Information, Interpretation, Executive, and Intelligence Bureaus; a "Look-Out Committee" that maintains a "Great Picture Gallery"; and Transportation, Food, Telegraph, and Police commissions. Each division employs amazing inventions and advanced science. The Look-Out Committee uses a combination telescope, stereoscope, and camera that can take perfect color photographs. (Color photography was then still crude and impractical.) The Information Bureau has a "wonderful" device that is "a piano, telephone and phonograph combined," as well as a "pipe organ" with only two reeds that can "play difficult music in three octaves" and "reproduce songs with or without words and can talk better than any talking machine." The body's circulatory system has an "oxidation laboratory" with a "pure food delivery system." (Pure food was another obsession of the Progressives, and the 1906 U.S. Pure Food and Drug Act was then relatively new.) The pancreas and liver also have laboratories. The digestive system manufactures "acids" in one factory and "alkaline fluids" in another. The muscles are engines; the heart is a "wonderful pumping station," so sophisticated that, even though it "has been studied by thousands of scientific men for a long time, . . . nobody knows how it manages to make itself go." "White blood corpuscles" serve as policemen who "arrest invaders" and sacrifice "their lives for the good of the community."[12] In whole and part, then, the body is a moral, organizational, and technological marvel.

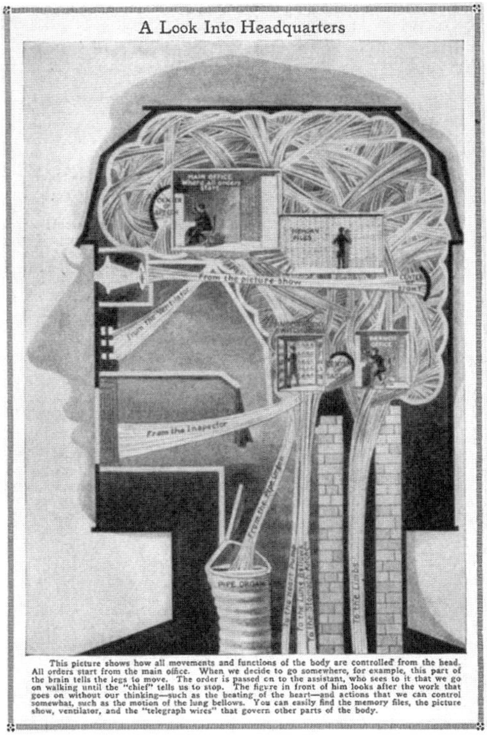

FIGURE 2.6. *"A Look into Headquarters,"* Pictured Knowledge *(1917), 266. This source version has a staff of four in the headquarters, one less than* Wunder in Uns *(1923), pl. 13. (See Plate 2.) National Library of Medicine.*

It gets pretty elaborate. The text of the essay has a metaphorical exuberance that far exceeds anything visualized in the pictures, as striking as they are. Like Kahn's "Oxygen Cycle," it addresses readers in the second person in present tense to encourage the reader's identification with "this marvelous community": "Did you know you are 'it'? . . . This marvelous community is your own body."[13] The argument is a strange brew of modernization oratory and Sunday school didacticism:

> When you think of the skill it takes to originate a camera, or devise a telephone; to manufacture acids for a particular chemical; or the ability necessary to govern a city, you will appreciate the marvel of the human body. . . . Here groups of cells specialize and become expert each in its own particular kind of work and are able to their work better than man himself has ever been able to do it. Surely the author and creator of this marvelous community [the human body] [has] all knowledge, all wisdom and all power.[14]

God is the master bureaucrat and technologist. The human body is a technoscientific revelation: a scripture of modernity.

VISUAL EDUCATION AND *PSYCHOTECHNIK*

A brief note at the end of "The Bodies We Live In" marks the novelty of the pictorial approach: "The illustrations form so important a part of this unique treatment of a difficult school subject that it is deemed proper to call attention to the fact."[15] Body metaphors and similes—the body as a house, kingdom, economy, machine—had for centuries appeared in texts dealing with medicine and health, but those metaphors were rendered in words, not images. In the print tradition of medical illustration, inaugurated in 1543 with publication of Andreas Vesalius's *De humani corporis fabrica,* illustrations display the anatomical structures of the body, showing how things look (albeit idealized or cleaned up for the sake of legibility). The images might be theatrical and allusive or might make reference to human mortality or a scene from Christian martyrology or Greek mythology, but they were not primarily metaphorical or narrative. According to art historian Martin Kemp, "illustrations reigned supreme for description, while the text remained best adapted to explaining how something worked."[16] That same division of labor, which prevailed in the time of Leonardo da Vinci, still prevailed in the nineteenth and early twentieth centuries. In contrast, the pictures in "The Bodies We Live In" used metaphor and sequential illustration to explain how things work.

The other entries in *Pictured Knowledge* also had lively monthly magazine-style layouts with photographs and illustrations (both commissioned and borrowed from other sources). Some of these also featured halftone cross-sectional drawings that show interior processes and explain how they function, including cutaways of a coal mine, a cotton mill, and an automobile engine. However, there was no metaphor in them. They dealt with the external world, not the body of the reader. "The Bodies We Live In" thus had a unique status within *Pictured Knowledge.* It was an exemplary article in an exemplary encyclopedia that aspired to serve as a model of the new approach to pedagogy: "visual education."

Although that term (along with "visual instruction") only came into widespread use in the early 1900s, a movement dedicated to the transformation of classroom instruction through the showing of objects and pictures had already been growing for decades in both Germany and the United States.[17] Its historians mostly focus on the new pedagogical enthusiasm for movies (and to a lesser degree stereograph cards and projected "lantern" slides), but there was also great enthusiasm for printed illustrations and illustrated texts. The invention and continuous improvement of halftone printing, photogravure, and photolithography, along with the invention of new inks and dyes, as well as new methods of page preparation, enabled publishers to cheaply print and manipulate photographic images, line art, and type, and to interweave them in new ways.[18] Enterprising authors, editors, and publishers used these new techniques to gain new readers for their publications, a mass audience. As the trend intensified in the twentieth century, the look of print changed radically and continuously.

New techniques of picture production, page layout, and product labeling emerged in tandem with ever more aggressive marketing tactics. These in turn fostered further innovations in image and print production. There was, historian Neil Harris argues, "an iconographical revolution of the first order": "The single generation . . . living between 1885 and 1910"—in America and Europe—"went through an experience of visual reorientation."[19] The visual world was being remade, and to many observers, that change signified the change of the century and the coming of a new age: modernity. New media, as well as new media genres and locations, proliferated. There were many leading edges: the educational and commercial exhibition, the store window display, the photographic print, the slide lecture, the billboard, the classroom chart, the wall poster, the movies, the electric light display, the illustrated mass-circulation periodical—along with a proliferation of labeled, branded, packaged, mass-produced consumer goods and mass spectacles. There emerged a new visual vocabulary of captioned photographs, illustrations, cartoons, models, fancy typefaces, logos, headlines, diagrams, decorative lines and borders, and graphs, a churning whirl of juxtaposed words, images, and abstract forms.[20] And there were new groups of workers who were recruited to invent and orchestrate the effects: art directors, advertising men, illustrators, and layout artists—the new professions and trades of early twentieth-century media culture.[21]

In that moment, the emergent advertising profession allied itself with academic psychologists who were inventing the field of applied psychology (in German, *angewandte Psychologie*, *praktische Psychologie*, or *Psychotechnik*).[22] These psychotechnicians advocated theories and methods that, they claimed, could shape the behavior of individuals and groups in mass society: workers, consumers, students, voters, jurors, housewives, and criminals. Steeped in the rhetoric of modernization, they wrote about "scientific" and "efficient" ways to achieve the instrumental control of human subjects through the manipulation of "instincts and "drives" by "suggestive" visual devices, displays, and textual appeals.[23] They offered their services to advertising agencies and manufacturers who wanted to brand themselves as modern and who wanted to use modern methods to lure consumers into purchasing products that in turn helped them to be modern.[24]

At the same time, a cadre of ambitious professional educators was emerging in tandem with newly forming and reforming state and municipal education bureaus, teaching colleges, and university education departments. There was a simultaneous expansion and reform of primary, secondary, and college-level education.[25] In their classrooms, professional journals, associations, conferences, and mass media articles, the new educators defined themselves in opposition to the methods of the past. With one eye on what mass media and advertisers were doing, they argued for a pedagogy based on commercial visual practices and recent psychological theories, experimentation with different methods of teaching and learning materials in the classroom and the laboratory, and close observation of students and teachers. A good part of the new agenda called for teachers to take students out of the classroom to museums or on field trips, where nontextual materials could be experienced firsthand. It also called for visual materials—printed pictures and blackboard drawings, but also motion pictures, lantern slides, specimens, models, and plaster casts—to be brought into the classroom, to make the classroom into a visual learning environment.[26]

Iconophilia ran rampant. Addressing a 1911 convention of Sunday-school teachers in Chicago, Bertha Lainé, a Canadian teacher, recited the new catechism: "'An ounce of pictures is worth a ton of talk.' Eye-teaching through pictures, objects and blackboard teaching is indispensable."[27] Aided and abetted by business enterprises—the term "visual education" first appeared in a 1906 catalog of a stereographic card publisher—reformers at every level sought to use the new techniques of image production and display pioneered in commercial publishing to remake classroom and textbook pedagogy.[28] For models, they looked to heavily illustrated mass-circulation magazines and Sunday newspapers—both articles and advertisements—and to children's magazines like *St. Nicholas.* They also looked to mass-circulation educational magazines like *Popular Mechanics* and *Scientific American.* Perhaps the most impressive model of all was *Popular Science Monthly,* which gained a readership of hundreds of thousands almost overnight when Waldemar Kaempffert became its editor in 1916. Under the previous editor, *Popular Science* was a well-respected, lightly illustrated small-circulation magazine. Under Kaempffert, it became aggressively visual. Hyperactive clusters of stand-on-your-head action-oriented halftone photos and drawings filled nearly every page.

In the 1910s and 1920s, encyclopedia publishers (many of them located in Chicago) tried to capitalize on the trend with a new generation of illustrated encyclopedias: *The How and Why Library, The Human Interest Library, The Book of Knowledge, The Wonder Book of Knowledge, The World Book Encyclopedia, The Volume Encyclopedia,* and *Compton's Pictured Encyclopedia.* The most articulate of these was *Pictured Knowledge* (1916 and many editions thereafter), subtitled *The New Method of Visual Instruction Applied to All School Subjects,* a multivolume encyclopedia "quite as much a work for adults as for children."[29]

Conceived as a showpiece for the visual education movement, *Pictured Knowledge* featured articles in a reader-friendly magazine format, arranged topically rather than alphabetically (as in the style of the authoritatively imposing *Encyclopædia Britannica*), each equipped with magazine-like headlines, page layouts, and captioned illustrations.[30] Its editors were Calvin N. Kendall (1858–1921), then New Jersey's state commissioner of education, and Eleanor Stackhouse Atkinson (1863–1942), a schoolteacher turned writer of newspaper articles, romances, educational books, and fiction for children, as well as author of *The How and Why Library* (1913), a one-volume illustrated children's encyclopedia. Its "director of visual instruction" was Alfred Willis Abrams (1866–1938), chief of visual instruction of the New York State Department of Education, which was then celebrated for the massive collection of prints and projectable lantern slides it distributed to

FIGURE 2.7. *A two-color whirl of pictorial and typographic effects.* New York World Sunday Magazine, *10 May 1911, 10–11. American Newspaper Repository, Duke University.*

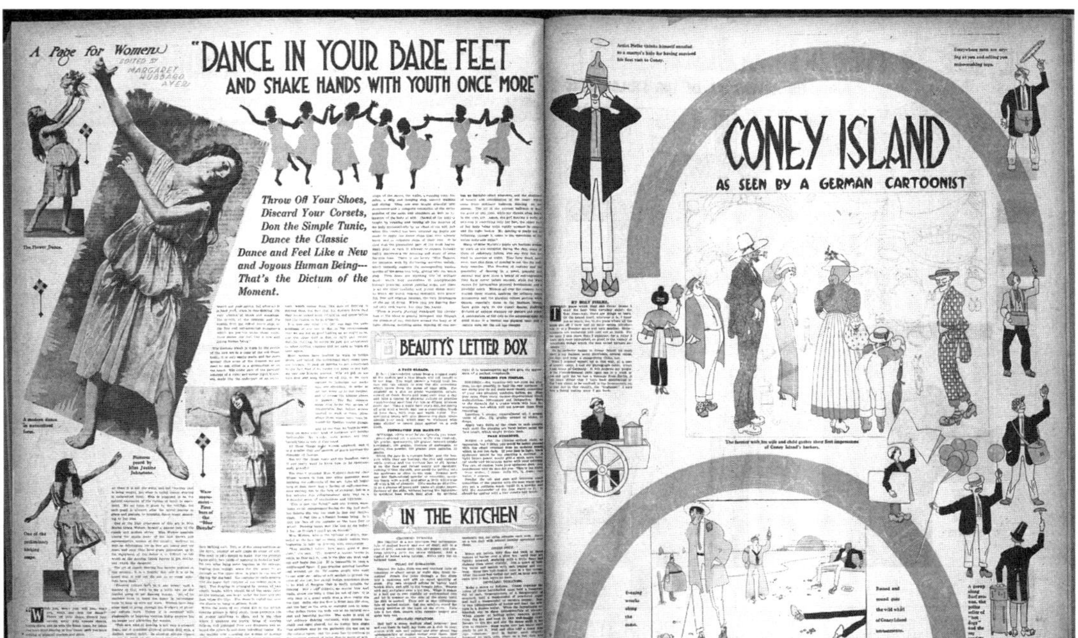

public schools.[31] Art editor Seymour Jones, who did some of the illustrations, was a veteran of Chicago commercial publishing. G. Stanley Hall, then the most prominent American psychologist, bestowed his blessing in a short preamble. Contributors included star Progressive reformers, officials, and academicians such as Luther Gulick, John Burroughs, David Starr Jordan, Josephine Baker, Charles Bolduan, James J. West, Gifford Pinchot, A. A. Hopkins, Leonard Wood, and Anna Comstock—names familiar to readers of mass-circulation newspapers and magazines as well as to professional journals.

The articles—on "industry," geography, history, literature, and nature—were chatty and richly illustrated with photographs, line drawings, and black-and-white halftone reproductions of paintings. These were obtained from photo bureaus, museums, newspapers, and magazines as well as the promotion departments of industrial companies and associations, along with art made by commercial illustrators commissioned especially for the project. The idea was to give the encyclopedia the look and feel of illustrated newspapers, monthly magazines, children's books, exhibitions, and movies (which other encyclopedias such as *The Book of Knowledge* were also trying to do), but also to theorize and propagate visuality. G. Stanley Hall's preamble stressed the power of modern image-making practices to mediate between reader and the natural and human world, to transmit knowledge and transform the reader:

> The limit and range of what pictures can do is steadily increasing. One can know many of the wonders of the vast new world of the microscope and telescope without ever looking through an eye glass. It would be a curious question . . . to work out how far an education based upon a wise selection and proper gradation of pictures might today be carried out without the ability to read. . . . Seeing is not only believing, but understanding, and a single judicious picture or blackboard drawing often tells in a moment what it would take paragraphs to describe, if . . . words could ever give it at all.[32]

Like other twentieth-century multivolume encyclopedias, *Pictured Knowledge* was sold mainly door to door. It was marketed in a variety of formats: as a complete set or in installments, in cheap and deluxe editions, purchased up front or on a layaway plan. We can imagine its traveling salesmen working hard to get a foot in the door and then quickly making his pitch for wary customers.[33] The selling points were set out in a short introductory précis, "What Is 'Pictured Knowledge'?" (unsigned but probably written by editor Calvin Kendall). Addressed to school book selection committees, librarians, teachers, and principals, as well as to self-improving parents of schoolchildren, it takes the form of a visual instructionist manifesto. Most schoolbooks, it argues, do not hold the attention of students. Children will only be attracted to materials that have "color and motion"; they resist "dry, dreary, dead" texts that are mere "catalogs of facts." The "old idea" was to "fill the child's mind full of an enormous mass of unrelated facts," a kind of coercive but ineffectual "mental discipline." In contrast, the "new idea" offers "essential, useful things logically related, . . . so presented as to be attractive to the child, to make him think and keep his curiosity alive."[34] Directly or indirectly, the introduction links visuality to modernity through repeated references to the modern age, newness, new visual technologies, "experts," and that hobbyhorse of early twentieth-century modernization discourse, efficiency. Visuality is efficient. On this point, the introduction leans on the authority of venerable American academic psychologist G. Stanley Hall (1844–1924): "Whenever . . . we can substitute a picture for mere words, the concrete for the abstract, we are doing a real work of mental economy of great value in this age of strain and fag of brain and nerve."[35]

Hall had long been enamored of "the graphic method." After a trip to Europe in the late 1870s, he marveled how in Germany "the lecture room" had been converted "into a sort of theatre where the graphic charts are the scenery, changed daily with the theme."[36] Back then Hall was only calling for the application of the "graphic method" to secondary and university education. Decades later, in *Pictured Knowledge,* he called for an overhaul of the entire educational system, from primary school on up. Visual education, he argued, would help students learn faster with less effort, and to understand and retain more. This was urgent because in the expanded curriculum of modern subjects, there was more to learn: more astronomy, chemistry, physics, geology, geography, history, literature, mathematics, physiology, hygiene, industrial technology. This expansion of knowledge converged with the encyclopedist ideal: the modern well-educated person should be encyclopediated, equipped with knowledge of nearly everything.

But how to get there? Modern children, like Kracauer's salaried masses, are easily distracted. "Not only educators but businessmen everywhere have felt that something should be done to capture the child's interest." The "use of pictures as substitutes for realities" appeals to "the vivid imagination of a child," for whom "it constitutes an experience"; images are a proxy of three-dimensional reality, a technology of pedagogy.[37] Here Hall echoed Johann Pestalozzi, G. F. Herbart, and Friedrich Junge, nineteenth-century pedagogical reformers who criticized text-based learning and who called for a pedagogy based on field trips and "object lessons" brought from the field into the classroom.[38]

But "pictured knowledge" is not only a matter of profuse illustration. The text surrounding the images must use "word pictures,"[39] a style of writing that vividly conjures up a scene. The illustrations must use textual labels and be set amid blocks of text. Even if *Pictured Knowledge* is a call for a visual pedagogy, it never relies solely on pictures. Its pictures are captioned below the frame (sometimes quite extensively), and words are incorporated inside the frame of the illustration, which is placed within blocks of text. An effective pedagogy must feature both words in pictures and pictures in words, image-laden texts and text-laden images that refer back to each other, a contrapuntal mirror *ekphrasis*.[40]

The turn to an image-based pedagogy has an evolutionary premise: ontogeny recapitulates phylogeny. "The child is all eyes. . . . He wishes to look into everything which moves and stirs. He loves to see color and note form. Gradually he begins to know and to classify objects, as his perceptions are trained."[41] In child development, as in the advance from monkey to human and from savagery to civilization, the visual field develops before language and is the basis on which language and higher ideational thought develops.[42] "Pictured knowledge" should follow this evolutionary/developmental trajectory: It "embraces [the] modern idea of starting from the child's level and building up."[43] In so doing, reformist educational methods, advertising, the entertainment industry, and government propaganda all have a shared assumption: mass humanity is morally, intellectually, and emotionally a child. The same techniques that advertisers, publishers, and filmmakers use to charm mass audiences can be used to hold the attention of children (and vice versa): "Pictures represent things much better than words alone and are a more accurate and vivid means of expressing ideas. It is for these reasons that pictures have come to be used so largely used not only in schools but in all kinds of publications and in commercial enterprises." "So widely has this truth come to be recognized that education has been defined as 'image growth.'"[44]

University of Chicago education professor Wilbur S. Jackman's phrase "image growth" was then much in vogue in pedagogical theory. According to Jackman (1855–1907), from "making, modeling, painting, drawing, written language . . . , the image emerges from vagueness towards clearness, and finally becomes a definite product in consciousness."[45] The mind is both the terminus and starting point of print reproduction: printed images end up as images imprinted on the brains of readers, where they can be collected until recalled by the memory and put back into circulation as speech, writing, or illustration. Knowledge of a particular subject is achieved by the accumulation of multiple images from visual sources, tactile experiences, and visualizing language, which gives thought precision, density, and a kind of perspectival depth and volume. Going one step further, *Pictured Knowledge* argues that "Pictures" can be used "as substitutes for realities for those who can not visit the scene." Pictorial illustration (in combination with vivid textual description, word pictures) stands in for material reality; it is a virtual field trip. In the "vivid imagination of a child, it constitutes an experience rather than a description."[46] Visual education is experiential learning.

HOW TO READ PICTURES: THEORIZING AND DISCIPLINING THE IMAGE

After this introduction comes a jumble of illustrated articles on the history of the United States, insects, Jane Addams and Hull House, the Seven Wonders of the World, wind and weather, famous inventors, modern industries, famous masterpieces like the Winged Victory of Samothrace, and so on. Allusions to the visual instruction method recur throughout. The geography section is titled "The Earth as a Picture Book." The section on the "book publishing industry" is titled "How Pictured Knowledge Was Taught to Talk" (1863). A later 1937 edition adds a subsection on "The Art of Reading a Picture."[47] The title page of each section comes equipped with a decorative border of hand-drawn line art that visually synopsizes the contents and reinforces the theme of the article and the overall theme of graphic presentation.

There is a lot on manufacturing and technology: articles on how rubber is made, the building of the Panama Canal, the telephone, coal mining, steel—and another technology of modernity: "visual instruction." "The Use of Pictures in Teaching" (which is addressed to both children and teachers) is simultaneously a how-to article and an inventory of arguments for visuality.[48] Under the heading "Visual Instruction," it presents a series of physiological, psychological, pedagogical, and historical briefs for the centrality of the visual and then gives lessons on how to study things visually.

The section begins by positing that "the mind comes to know the world chiefly through the sense of sight. It is estimated that eighty per cent of our primary notions about the things of the world come through the eye": "Not only do we acquire so many ideas through seeing, but ideas thus gained seem much more real to us . . . and also more exact and vivid."[49] Pictures, it claims, give a "more clear and definite" impression of material phenomena: "Until you have a clear mental picture of a thing you have not really observed it."[50] Pictures "are the most direct aid to understanding": "Words, whether spoken, written or printed, are only symbols of ideas. They have no meaning until real mental pictures are associated with them. . . . It is for these reasons that pictures have come to be so largely used not only in schools, but in all kinds of publications and in commercial enterprises."[51] Innate human receptivity to images accounts for "the development of photography and the invention of various processes of printing pictures" in the modern era. At the same time, it also accounts for the development of drawing and painting in antiquity by "the early Egyptians," the use of pictographic sign language by the American Indians, and the making of "the first illustrated text-book," Comenius's *Orbis Pictus* (1657).[52]

But even with the human predisposition to receive communicative images, a naive or casual glance is inadequate. The student must be trained to look productively:

> The eye is merely an optical instrument that brings light from an object to the brain. The eye presents to the mind much that is very interesting, but the mind must give close attention to the various features of the object and must react upon them. The aim should be to form a correct mental picture corresponding to the thing or the picture presented to the eye. . . . It is necessary to hold the attention to the object for some time and to put forth some effort to note the several parts shown, their size, position, color and arrangement.[53]

Students must therefore learn to methodically observe a picture, to make "intelligent observations," much like a scientist observes nature. Students can analyze a picture by breaking it down into its constituent parts and then "think about what they see": "You cannot at first see a picture as a whole. You must begin with some part of it. When you understand this part, you may pass on to another." Seeing has to be systematic: "Learn to observe a picture in an orderly way and to bring your observations together into some significant groups. In this way you can acquire the ability to think."[54]

Pictures, then, cannot stand on their own. One must study them in conjunction with a textual apparatus of captions and surrounding sentences:

> Pictures are attractive. How interesting they are! There is a temptation to turn over the pages to find others. But one should not be content merely to look at pictures one after another in a superficial manner. First you should examine them closely to note just what they show, and when you have interpreted them . . . with the knowledge and training you have, then you should read the text for suggestions and explanations. You will, perhaps, be surprised . . . to find how much more there is in the picture than at first you thought.[55]

To reinforce this lesson, the section then presents a series of "Exercises in Observing Pictures," and asks the reader to analyze some halftone photographs of scenes and paintings:

> Test yourself by the accompanying pictures to learn how . . . to observe and interpret them. In each case examine the picture carefully before you read what is said about it. . . . Note the several parts and the form, position and size of each. Observe how the parts are related to each other. See if you can tell what useful purpose is served by the different things shown.

> Try to explain the meaning of everything you see. Then read the text immediately following the picture. When you have done this, reading very thoughtfully, examine the picture again.[56]

As the exercises continue, a certain slippage becomes evident. Some pictures are not "attractive" or "interesting"; they are as dull as the old text-based pedagogy that the visual instructionists condemn. "Next is another picture"—a pedestrian halftone photograph of "Clay and Sand Deposits, at Roseton, N.Y."—"that is very instructive if you read it rightly." Such pictures, the author argues, are still full of meaning and are more efficient than words—a kind of information compression. However, they are evidentiary, not interpretive or performative or powerful in any affective way. They don't work to recruit the student's gaze. The student must be told to study them and taught to read and analyze them intensively to mine their content. Figure 2.8 shows what appears to be some kind of quarry with three horses pulling wagons lined up near some hard-to-distinguish square structures. The text admonishes the reader, "Look first to note the different parts of the picture. Remember that objects, form, structure and relative position are indicated in a picture by light and shade and by lines of different character and direction. . . . Now try to interpret the meaning of this picture."[57] The student is instructed to use the same formalist approach to the photographic object that would govern art history and literary criticism for most of the twentieth century.

The paragraphs that follow discuss the natural history of the deposits, including how they were formed by glaciers and are "now the basis of a great brick making industry"—information that would be almost impossible to glean from the grainy halftone. The author comments, "What a lot of information is thus to be gotten from a very simple picture that does not at first attract much attention! Those who have not learned how to observe would probably pass such a picture by with a mere glance of casual interest."[58]

But enhancement of a student's powers of observation is not the only benefit of the training. In acquiring the habit of close and systematic reading of pictures, the author argues, students learn to think critically:

> Pictures can be used effectively to challenge thought. . . . Teachers often complain that their pupils do not think. How are they to learn to think, when the school exercises consist very largely of giving out information through words? Very often these words produce no clear mental picture. . . . The ability to reason is acquired, not so much by repeating reasons given by others, as by practice in reflecting on what is seen. Avoid, therefore, looking at pictures in an aimless way. Every good picture presents a problem in thinking.[59]

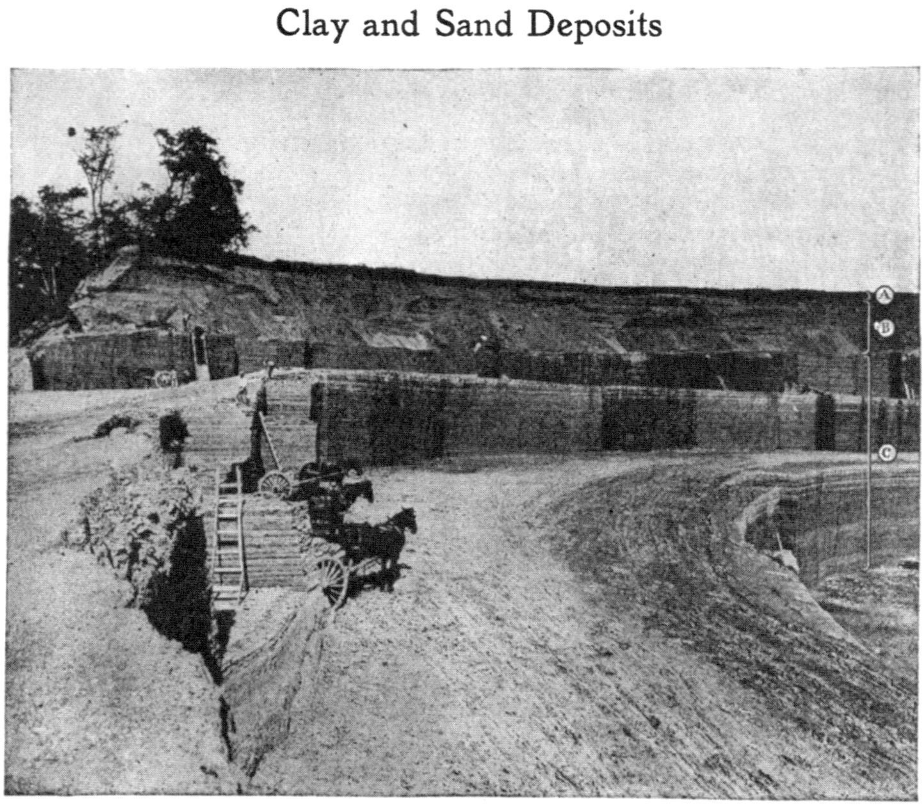

FIGURE 2.8. *"Clay and Sand Deposits,"* Pictured Knowledge *(1917), 496. National Library of Medicine.*

The passage betrays some anxiety: the worry that the viewer will merely browse the volume, merely window-shopping among the images in the fashion of the typical magazine reader—or worse, that the reader may actively resist the effort to impart knowledge and instill moral discipline. It is the reader's habit to take pleasure in pleasurable images and to be entertained; this opens the door for manipulation from teachers, governments, and advertisers. But not every image is pleasurable, and even pleasurable images can sometimes fail to capture the viewer.

In all of this, the relationship between image and word is unstable. *Pictured Knowledge*'s emphasis on the image is a protest against logocentrism in academic instruction.[60] The encyclopedia's starting premise is precisely that "aimless looking," images, mute and unnarrated, precede intensive looking, language, and complex thinking—and that essential characteristic of the image is useful. The image recruits the eyes of the viewer. Publishers, advertisers, war propagandists, and filmmakers were daily demonstrating their ability to use this essential characteristic of visuality. Educators, following their example, should also use images to manipulate the malleable pupil: "Pictures . . . have a molding influence on children both by the information they convey and by the feelings they arouse."[61] New technologies that yoke the viewer to the visual (and then the visual to the textual) can thus improve the uneducated, undomesticated child as well as its phylogenetic equivalent: the masses. "In the beginning was the Word" is replaced by "In the beginning was the Image."

3

OCULARCENTRIC!

CONCEPTUAL ILLUSTRATION AT WORK IN THE "GREAT LOOP"

> The framing of life, the need that a picture . . . remain in its frame was over. . . . Now pictures commenced to want to leave their frames.
>
> —GERTRUDE STEIN, *PICASSO*, 1938

> THE IMAGE IS NOT an idea. It is a radiant node or cluster . . . , a VORTEX, from which, and through which, and into which, ideas are constantly rushing.
>
> —EZRA POUND, "VORTICISM," *BLAST*, JUNE 1914

THE NEW PICTORIALISM IN GERMANY

Pictured Knowledge did double duty as both an image-laden textbook and a modernizing manifesto for image work. But it didn't revolutionize pedagogy, and it didn't entirely deliver the goods. Although its section on building character through imagery argues that pictures can contribute to moral, intellectual, and cultural improvement—pictures can "challenge thought" and "make you think"—the end result disappoints.[1] Readers are left with the standard curriculum: a pictorial celebration of progress combined with an uncritical affirmation of the status quo. The new-picture pedagogy doesn't challenge the verities of the pulpit, parlor, legislative chamber, or classroom. Nowhere does the visual instructionist program come close to effecting (or even imagining) any kind of "scopic regime" change—a transformation of the dominant practices of image creation, deployment, and reception that fundamentally reshapes the subjectivity of the viewer.[2] Which is to say that even though the visual instructionists aimed to reengineer the moral character of their student and working-class readers, their founding assumption was that the viewer, the viewed, and viewing are all stable entities.

Of all the sections of *Pictured Knowledge,* only "The Bodies We Live In" exploits some of the affective and conceptual possibilities of image practice as a visual vocabulary of the self in modernity. But even "The Bodies We Live In" falls short. It connects the human body to a utopian (though not particularly radical) social vision: the state-building, rationalist, moral reform agenda of American progressivism. Even there the social vision mainly takes place in textual metaphors—word pictures that far exceed the metaphors presented in the illustrations.

The opposite came to prevail in the works of Fritz Kahn. Kahn's images exceed the text. Affectively, methodologically, scientifically, quantitatively, and aesthetically, they went way beyond anything published in America. Inspired by advertising, avant-garde modernism, and mass entertainment, his visual antics, even reduced to the scale of book and magazine pages, dazzled and troubled his large readership.

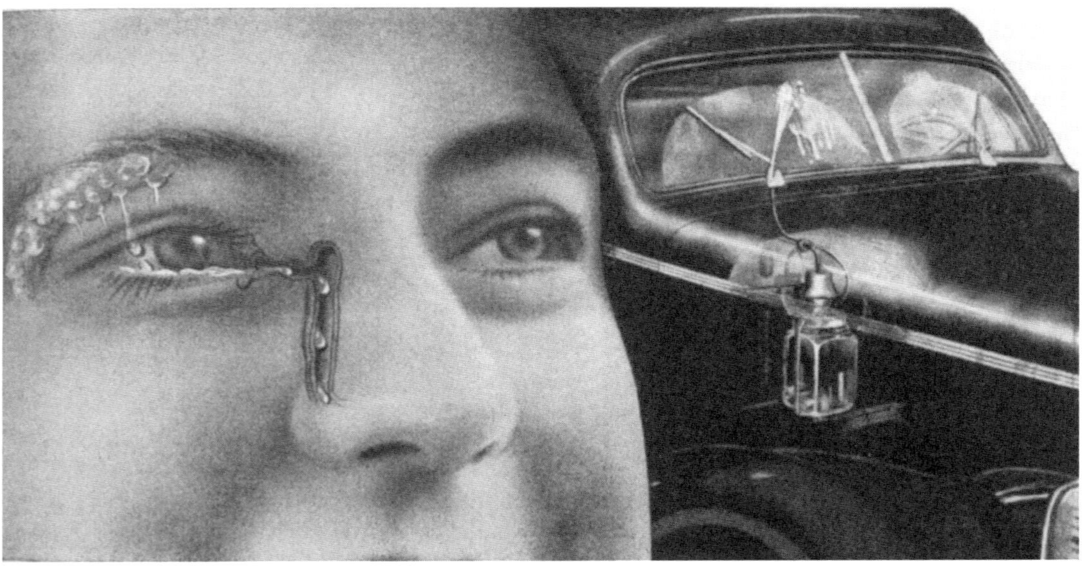

FIGURE 3.1. *"The Lid and Tear Apparatus Is Analogous to a Windshield Wiper with an Automatic Arrangement for Spraying Water on the Windshield,"* Der Mensch Gesund und Krank *(1939), 2:313. Artist uncredited. National Library of Medicine. This photomontage, rendered in the seamless style of John Heartfield, compares the panes of the divided car windshield to eyes and the wiper fluid to tears. The metallic surface of the automobile and the smooth skin of the face glow in parallel with the shiny luster of industrial modernism. The eyes (and windshield) gaze into the distance, as if a bright, modern future beckons—a facial posture typically seen in Soviet and National Socialist iconography.*

That odd suspension of mixed emotions had an instrumental purpose, and something else. Kahn invited the public to see the modern world through modern eyes, with double or triple vision: to be scientized, maybe burdened, but also seduced. His illustrations asked readers to take a joyride in the industrial body the way you would take a ride in a shiny new car. (Indeed, cars show up in many of the illustrations.)

Mostly Kahn's images refer back to the human body, the body of the reader. This fostered the same empathic identification that advertisers, mass-market publishers, and the motion picture industry mobilized: a habitual reflexivity. His pictures were mirrors in a hall of mirrors, a view of views. Many of the best ones refer back to the physiology and subjective experience of seeing in modernity—readers seeing themselves seeing.

Kahn, of course, was a commercial author who wrote with an eye on the market. He was not a professional pedagogue, and up until 1933, he lived in Germany, not America. His books and articles were tailored to the exigencies of interwar German politics and Weimar culture, and to the opportunities they presented. It was a milieu stocked with a bountiful storeroom of visual strategies to draw on, strategies that were perfectly pitched to an ongoing atmosphere of crisis and transformation. In that fertile environment, Kahn and his cadre of illustrators were inspired to invent many new kinds of images that invoked modernity in content, style, and function—and they implicitly critiqued the style of illustration offered up in contemporary German popular science and health.

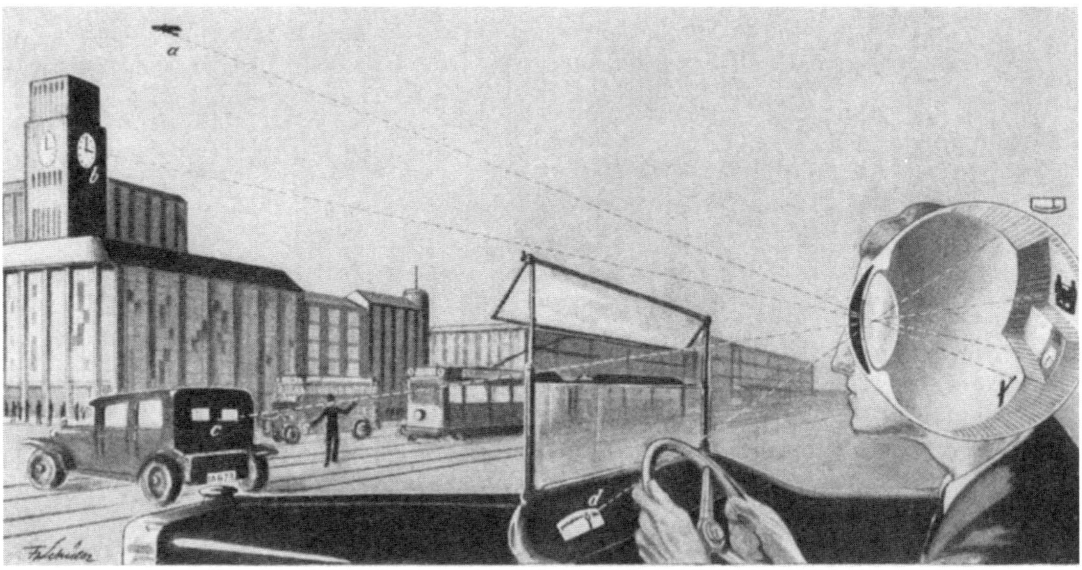

FIGURE 3.2. *"Adaptation Mechanisms in the Animal Kingdom III,"* Das Leben des Menschen *(1931), 5:14. Artist: Fritz Schüler. National Library of Medicine. A man driving a Model T serves to demonstrate the optics of distance vision in a Weimar landscape of modern buildings, automobiles, and electric streetcars. Copyright Kosmos Verlag, Stuttgart.*

Kahn and his artists were also favored in other ways. In America, commercial illustrators (such as those who worked on *Pictured Knowledge*) labored under the heavy hand of an art director, editor, or client.[3] Given an assignment, the artist was expected to follow detailed instructions. It was a given that images had to have a modern look, even if they showed historical scenes, but (apart from high-end magazine art) most commercial illustrations had no direct relation to any articulated modernist aesthetic or movement. Journeymen illustrators (many of them women) were given little freedom to play or experiment.[4] In everyday newspaper, magazine, and advertising illustration, and especially in popular scientific illustration, the artist was expected to produce a commonsense rendering that ranged from artful to competent to crude but that rarely strayed from the argument or narrative it was commissioned to support.

Such illustrations assumed that viewer and artist shared the same everyday visual experience of modernity, an approach that historian T. Jackson Lears provocatively terms "stupid realism."[5] Sentimental, dramatic, coy, silly, or woodenly didactic, the realistic style of commercial illustration results in pictures that disrupt nothing. The viewer (and artist) need exercise only the most reflexive interpretive procedures. The conventional wisdom in American advertising and publishing was that modern art was pretentious, undemocratic, illegible, and ineffective—too fancy for popular taste. The average reader, it was thought, was incapable of textual and graphic interpretation, only responsive to some form of direct address. Advertisers, editors, and publishers judged the merits of art on how much it seemed to stimulate sales. Illustrators were regarded as instruments (variably compliant, variably skilled) through which an assignment or campaign was to be implemented.

In Germany, artists and designers had a different status. They tended to be more aesthetically sophisticated and exercised more creative control. There was a proud tradition of independence, an esprit de corps fostered by high-end commercial art and design magazines such as *Gebrauchsgraphik* (1924–44) and professional societies such as the Bund deutscher Gebrauchsgraphiker (League of German Commercial Artists) and Verein der Plakatfreunde (Society for Friends of the Poster). Such organizations bound artists together with a collective ethos of artistic excellence and feelings of cultural importance. There was a sense that the artist played a vital role in advancing the universalist modernizing project that went from the ancient Greeks to the Renaissance to the Enlightenment, right up to the present day. German commercial art valorized the techniques and principles of both academic fine art *and* modernist critiques of the academy, such as the Bauhaus program. The artist could serve as a high priest of the cult of progress, tradition, or both.[6]

To some critics, this was not a good thing. Hanns Kropff, a prominent advertising consultant, theorist, and university lecturer, complained that German artists "seldom concern themselves over whether their work is good from an advertising standpoint":

> They draw or paint a design and then fill the precious space with it. And the odd thing is that the commissioning firm accepts and publicizes the design simply because the artist says it is good and because it was made by *him*. Advertising is produced not for the person paying for it, but rather for the reputation of the artist.[7]

Kropff called for greater managerial control over art and design, similar to what was practiced in America, and condemned German commercial artists for taking on airs. He was not completely wrong. The artist's artwork was often a self-aggrandizing "advertisement for myself."

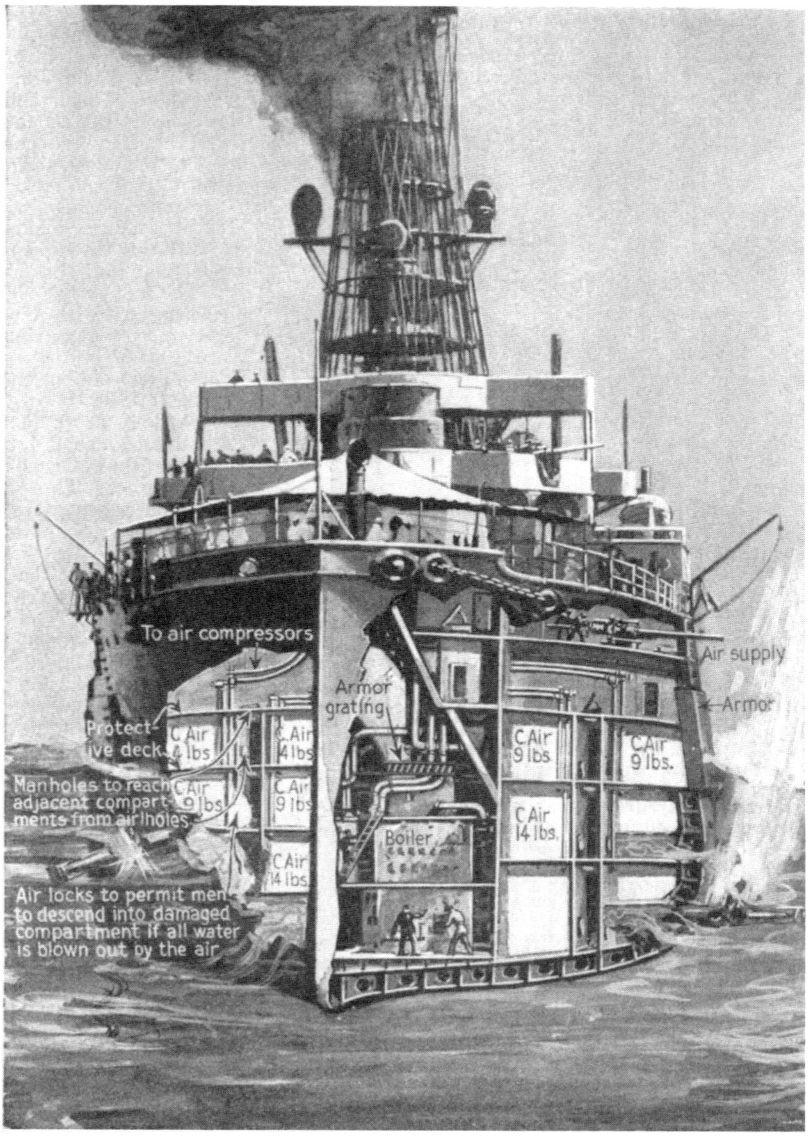

FIGURE 3.3. *"The Air-Lock Principle Applied to the Battleship,"* Popular Science Monthly *91, no. 6 (December 1917): 887. Artist uncredited. A cutaway diagram reveals an interior divided into boxes with equipment, labels, caption, and little men doing work—the same arrangement that appears in* Pictured Knowledge, Wunder in Uns, *and "Der Mensch als Industriepalast." The aesthetic: "stupid realism"?*

Many artists cultivated a distinctively individual style, which served as a signature (regardless of whether they actually signed their work). But that was arguably a good thing: it created a rich visual vocabulary for advertisers to make use of; it broke down the distinction between avant-garde aesthetics and commercial art for the greater benefit of both domains.[8] While artists and designers used their commissions to create a brand name for themselves through displays of ingenuity and craftsmanship, they also sought to delight themselves and their colleagues—to be artful *and* create images that strongly attracted and pleasured viewers. In many ways, that way of doing things was perfectly adapted to the media environment: in a world crowded with images (and image makers), the first job of an image is to get the viewer's attention. German artists, like their employers, were well aware from experience that their images were seen in a saturated visual environment where no single image or phrase could register anything more than a momentary impression. To engineer subjectivity, the publisher, author, advertiser, filmmaker, or artist had to deploy images, themes, and tropes in barrages of disruptive novelty using jesting wit, distinctive style, and arresting beauty.

Rudi Feld, celebrated poster artist as well as art director and set designer for UFA (the largest German film studio), understood this imperative in physiological terms as an innate characteristic of human subjectivity (really the premise of Kahn's entire oeuvre). Agency regresses from person to body part: "All eyeballs are sensitive to new entities; the cerebral cortexes give the order to stop" and pay attention. "Because of the unusual, curiosity sets in."[9]

For H. K. Frenzel, editor of the elite graphic design journal *Gebrauchsgraphik,* the condition of modernity—the accelerating, shocking proliferation of events and experiences—was the most important thing: "Sensational times require unusual methods" (Plate 7). In a 1931 article on the advertising firm of Werbekraft and the design team of Plucer and Elfer, Frenzel argued that "all the interest of the reading public is centered around . . . sensational news, consequently an advertisement of ordinary standards stands little chance of attracting attention. The Werbekraft GmbH have carried out experiments with sensational advertising to combat sensational news."[10] They were not the only ones. Influenced by avant-gardist innovations (and sometimes joined by avant-gardists who were not above working in mass-circulation and high-end media), commercial artists and designers developed idiosyncratic approaches. They fashioned bizarre, almost hallucinatory images set in compositions that flamboyantly occupied the entire page or picture frame, with little or no explanatory text to distract the eye. Their illustrations, designs, and headlines directly hailed the viewer, a mode of visual address that sought to outshout the competition—a "yelling" style.[11] American advertising executive J. Murray Allison described the European approach as "a blow between the eyes, as hard as possible to deliver it, through the medium of printer's ink."[12] Such images made use of bold and simple color, provocative and theatrical poses, amusing juxtapositions of image and image, and image and text. Often they featured an attractive young woman, smiling, dancing, talking, or in some other way displaying herself, as in Kahn's "Sunbath!" (Figure 3.4) and a related promotional advertisement (Plate 8), both likely produced by Plucer and Elfer.[13]

Feld and Frenzel called for a confrontational approach, but images could also charm, even hypnotize, the viewer. Freed from the obligation to literally illustrate a textual narrative or make an advertising

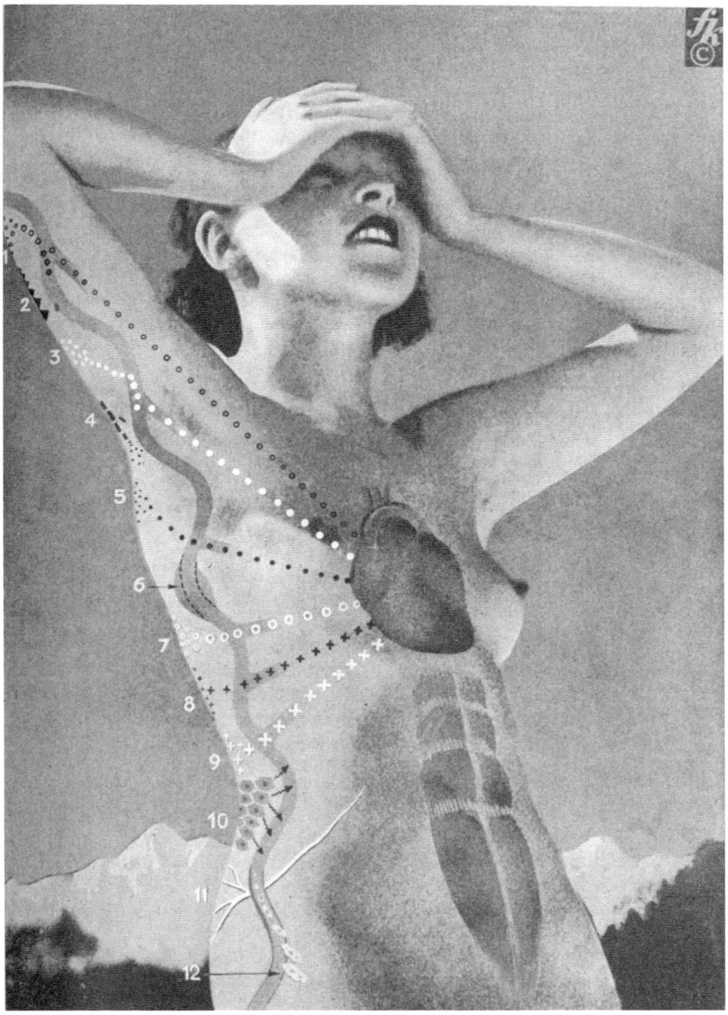

FIGURE 3.4. *"Sunbath! The Twelve Now Well-Known Effects of Sunlight on the Skin,"* Der Mensch Gesund und Krank *(1939), 2:253. Artist uncredited, probably Plucer and [Arpad] Elfer. National Library of Medicine. An alluring nude woman basks in sunlight against an alpine landscape. As the rays shine on her skin, they are converted into dotted lines of energy that seem to caress her breast and torso. The image is reminiscent of the borderline erotic photographs that were featured in contemporary nudist health* (Nacktkultur) *publications. A two-color red-and-black version also appears on the dust jacket of volume 2 of* Der Mensch Gesund und Krank *and in a promotional advertisement (Plate 8).*

pitch, German artists and designers had the license to be suggestive, allusive, poetic, even glyphic. One way to do that was to use the visual tropes of hypnotism: circles, pendulums, spirals, and the penetrating gaze. Like hypnosis, the image would act on the viewer's primal unconscious desires and appeal to the irrational mind. This conviction was grounded as much in the artists' own visual experience and aesthetic fashions as it was in advertising psychology, Freudian psychoanalysis, and marketing theory. The viewer's subconscious served as a target for modernist visual production up and down the line.

Hans Schleger, a modernist graphic designer who worked in New York and Berlin in the 1920s and early 1930s, explained the psychology of image production in this way: "The essential driving forces . . . always remain the same: hunger, thirst, sex, ambition and self-preservation. Effective advertising must concentrate upon these urges. It must loosen up the brakes that set too great a check upon human actions and

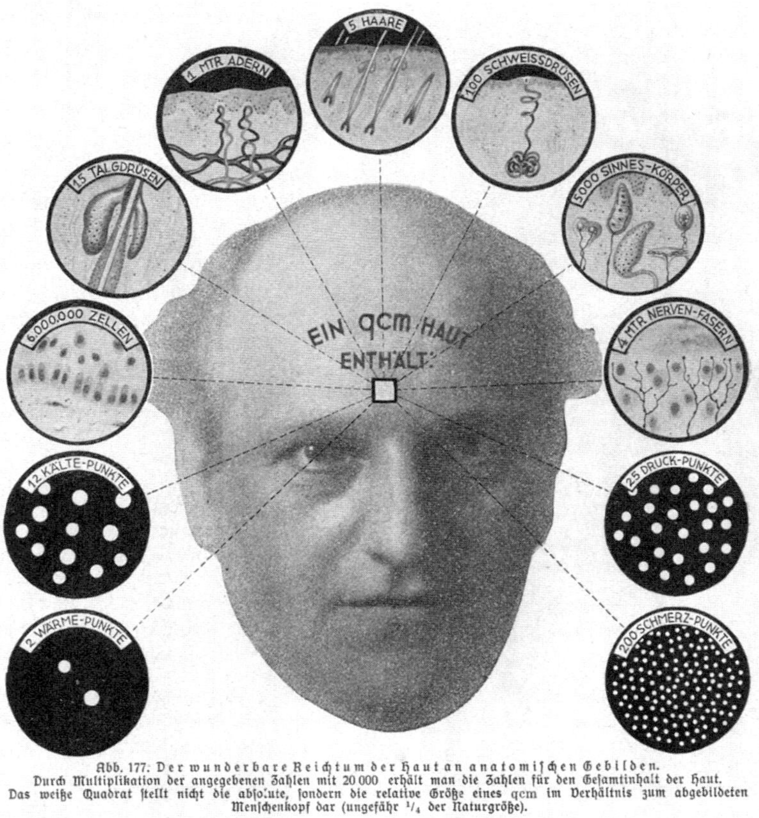

FIGURE 3.5. *A Dadaist head mesmerically stares at the viewer in an illustration that teaches a lesson on sensory perception. "The Wonderful Richness of the Anatomical Features of the Human Skin,"* Das Leben des Menschen *(1929), 4:249. Artist uncredited. National Library of Medicine. The inscription on the forehead reads, "One cubic centimeter of skin contains . . ." The caption is discreetly set in small type at the bottom of the figure so as not to distract from the overall composition.*

human will."[14] In Schleger's account, the viewer of images is a pure abstraction. But that abstraction implicitly refers back to the individual in mass society who was the subject of contemporary sociological analysis and literary, cinematic, and theatrical depiction. In a 1921 issue of the *Berliner Illustrierte Zeitung*, an anonymous critic went further: "What interests the masses of people are matters of sensory perception. The formation of their opinions proceeds from the visual appearance of life and its occurrences, not through intellectual consideration and speculation."[15]

If the masses existed in a borderline state where trance could be easily mobilized, then the individual in mass society was not much of an

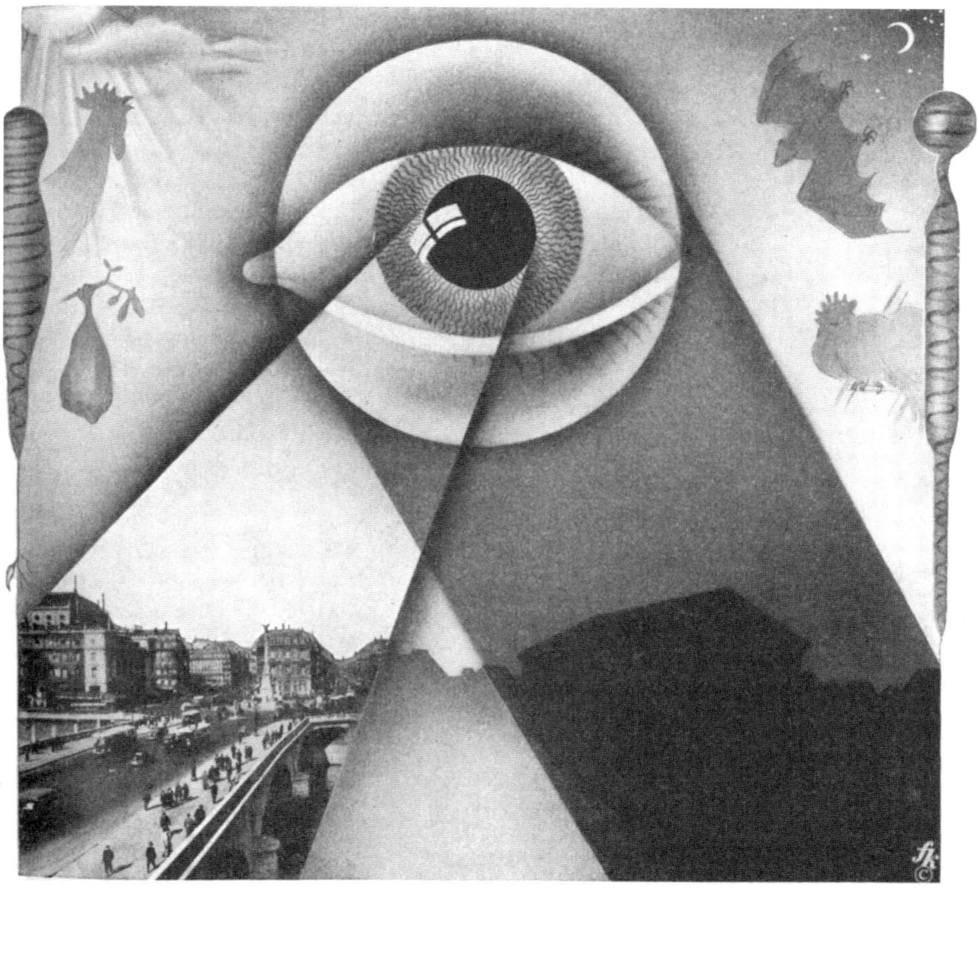

FIGURE 3.6. *"The Human Eye: A Combination Camera for Day and Night Vision,"* Der Mensch Gesund und Krank *(1939), 2:331. Artist uncredited. National Library of Medicine. Part Art Deco, part Dada, the illustration supports a lesson on the physiology of day and night vision (rods work at night; cones work during the day) that refers to the visual landscape of urban modernity (the scene is a photograph of Paris). The monocular eye fixes the reader. Lines from the eyeball oscillate between day and night, like a hypnotist's medallion. The open eye simultaneously alludes to the camera's iris, the hermetic eye of knowledge, the all-seeing eye of God, and the surveying eye that served as logo for the Dresden Health Exposition (1911) and its au courant successor, the Deutsches Hygiene-Museum (1931).*

individual. The boundaries of personhood were porous and open to invasion. The public was suggestible; it could be controlled through emotional appeals and striking imagery. Among European image producers, historian Victoria de Grazia argues, there was a sense that "the aesthetics of representation" must be "animated by strong symbolic and pictorial design traditions in order to activate latent desire." In a civilization of sleepwalkers, images should work in the register of dreams by directly addressing the unconscious.[16] True, the masses often resented attempts at manipulation or responded with indifference. Popular opinion was mystifying and hard to predict. Even so, professional and academic experts asserted that there was a psychological science to advertising, marketing, and political propaganda. Greater successes in

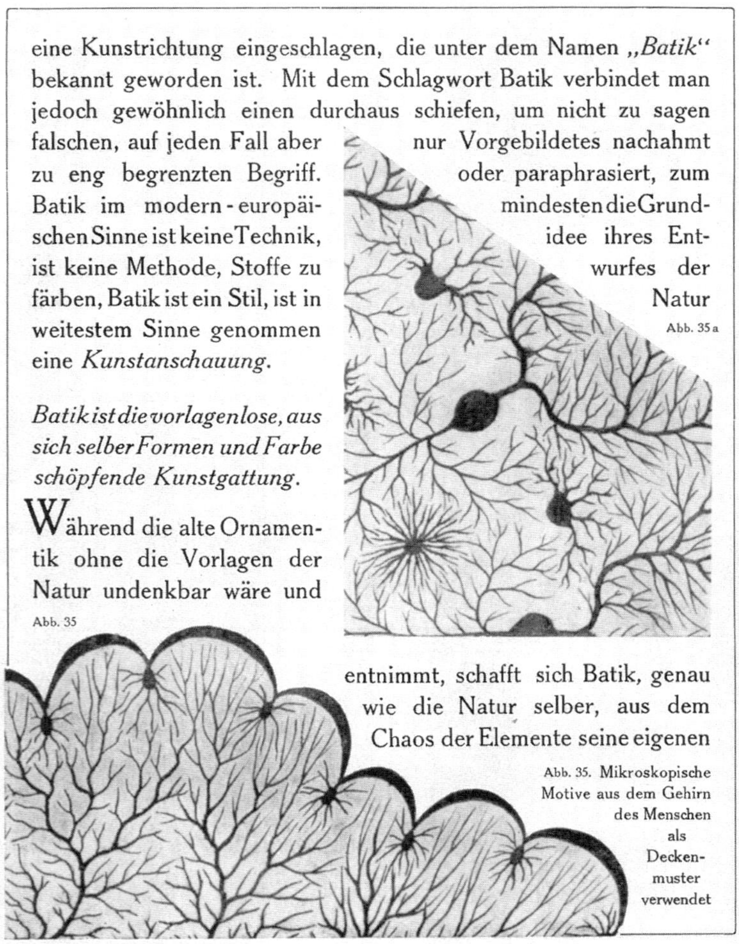

eine Kunstrichtung eingeschlagen, die unter dem Namen „*Batik*" bekannt geworden ist. Mit dem Schlagwort Batik verbindet man jedoch gewöhnlich einen durchaus schiefen, um nicht zu sagen falschen, auf jeden Fall aber zu eng begrenzten Begriff. Batik im modern-europäischen Sinne ist keine Technik, ist keine Methode, Stoffe zu färben, Batik ist ein Stil, ist in weitestem Sinne genommen eine *Kunstanschauung.*

Batik ist die vorlagenlose, aus sich selber Formen und Farbe schöpfende Kunstgattung.

Während die alte Ornamentik ohne die Vorlagen der Natur undenkbar wäre und nur Vorgebildetes nachahmt oder paraphrasiert, zum mindesten die Grundidee ihres Entwurfes der Natur entnimmt, schafft sich Batik, genau wie die Natur selber, aus dem Chaos der Elemente seine eigenen

Abb. 35a

Abb. 35

Abb. 35. Mikroskopische Motive aus dem Gehirn des Menschen als Deckenmuster verwendet

FIGURE 3.7. *"Microscopic Motif from the Human Brain Used as a Ceiling Pattern,"* Rhythmische Musterkunst der Natur *(1923), 38. National Library of Medicine. Copyright Kosmos Verlag, Stuttgart.*

social engineering of popular opinion and behavior would be achieved through experiment, research, and close analysis of trial-and-error experience. But it was also accepted that the generation of visual material required artistic intuition—and that required the utilization of artists, not just craftsmen and psychotechnicians.

At the same time, commercial illustration was not just an instrument of social engineering; the artist could offer it positively as *art.* Acting on behalf of his employers, the artist was a manipulator. But if the artist shared the viewers' social experiences and interpretive practices—and identified (socially, ideologically, aesthetically) with the mass viewer—then commercial art might have a democratic and progressive valence. It need not be only elitist or manipulative. As the bizarre image, pleasurable and affective, entered into the imaginative life of the viewer, maybe it worked to deepen the trance state of industrial dreamtime. But maybe it could also give the viewer a jolt—wake up the sleepwalking masses. If to be modern was to be emancipated, then the image could emancipate; it could work as liberation iconography.

Kahn's commitment to image work, while not overtly political, aligned more with this hopeful vision. Initially, that only signaled a commitment to the modern, but not modernism. Take, for example, his 1924 collaboration with designer Georg Piek-Patrik, *Rhythmische Musterkunst der Natur* (Rhythmic art patterns of nature). Influenced by Ernst Haeckel's late nineteenth-century bioaesthetics of organic life, *Rhythmische Musterkunst* followed the not terribly modern practice of taking patterns from nature for use in decorative art, furniture, and clothing design.[17] Its only innovation (and the reason for Kahn's participation) is that it derived its patterns from microscopy and anatomy—veins, cells, organs, tissues—rather than from the traditional vocabulary of leaves, flowers, branches, and feathers. (The project was ill conceived: the designs were uneven, with some downright ugly; and it went against the modernist rejection of elaborate detail and decoration, and its embrace of pared-down geometrical abstraction or raw primitivism.) Similarly, while the five-volume *Das Leben des Menschen* tried to do illustrated popular science in a modern way, it was a hodgepodge of visual material, styles, and approaches—

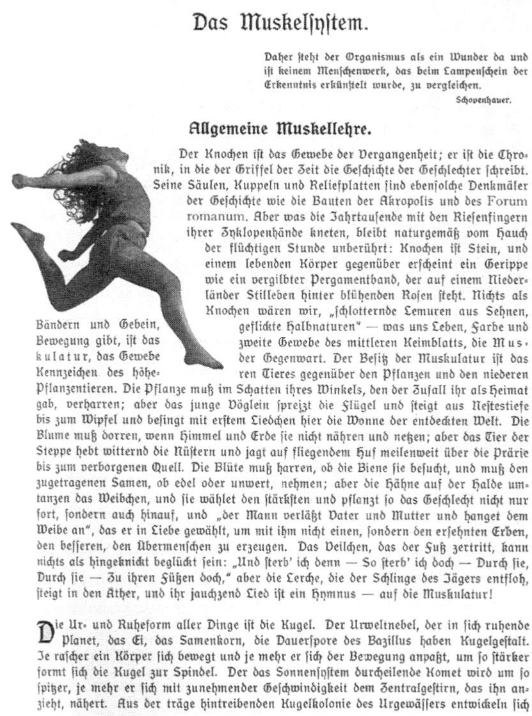

Das Muskelsystem.

Daher steht der Organismus als ein Wunder da und ist keinem Menschenwerk, das beim Lampenschein der Erkenntnis erkünstelt wurde, zu vergleichen.

Schopenhauer.

Allgemeine Muskellehre.

Der Knochen ist das Gewebe der Vergangenheit; er ist die Chronik, in die der Griffel der Zeit die Geschichte der Geschlechter schreibt. Seine Säulen, Kuppeln und Reliefplatten sind ebensolche Denkmäler der Geschichte wie die Bauten der Akropolis und des Forum romanum. Aber was die Jahrtausende mit den Riesenfingern ihrer Zyklopenhände kneten, bleibt naturgemäß vom Hauch der flüchtigen Stunde unberührt: Knochen ist Stein, und einem lebenden Körper gegenüber erscheint ein Gerippe wie ein vergilbter Pergamentband, der auf einem Niederländer Stilleben hinter blühenden Rosen steht. Nichts als Knochen wären wir, „schlotternde Lemuren aus Sehnen, Bändern und Gebein, geflickte Halbnaturen" — was uns Leben, Farbe und Bewegung gibt, ist das zweite Gewebe des mittleren Keimblatts, die Muskulatur, das Gewebe der Gegenwart. Der Besitz der Muskulatur ist das Kennzeichen des höheren Tieres gegenüber den Pflanzen und den niederen Pflanzentieren. Die Pflanze muß im Schatten ihres Winkels, den der Zufall ihr als Heimat gab, verharren; aber das junge Vöglein spreizt die Flügel und steigt aus Nestestiefe bis zum Wipfel und besingt mit erstem Liedchen hier die Wonne der entdeckten Welt. Die Blume muß dorren, wenn Himmel und Erde sie nicht nähren und netzen; aber das Tier der Steppe hebt witternd die Nüstern und jagt auf fliegendem Huf meilenweit über die Prärie bis zum verborgenen Quell. Die Blüte muß harren, ob die Biene sie besucht, und muß den zugetragenen Samen, ob edel oder unwert, nehmen; aber die Hähne auf der Halde umtanzen das Weibchen, und sie wählet den stärksten und pflanzt so das Geschlecht nicht nur fort, sondern auch hinauf, und „der Mann verläßt Vater und Mutter und hanget dem Weibe an", das er in Liebe gewählt, um mit ihm nicht einen, sondern den ersehnten Erben, den besseren, den Übermenschen zu erzeugen. Das Veilchen, das der Fuß zertritt, kann nichts als hingeknickt beglückt sein: „Und sterb' ich denn — So sterb' ich doch — Durch sie, Durch sie — Zu ihren Füßen doch," aber die Lerche, die der Schlinge des Jägers entfloh, steigt in den Äther, und ihr jauchzend Lied ist ein Hymnus — auf die Muskulatur!

Die Ur- und Ruheform aller Dinge ist die Kugel. Der Urweltnebel, der in sich ruhende Planet, das Ei, das Samenkorn, die Dauerspore des Bazillus haben Kugelgestalt. Je rascher ein Körper sich bewegt und je mehr er sich der Bewegung anpaßt, um so stärker formt sich die Kugel zur Spindel. Der das Sonnensystem durcheilende Komet wird um so spitzer, je mehr er sich mit zunehmender Geschwindigkeit dem Zentralgestirn, das ihn anzieht, nähert. Aus der träge hintreibenden Kugelkolonie des Urgewässers entwickeln sich

FIGURE 3.8. *Page layout with a silhouetted modern dancer, a figure of emancipated modernity.* Das Leben des Menschen *(2nd ed., 1927), 2:175. National Library of Medicine. Copyright Kosmos Verlag, Stuttgart.*

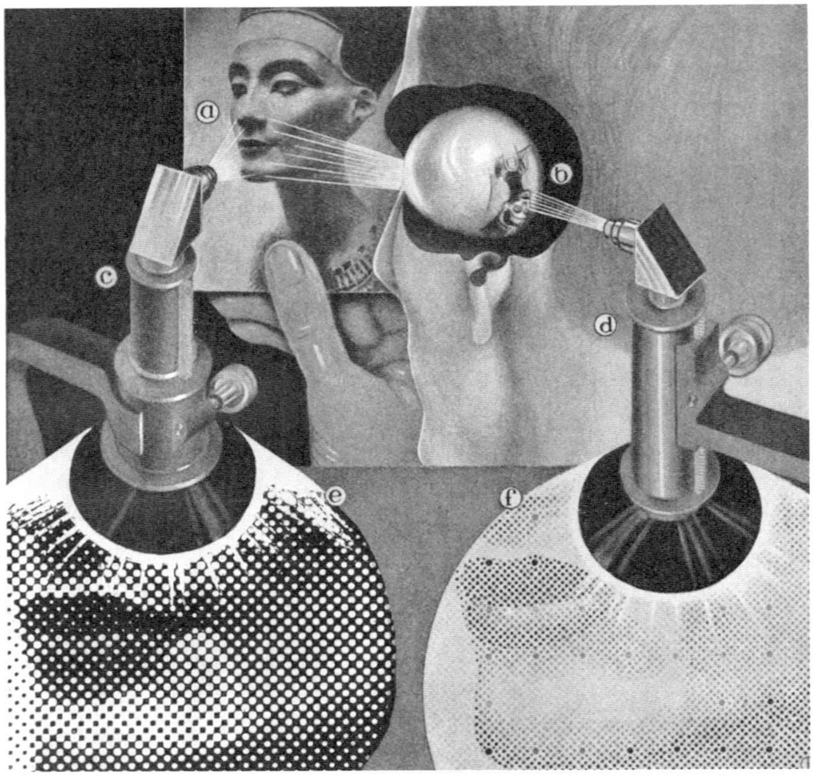

FIGURE 3.9. *"The Similarity between Halftone and Retinal Image,"* Das Leben des Menschen *(1931), 5:53. Artist: Roman Rechn. National Library of Medicine. The surrealist mixture of fingers, lips, eyeball, halftone textures, shiny devices, and radiant light creates an oddly dreamy, sensual effect. But why an ancient sculpture in a 1931 modernist illustration? The illustration was first conceived in the mid-1920s, when the bust of Nefertiti had recently been installed in Berlin's Egyptian museum, but publication of volume 5 of* Das Leben des Menschen *was delayed by the economic crisis. Rechn's masterpiece asserts that new regimes of science and scholarship incorporate antiquities into modernity, and so give us new ways to see old things.*

an image dump. Many of its pictures were artlessly drawn or were lifted from sources that were indifferent to modernism. These outnumbered the imaginative, finely executed modernist illustrations. The settings for the illustrations likewise had no modernist sensibility. The page layouts were haphazard; none of them showed any strong commitment to modernist design, perhaps because they had to adhere to Franckh's unadventurous house style. There were very few splash pages—a staple of contemporary mass-circulation magazine design—and the layouts paid little attention to negative space, nonrectangular frames, typographical composition, and other staples of modernist design.

This visual eclecticism was partly dictated by the resources available to Kahn and his publishers. Commissioned illustrations cost money. Splashy modernist layouts used lots of negative space, and that

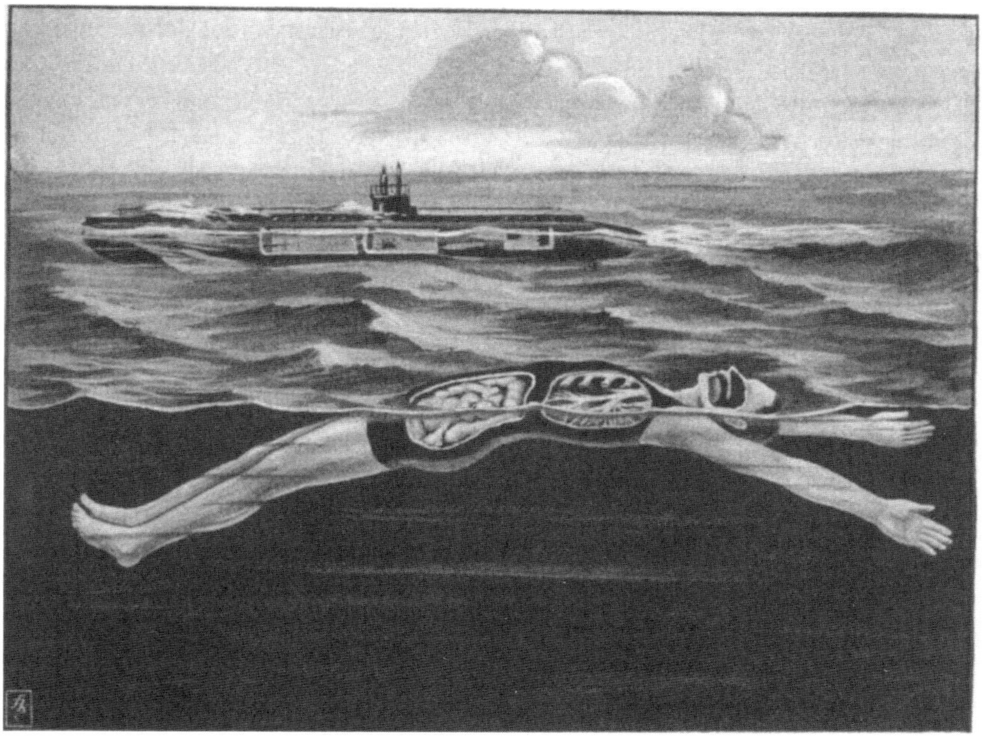

FIGURE 3.10. *"The human body does not sink in water—if it remains quietly relaxed. The head, the thorax, and the abdomen all contain gases and help to keep the body afloat,"* Der Mensch Gesund und Krank *(1939), 1:295. Artist uncredited. National Library of Medicine. A body is like a U-boat—a bizarre image perhaps influenced by the surrealism of René Magritte (who was himself trained as an advertising artist). The caption entirely omits mention of the visual metaphor, an indication of Kahn's growing confidence in the power of the image.*

would have increased page counts and required design expertise and care. We don't have records of the fees paid to the artists who did Kahn's illustrations or of the budgets that Kahn's publishers allotted him. We do know, however, that it was fast and cheap to take images from preexisting sources. (Carrying this forward into the late 1930s, as an exile from Hitler's Germany, Kahn's modus operandi was to borrow extensively from his own previously published works.) Some of Kahn's artists were sophisticated image makers; others were able craftsmen; still others were barely competent hacks. But pushed by Kahn's need to entertain the reader with novel visual metaphors, even his least modernist artists sometimes carried naive realism into the register of surrealism. Thus, while the schematic depiction of mind featured in "Der Mensch als Industriepalast" only had office space for Will, Reason, and Understanding—and none for dreams, imagination, or eros—we are induced to smile at what is left out. In many other illustrations in *Das Leben des Menschen* and *Der Mensch Gesund und*

Krank, unreason and sexual desire come right to the surface (Plate 1). This return of the repressed, while a practical method of mobilizing the reader's attention, was also pleasurable in ways that exceeded any instrumental purpose.

POPULAR MEDICAL ILLUSTRATION BEFORE KAHN

All of this was a far cry from the image practice of prewar popular health and science illustrated publications. Before 1914, popular books and articles on anatomy and physiology mainly presented views of anatomical structures, microorganisms, abnormalities, disease conditions, home health, and hygiene. They did not graphically attempt to contribute to the reader's experience of life in modernity except by invoking the rhetoric of anatomical objectivity or the literal depiction of other topics.

Scholars of visual culture do not have a good name for the pragmatic illustrationism that dominated medical publication in the nineteenth century. Art historian Martin Kemp calls it a "renunciation of stylishness" or "style-less style"—the deliberate "plain" opposition to style.[18] That was certainly part of it. But for popular texts, there was also a large admixture of commercial indifference to stylishness combined with journeyman incompetence or carelessness: no style. Or there was some tacit aggregate aesthetic, a commercial blend of things. While many popular medical publications used illustrations derived directly or indirectly from sources committed to antistyle technical standards of scientific illustration, others were taken from sources where there was some neoclassical or Romantic aesthetic, and still others were drawn from sources that had a tacit commercial aesthetic. No style, antistyle, and style could appear side by side in the same book. There was no consistency because, for economic reasons, publishers often reprinted or copied illustrations from previously published works rather than getting an artist to create something new. Some of these derivative illustrations maintained the integrity of a source illustration, but many were corrupted in a cascade of uncredited copies and copies of copies. Ultimately, this cumulative mixing of anonymized pictures from different sources undermined the very notion of style. The haphazard recycling of images and styles ended up making woodcut anatomical illustration into a genre of illustration that was the work of everyone and

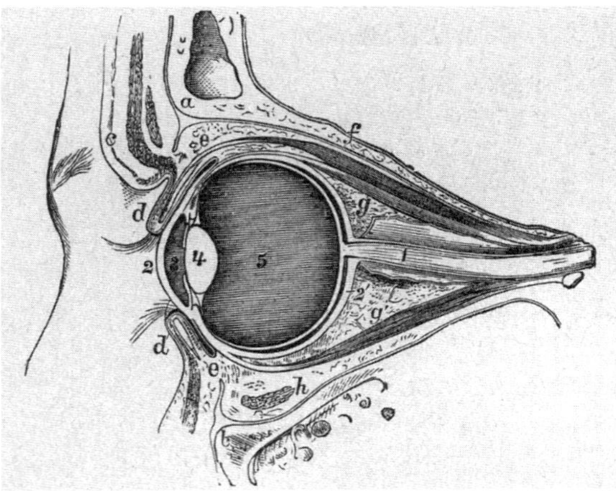

FIGURE 3.11. *Nineteenth-century technical anatomical engraving, recycled. "Vertical Cross Section of the Eye Socket," in Anna Fischer-Dückelmann's popular* Die Frau als Hausärztin: Ein ärztliches Nachschlagebuch der Gesundheitspflege . . . *(Stuttgart: Süddeutsches Verlags Institut, 1901), 36. National Library of Medicine.*

no one.[19] It also worked to give the anatomical illustration a certain generic authority, as if it came direct from the body, without any need to consider authorship or an artist's hand.

Mostly the images in nineteenth-century popular medical books and pamphlets earnestly depicted the structures of the body. Here and there, some pictures tried to visually entertain, to perform a kind of anatomical burlesque, but mostly they didn't. In the anatomies of the sixteenth and seventeenth centuries, the reader was treated to images of dancing cadavers and death figures.[20] Nineteenth-century popular medical books often bid for attention with images of pathology or racial exoticism (mixed with large doses of phrenology), but in their sections on anatomy and physiology, the torso and whole figure were usually modestly placed before the reader in static poses. Often the illustration focused on some entirely detached body part or system. Metaphor, process, and motion went unvisualized. In a popular medical book of the 1880s, for example, an illustration titled "Professor Brain's Telegraph" shows a standing black figure inscribed with a tree of white nerves. The metaphorical comparison of the nervous system to an electric telegraph system appears in the caption and accompanying text, but not in the illustration itself. In nineteenth-century medical books, metaphor was confined to words; there was a firewall between metaphor and image.[21] (For more on metaphor and the firewall, see chapter 6.)

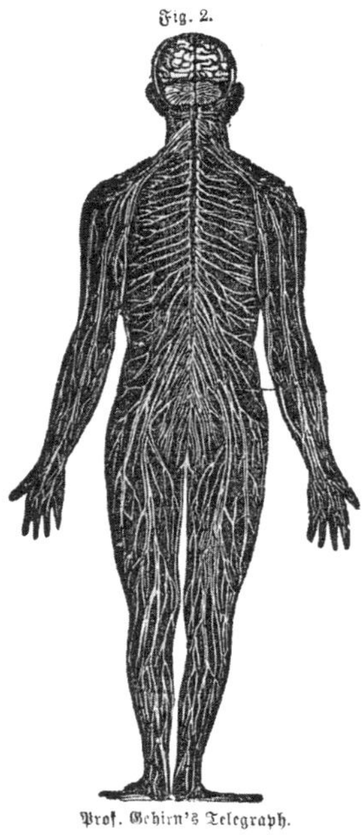

FIGURE 3.12. *"Prof. Brain's Telegraph," E. B. Foote,* Offene Volks-Sprache über das Menschensystem . . . und Gesunder Menschen-Verstand *(New York: Murray Hill, 1882), 29. National Library of Medicine. Artist and source uncredited. The image is a crude variation on an often-copied series of white-on-black diagrams of the venous, arterial, and nervous systems that goes back at least to the 1850s, if not earlier.*

WHODUNIT? KAHN AND HIS ARTISTS

By the early decades of the twentieth century, popular tastes were changing: the reading public for popular science was increasingly seeking out works that were modern and scientific. However, illustrations in popular medical texts lagged behind, were often of nineteenth-century or earlier vintage, and were recycled across national and linguistic borders from text to text. It was economical for publishers and authors—they could save time and money by repurposing old art—but by the 1920s such illustrations had grown stale. Nineteenth-century anatomical engravings looked old-fashioned in an industrial world that prized novelty, progress, and up-to-date science. Kahn thus promoted his approach to popular science by marking the distance between his work and "the anatomy of the 1880s."[22]

However, Kahn also had economic incentives to plunder the past. The sheer volume of illustrations in *Das Leben des Menschen* made it necessary: he just didn't have a large enough budget to hire artists to work up new drawings. So he liberally recycled from past volumes of *Kosmos* and predecessor works such as Hermann Dekker's *Lebensrätsel: Der Mensch biologisch dargestellt* (1906; The mystery of life: Man illustrated biologically). Dekker, Franckh's house expert on the human body until Kahn supplanted him, had also recycled quite a bit, and he perhaps served as a model before *Wunder in Uns* pulled Kahn in new directions.[23]

Even so, well into the 1930s, Kahn was still borrowing from nineteenth-century works. J.-B. M. Bourgery's showstopping anatomical/surgical lithographic atlas, published between 1831 and 1854, was a particular favorite. A few of the large-format full-page color plates (rendered brilliantly for Bourgery by artist N. H. Jacob) appeared in Kahn as much smaller black-and-white halftones and outfitted with added elements, such as pastiches that illustrated points that were entirely unrelated to the aims of Bourgery's original (Figures 3.13 and 4.19).[24] Of course, in many ways, Kahn's entire oeuvre was a pastiche of repurposed and newly created materials. He wrote in his own voice, but like the graphics, much of his text derives from predecessor and contemporary sources. Kahn was by nature an encyclopedist, not an original thinker or researcher. His authority depended on a show of familiarity

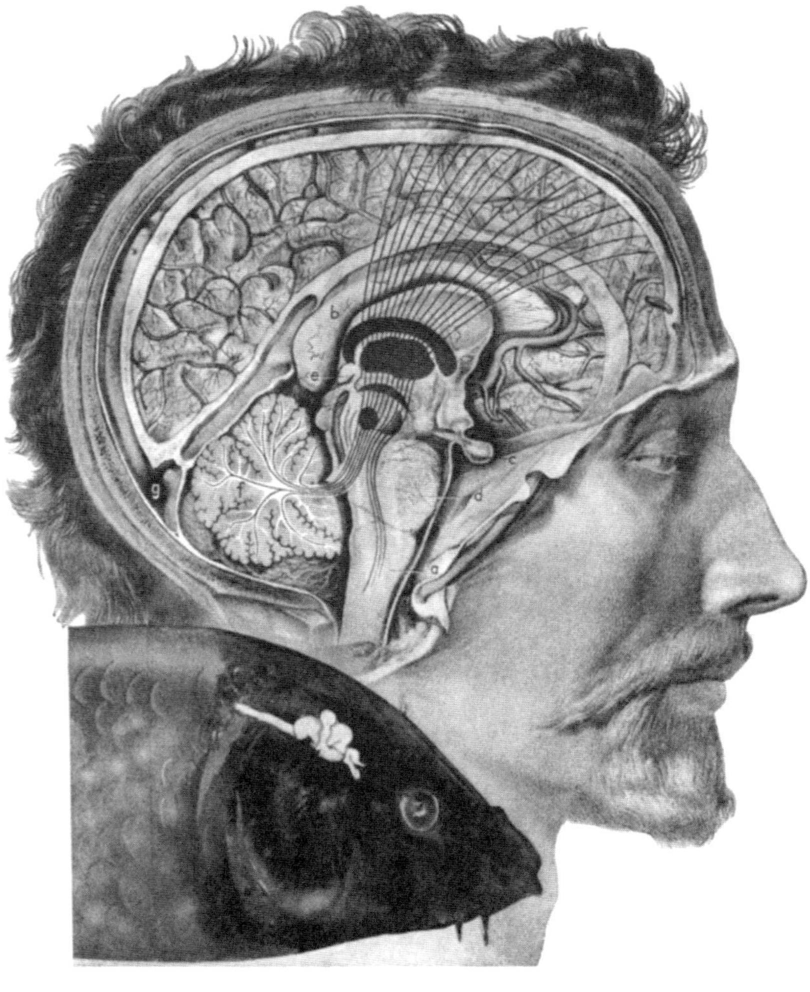

FIGURE 3.13. *A human figure from J.-B. M. Bourgery's mid-nineteenth-century anatomical atlas, repurposed to compare the brain size of man and fish.* Der Mensch Gesund und Krank *(1939), 2:171. National Library of Medicine. Kahn frequently raided Bourgery's atlas for reusable images.*

with the great discoveries of the past and current biomedical knowledge. He made especially sure to discuss the latest developments and include illustrations borrowed from recent scientific publications.

There was a progression to Kahn's image work, a learning curve. To advance within the Franckh stable, and to gain a wider readership, Kahn needed to differentiate himself from Dekker and other authors. His first career move in that direction was to outdo them with more varied, more lively illustrations, in greater quantities. In later publications, as he gained his footing, the images become more imaginative, more metaphorical, more allusive, more various, and more modern in content or style.[25] But there was always an uneven mixture of things. "Der Mensch als Industriepalast," which first appeared in 1926, in volume 3 of *Das Leben des Menschen,* has a sparkling modernist aesthetic, as do some other illustrations in the same volume, but other images are clumsy and innocent of modernism (including some industrial profiles that look like tryouts for "Der Mensch als Industriepalast"), and still other images are positively dull. Dissatisfied with some of the artwork, Kahn took the trouble to have his artists redo them in later editions of *Das Leben des Menschen* to make them more entertaining, more scientific, more modernist.[26]

For the two-volume *Der Mensch Gesund und Krank* (1939), he went further: he revised many illustrations that had previously appeared in *Das Leben des Menschen* and added many new ones. The newer book was more modernist in design and layout, and more unified in conception and execution. This was partly the result of challenging circumstances: Kahn was forced to leave behind the original camera-ready art when the Nazis seized control of Franckh. Moving over to the Swiss-based Albert Müller publishing house for *Der Mensch Gesund und Krank,* Kahn was forced to work up new illustrations, using new (uncredited) artists, basing many of the new pictures on photographs of the original art or printed illustrations. This degraded picture quality, but it also gave Kahn the opportunity to revise and reconceive the images. At the same time, the new publication was an opportunity to add many entirely new illustrations.[27]

Through the 1920s and 1930s, a convergence of forces pushed and pulled Kahn, the modernizer, ever further in the direction of aesthetic modernism. There was a growing market for books that aligned with the modern, as the attention and acclaim for the modernism of "Der Mensch as Industriepalast" made clear. There was also a need for brand distinction as Kahn built a new career for himself outside Germany. But most of all, modernism came to Kahn through the employment of artists who brought with them their own unique aesthetic commitments—artists who came to be vital collaborators in the works issued under Kahn's name.

Little is known about their lives and careers—in many cases not even their birth and death dates—and little is known about the details of their working relationships with Kahn. On the Internet, their trails,

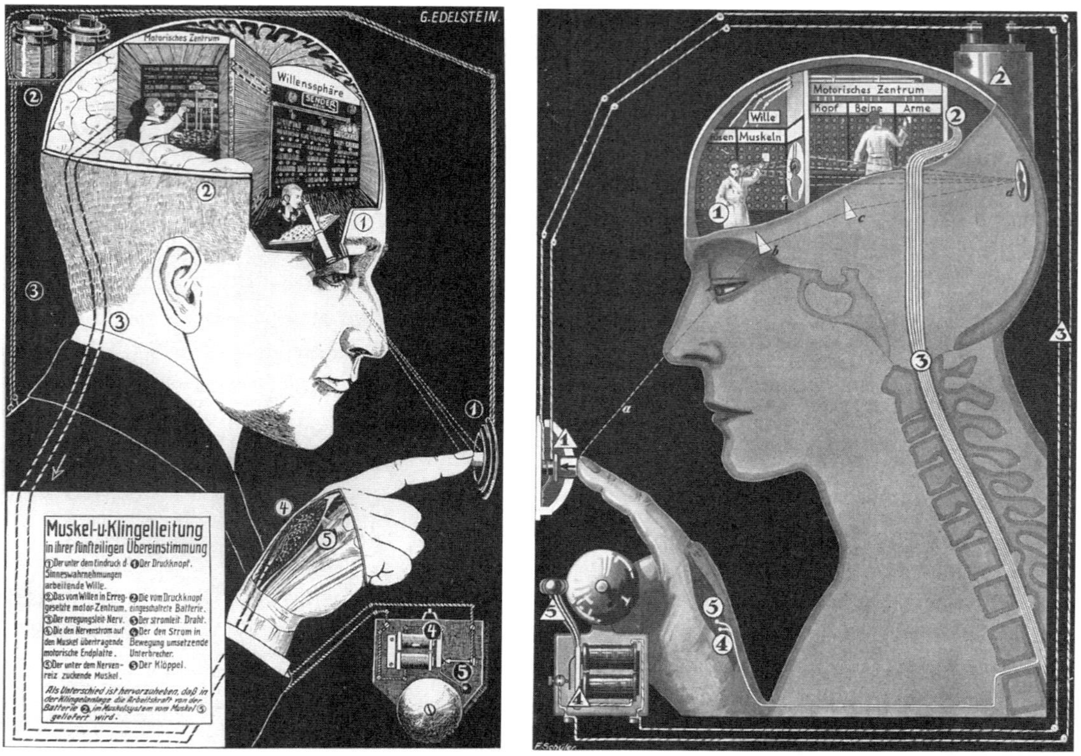

FIGURE 3.14. *Two versions of "Muscle and Doorbell Circuits Compared in Five Ways,"* Das Leben des Menschen, *volume 2. (a) First edition (1924), pl. 17: modern but not modernist. Artist: G. Edelstein. National Library of Medicine. (b) A modernist 1927 version, pl. 13. Artist: F. Schüler. National Library of Medicine. The 1924 Edelstein version was also marketed as a small 44 × 84 cm poster. Copyright Kosmos Verlag, Stuttgart.*

if present at all, quickly evaporate, and much archival work remains to be done.[28] Some of the artists had a season of recognition; others struggled.[29] Some undoubtedly lost careers, or lives, to the economic turmoil of the times, the repression of the Third Reich, or the devastation of World War II. However, even at a glance, it is apparent that Roman Rechn, Alwin Freund-Beliani, Otmar Trester—Kahn's most sophisticated artist-collaborators—needed no instruction to supply Kahn with modernism. Instead, they led Kahn, who was never committed to modernism or any particular aesthetic. In a 1936 letter to Kahn's publisher, Roman Rechn (1892–1945), perhaps the most modernist of them all, insists on the unique contributions of the artist and describes the work process:

> On the question of the authorship of "my" images in "Leben des Menschen," I can only say this: from first to last, they were designed and . . . structured using my ideas. Be it the plates "The spinal cord," "Hair" or the many others from the fourth and fifth volume.

> So too with . . . Fritz Schüler: "his" plates and images were entirely independently, freely, creatively conceived and made, from "Man as Industrial Palace" to the "Bloodstream." Dr. Kahn gave us only the subject and acted as a sharp critic. . . . He was not even able to make the most primitive sketch on paper. . . . The execution of a drawing often lasted for weeks, but always new ideas came when working through the problem. . . . No!—Dr. Kahn gave us the topic—but we had the ideas.[30]

Kahn gave a different account of the work process. A laudatory 1943 *Life* magazine article, based on interviews with Kahn, reported that the "drawings, photo-montages and diagrams," although "rendered by various artists, . . . were roughed out by Dr. Kahn to the last detail."[31] In exile (and against the Nazi expropriation of his images), Kahn had a vested interest in asserting his intellectual authorship and property rights. Transplanted to America, he needed to be recognized as the "gifted inventor" (as the *New York Mirror* proclaimed him) of the conceptual scientific image and the custodian of the popular medical imagination.[32]

Whatever the truth of the matter, Kahn's eclecticism—and, if we can believe Rechn, the latitude Kahn granted or ceded to his artists to do things in their own various ways—was unlike the working method of Otto Neurath, his celebrated contemporary. Neurath's ISOTYPE pictographic system, made in collaboration with designers Erwin Bernath and Gerd Arntz, was a grand philosophical project to universalize and modernize medical, demographic, and economic knowledge in the form of statistical and cartographic diagrams. The aim was to create a descriptive vocabulary that both explained and designed modern life. With its famously minimalist homuncular icons, ISOTYPE was a total system.[33] In contrast, Kahn had no system. Even taken individually, his books do not aspire to be a *Gesamtkunstwerk,* and they do not envision life in modernity as a *Gesamtkunstwerk.* In keeping with a fragmented, episodic work process that involved many different hands, used many different industrial metaphors, and cobbled together a jumble of many different medical and scientific findings, Kahn had no coherent design for living in any register.

"A GREAT LOOP": THE PICTORIALITY OF VISION AND VISIONS OF PICTORIALITY

But Kahn did have a consistent theme: visuality as a modernizing practice. In early publications, illustrations and text are fairly evenly matched, but over the course of the 1920s, the visuals ascend in number and scope. By the late 1930s, the proportion of words to images markedly diminishes; the emphasis on striking images grows

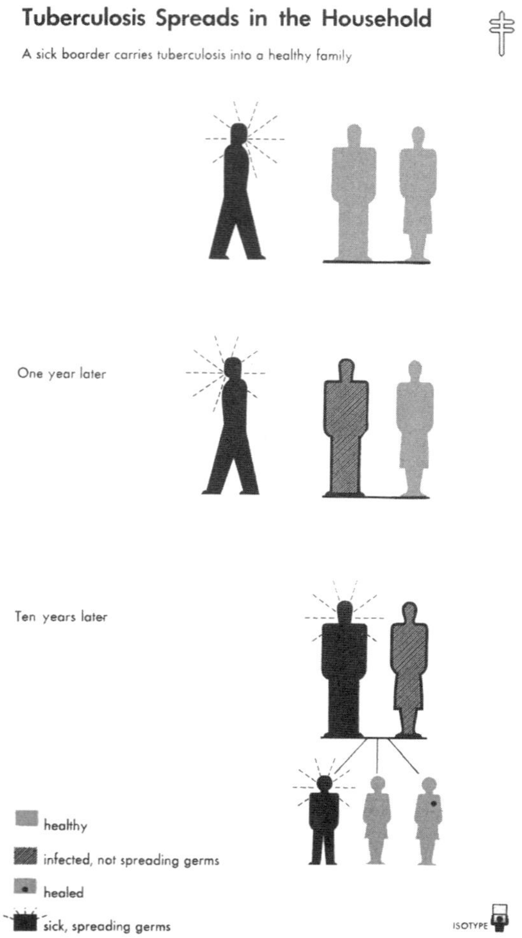

FIGURE 3.15. *Otto Neurath and H. E. Kleinschmidt,* Health Education by ISOTYPE *(New York: American Public Health Association, 1939), 21. National Library of Medicine.*

correspondingly. This increase did not emanate from the dictates of any manifesto or movement. Instead, it came from experience—the experience of picture proliferation in everyday life, the experience of consumer demand for texts with illustrations, and the experience of competition with other popular science authors and publications. All of these pushed Kahn and his publishers to become increasingly attuned to the market. Like any good mass marketers, they endeavored to give the people what they want. And what the people wanted were images and more images.

Kahn added something extra to all of that: images of people watching images, and images of the optical, physiological, and experiential elements of image watching. "We have moved in a great loop from vision to pictures to the pictoriality of vision to the visuality of pictures . . . and . . . round and round again," says art historian Whitney Davis, writing about Erwin Panofsky's concept of pictoriality. In the 1930s, roughly contemporary with Kahn's most productive period, Panofsky suggested that it had become necessary to distinguish visuality—our experience of viewing—from pictoriality. Over the centuries, from the Middle Ages to the Renaissance to the Enlightenment to the present, new picture-making and -viewing practices had developed, that we, the inheritors of those earlier traditions and inhabitants of our own moment, confuse with seeing itself.[34] We become so accustomed to seeing pictorial representations of the world that we see the world itself through the conventions of representation, whatever those might be in any particular time and place. Panofsky, Aby Warburg, Ernst Cassirer, Rudolf Arnheim, and Ernst Gombrich, who were then constituting themselves as the new priesthood of art interpretation, and whom we regard as the great mid-twentieth-century art-interpreting moderns, came of age in an oxygen-rich visual environment. Pictures were all around them—all sorts of images, not just works of great art. Whether hanging as a picture on a wall, appearing as a rectangular illustration inset into a page of text, or framed by the edges of a printed page or movie screen, these images came bundled with a certain "period eye" (historian Michael Baxandall's indispensable phrase)—the same pictorializing period eye that belonged to Kahn and his artists and readers.[35] Within that cognitive field, Kahn's illustrations participated in the "great loop" and commented on it (all those pictures that explain the mechanics of vision!). But it also deranged it. By claiming to demonstrate timeless scientific truths of the body while staging those truths in modern, up-to-date ways, Kahn's picture shows inevitably became dated as fashions, technologies, aesthetics, and the latest scientific certainties became obsolete, superseded by newer versions of modernity.

There were other disparities. Kahn could not fully control his artists or his own visual imagination, so his illustrations were sometimes at variance with the text.[36] The inconsistency occasionally occurs in the image content—an illustration makes a point unsupported by the text—but it more often occurs in metaphor, tone, or aesthetic. The illustrations run away from Kahn; they have a visual poetry that exceeds his written words. None of this seems to have much troubled Kahn's readers. Given the encyclopedism of his publications, obviously many readers did not read the multivolume works in linear fashion, from cover to cover. Instead, they browsed, selectively wandering over to illustrations and topics that caught the eye and ignoring large portions of written text.[37] Over time, Kahn increasingly responded to that. His career depended on generating illustrations (of science and modernity) that catered to the public's hunger for pleasurable images, the all-consuming period eye of mid-twentieth-century pictorialism.

4

VARIETY SHOW

THE STUDIO OF KAHN AND ITS VISUAL DEVICES

THE STUDIO OF KAHN

Driven by market imperatives to make images that attracted and held readers, the studio of Kahn devoured whatever visual devices were close to hand and cooked up many new ones: the body in modernity and modernity in the body; body factories; the bodyscape; the global body; body dynamism; photomontagery; surrealism; the chemical universe; the physiological advertisement; the dramatized statistic; body architecture and mechanics; the schematic body; the aestheticized diagram; organicism; the evolutionary body. Today, while we don't usually attach names to them, they are completely familiar. They turn up in magazine articles, advertisements, motion pictures, television, children's books, museum exhibitions, videos, and websites. But to the readers and viewers of the 1920s, 1930s, and 1940s, the images were freshly hatched, extraordinary, harbingers of the modern age. No one had ever seen anything like them.

Kahn was an idea man. He could sketch out rough drawings, but was not a trained artist. For finished, camera-ready art, he relied on hired hands. *Das Leben des Menschen* used at least twenty-five artists, probably more.[1] We can't be precise because many of them were never credited. But some did sign their pieces or had a distinctive style that permits an attribution. Kahn relied on specific artists for certain types of illustrations. For synoptic illustrations, he used Roman Rechn, who designed polished compositions, full of smooth surfaces and pictures within pictures, that commented on the prolific visuality of modernity; Rechn did four of the five dust jackets for *Das Leben des Menschen* and some of the finest interior illustrations. For diagrams, Kahn favored Otmar Trester, who reconceptualized the diagram as modern art; his flowcharts and schematics were abstract designs built out of elegant, ruled pen-and-ink lines, geometrical shapes, and perfect sans-serif lettering in the mode of Bauhaus. Arthur Schmitson, a veteran medical illustrator, took on the chores of expert technical anatomical illustration, but he also drew Jugendstil fantasy journeys in the human interior, with overgrown bodyscapes populated by fairies and fantastic creatures. Alwin Freund-Beliani combined exquisite draftsmanship with a stylish command of modernist graphics and a designer's sense of negative space. Fritz Schüler and Georg Helbig were Kahn's workhorses; they took on a staggering variety of assignments with more than serviceable journeyman skill, visualizing every metaphor thrown their way. Their illustrations, always in some way modern but mostly not modernist, entertained readers with clumsy stock figures and scenes of everyday life, oddly juxtaposed, situated, or scaled—quotidian realism as surrealism.

This is not to say that the work of Kahn's artists was entirely original. They didn't invent the before-and-after, the sequential diagram, or the cutaway. Those visual devices had been under development for centuries.[2] But the modernizing inflections—the love affair with technology, the urban, industrial, and domestic settings, the eclectic modernism, the visualization of metaphor, the theatricality of visual narrative—made something entirely new. Their images are full of playful and suggestive detail; they teem with cultural and aesthetic references; they overflow with meaning. Kahn attempted to control that, to some degree, through textual commentary, and by supplying labels and captions keyed to the image. Still, for contemporary readers, and for us today, the pictures "leave their frames" and call out for exegesis.[3] Bowing to that demand, this chapter extracts a selection of them for analysis, focusing on genres, tropes, style, iconology, and allusions. We will also consider the distinctive approaches and contributions of the artists.

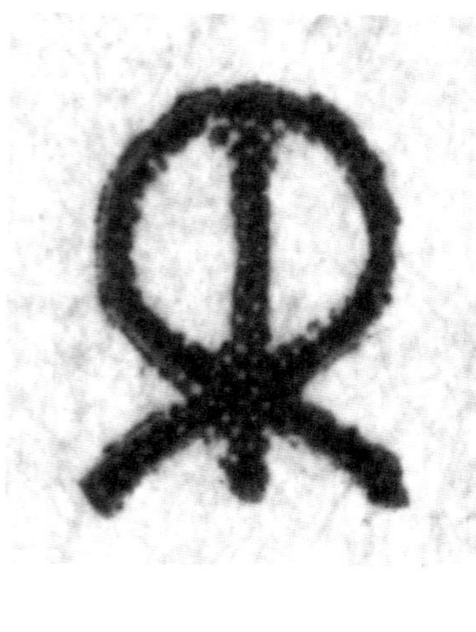

FIGURE 4.1. *Roman Rechn signed most of his illustrations with this double R logo (here taken from "The Iris-Key,"* Das Leben des Menschen *[1929], 4:129 [Figure 4.16 in this chapter]). National Library of Medicine. Copyright Kosmos Verlag, Stuttgart.*

A caveat. This chapter's close reading of particular illustrations may be a bit misleading. With the exception of "Der Mensch als Industriepalast" (and a few smaller monochromatic posters), readers encountered Kahn's images in newspapers, magazines, pamphlets, and small octavo-size books, mostly in grayish black-and-white halftones, leavened by an occasional two- or four-color plate. They were images in bulk—an illustration on nearly every page, sometimes more than one—surrounded by, and in dialogue with, text. The structure of Kahn's illustrated publications converged with and mimicked the structure of illustrated advertising. Like contemporary advertisements, each picture had to make a pitch, not to sell a product but to teach a lesson or illustrate a point. And like an advertisement, many of the illustrations have a visual hook (sometimes more than one) to catch the reader's eye. This was a visual practice that assumes that readers will approach the material with casual randomness. Some images, some sections, will be heeded and others ignored, unlike the linear immersive reading that readers of novels indulge in.

Even so, Kahn's artists often took great care with even seemingly minor assignments. The process of making camera-ready art required them to make illustrations that were much larger than what would ultimately appear in print, a method that encouraged added detail and polish. At the same time, their work was separate from the work of the layout men who arranged the pictures on the page for publication, and to some degree separate from Kahn himself. All of that worked to increase the tendency of the finished artwork to express something more than its stated purpose, to instigate a swirl of meanings that neither Kahn nor the artist (nor us today) could fully control or understand.

A final point: Kahn's artists toiled in the cultural margins. Some lacked talent, others were gifted, but all were working commercial artists. They were perhaps respected, even admired, by colleagues, but they were never celebrated. The work they did with Kahn, as influential as it turned out to be, and in many cases their career best, did not win them acclaim. Schmitson, Rechn, Trester, Freund-Beliani, and Schüler never showed their work in galleries or got noticed in prestigious publications such as *Gebrauchsgraphik*. They did not get taken up as the subject of articles or monographs.[4] Even if their signatures appeared (as was sometimes the case in *Das Leben des Menschen,* but not at all in *Der Mensch Gesund und Krank* and later works), they went unrecognized by readers and critics. Kahn fared better; he got all the credit. Kahn was the brand name.[5] Many readers, if they gave it any thought at all, took Kahn to be the artist. (To keep things in perspective, we should note that Kahn achieved only minor celebrity in Germany. Transplanted to America, he never ascended to the star status of scientific and cultural refugees, such as Hannah Arendt, who arrived in the same wave.) With that in mind, in this chapter and elsewhere, I have tried to use the wordy and sometimes grammatically difficult "Kahn and his artists." But in some places, as a kind of shorthand, I just use

"Kahn" as the subject of the sentence—"Kahn did this or that." If that has the unfortunate effect of making the artists again disappear, I apologize and ask readers to keep them in mind. In every case, Kahn's pictures were collaborations.

That said, our gallery tour commences.

THE BODY IN MODERNITY, MODERNITY IN THE BODY

"THE PRINCIPAL WORK OF THE HEART"

Kahn's most distinctive move was to stage modernity within the body. The body, or its systems or parts, could be figured as skyscraper, automobile, combustion engine, suspension bridge, conveyor belt, movie camera, radio transmitter, phonograph, pump, piston, plumbing, or electrical circuit. The human body could contain just about any feature of the industrial world. But just as often Kahn showed the body or body part *in* modernity: in the city of skyscrapers, elevators, omnibuses, and electric lights; next to airplanes, diesel trains, cars, and steamships; inside factories or next to radios, telephones, phonographs, and telegraphs. Sometimes both metatropes appear in the frame of a single illustration: modernity is in the body, which is in modernity. The body has the architecture of industrial engineering and construction, is the site of industrial technologies and production, is surrounded by modern people, buildings, machines, and machine-made goods. By proximity and physiology, the body and the industrial world become intermingled, equivalent, identical.

The assignment of so many semiotic and pedagogical tasks to a single illustration inevitably has a mimetic effect: it says that modernity is busy, crowded with actions and signifiers. Take, for example, "The Principal Work of the Heart," a physiological cityscape (from the hand of an unidentified artist) that makes three arguments. First, to the left of the image, the heart is figured as an elevator, continuously circulating blood around the body in the same way that elevators circulate people in buildings. Second, to the right, the heart is depicted as a dynamic machine pump that, over a day, is capable of filling metallic industrial tanker trucks. Third (center), over a lifetime of seventy years, the heart pumps enough blood to fill up most of the volume of a massive skyscraper. The illustration is recursive: the human being is like an elevator building, which is like a human being. Further, it refers to something real: the New York Telephone Building, a much-admired thirty-two-story stepped skyscraper, one of the city's first (completed in 1926), heralded as an exemplar of modern architecture and industrial civilization. (The source for Kahn's illustration was a halftone photograph of the New York Telephone Building that appears as the frontispiece of the 1931 English translation of Le Corbusier's modernist manifesto *Towards a New Architecture.*[6])

One is reminded of Joseph Roth's 1922 nearly delirious reaction to the announcement of a plan to construct Berlin's first skyscraper:

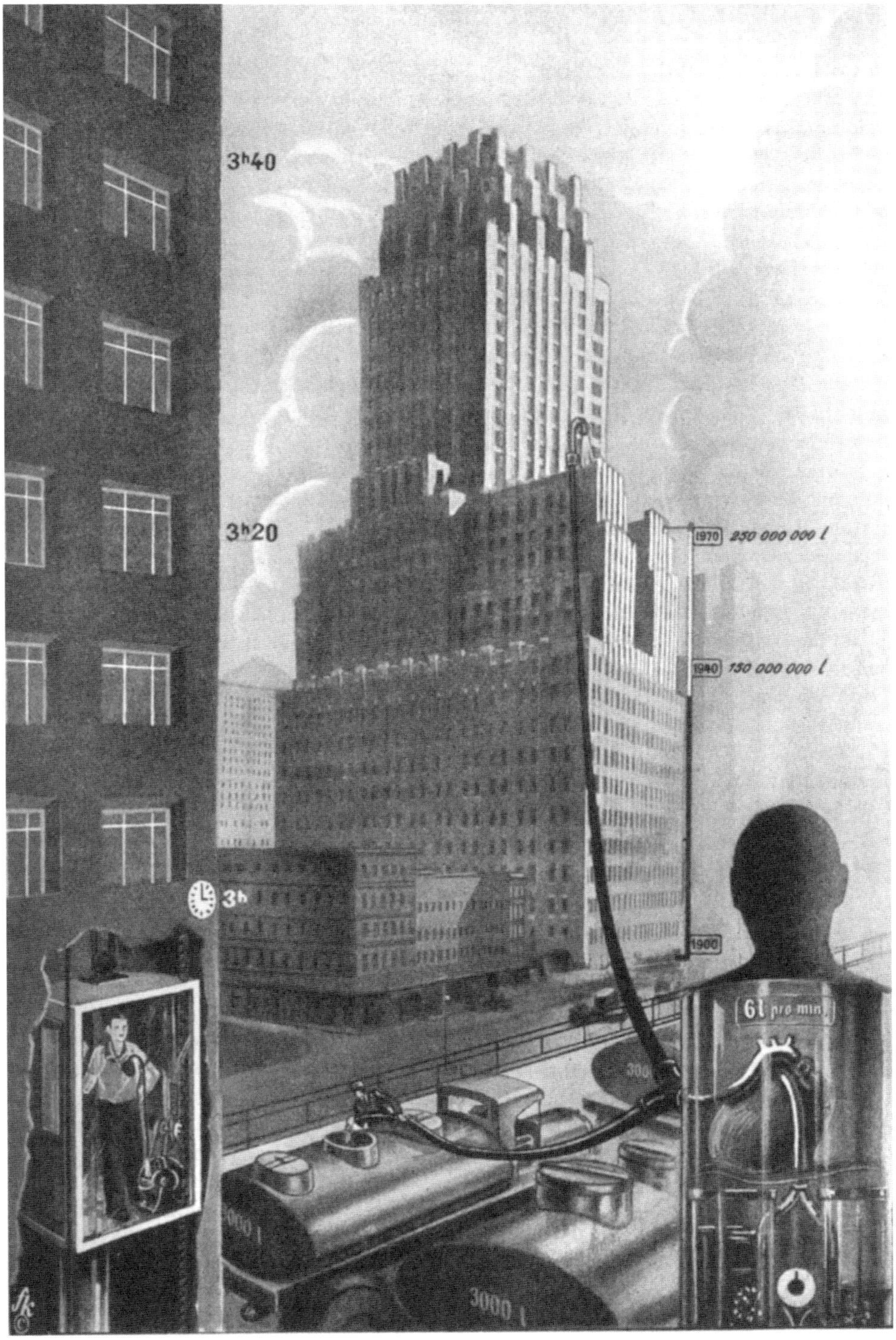

FIGURE 4.2. *Three lessons in one illustration. "The Principal Work of the Heart,"* Der Mensch Gesund und Krank *(1939), 1:193. Artist uncredited. National Library of Medicine. "In the course of a day the heart is theoretically capable of filling three tank cars with ten thousand liters of blood, . . . in seventy years it pumps 250 million liters of blood, a quantity sufficient to fill the cubic area occupied by a modern skyscraper."*

> The skyscraper stands at the summit of technical development. It has already overthrown the cold sobriety of "construction" and has begun to approach the romance of nature. . . . We will make ourselves comfortable among the clouds. . . . It will be a sort of return of the evolved human to the primordial forces of nature. This is an important juncture. . . . The building of the first skyscraper is a historical turning point. Whenever I look at pictures of New York, I am filled with awe at the omnipotence of human technology.[7]

In the same vein, Kahn's illustration stages the skyscraper as both a historical turning point and a modern sublime: an industrial landscape to match the monumental natural landscapes of Albert Bierstadt and other nineteenth-century romantics. The viewer is invited to identify with the silhouette figure, an evocative shadow body, who stands closest to the reader in the picture plane and who views the scene in the same orientation as the reader. The skyscraper, theatrically lit (bathed on one side by the light of a grimy sunset), looms in front of a backdrop of steamy, smoggy clouds. We are carried into the domain of the industrial gothic of Fritz Lang's *Metropolis* (1927).

Although connected by a gleaming pump fixture to nearby tanks and the distant skyscraper, the armless silhouette figure appears as a pictorial cipher, utterly estranged from everything, a surrealistic artifact that far exceeds the didactic lesson that instigated it. A mood of anomie dominates, but as in many Kahn illustrations, there is no consistency. A small, disparate element in the lower left corner of the image breaks the industrial reverie: a smiling homunculus in an elevator, whose exposed heart is hooked up to what appears to be a motor. That little ensemble of winking details must have been overlooked by many readers. As in so many of Kahn's images, there is a too-muchness about the illustration: too many meanings and figures packed into a black-and-white illustration that only takes up most of a page in an average-sized book, not a wall in a museum gallery or a movie projected on the big screen. This works to reduce and domesticate the sublime industrial monumentality of the scene: the medium of book illustration shrinks it down to a friendly pocket size.

The image is one of many where Kahn shows a silhouette figure in a modern setting. Historically, silhouettes mainly did the work of shadow portraiture, a genre that showed off a distinctive social type or individual profile (sometimes in the mode of caricature). In contrast, the silhouette in Kahn turns the human figure into a blank, scrubs away any historical, morphological, cultural, or individual specificity. It is a universalizing modernizing tactic. Otto Neurath took it further, making a universal picture language out of generic mass-produced homuncular silhouettes. On Neurath's pages, standardized icon silhouettes line up in ranks to make parades of statistical data. They are a mirror image of the mass rallies and military extravaganzas of National Socialism, but without the grandiosity. (Maybe the patterned choreography of the Tiller Girls and Busby Berkeley is a better comparison.) On an exhibition wall or in the pages of a book or magazine, Neurath's little multiples don't carry any emotional charge. Instead, they line up to charm the reader, as small figures in books and signs are wont to do.

Kahn sometimes worked the same iconographic territory (possibly under Neurath's influence), but mostly his silhouettes are bit players in a pictorial Tower of Babel. They don't all obey the same visual grammar, or even speak the same dialect. They work in all kinds of locations, in groups or solo, sometimes in the negative space of the page (where they too may be induced to perform a statistical dance), but more often in the body or factory or city.

"THE BLIND SPOT"

In an essay on visuality and the city in modernity, art historian John Tagg comments, "The city was ground for modernist representation because . . . it was already text," and not just text, "a compendium of texts." If we substitute "image" for "text," or add image and text together (Kahn's books are really compendia of image/texts), then we can use Tagg's comment to describe the city that received representation in Kahn—and also the illustrations that came out of Kahn's compounding pictorial strategy:

> A compendium of *images*; one construction impacting on another, in a process of accumulation and condensation that replicated the culture, political and economic concentration on which the power of the city had been founded. . . . Through an immense accumulation of capital, this city took form not only as the site, but also the *object*, of an emphatically visual consumption, articulating and articulated in new subjective experiences of exhilaration and alienation, pleasure and fear, mobility and confinement, expansiveness and fragmentation. . . . The city was . . . both a complex of representations and the place of circulation of representations; the effects of the one always articulating into and reworking the other, and constituting the working material for a range of ambivalent modernisms.[8]

"The Blind Spot" takes up that inversion of site (sight) and object. Within two corresponding circles—the eye and eye view—artist Alwin Freund-Beliani makes a cityscape of rectangles and cubes, which expand vertically and stretch horizontally. We see skyscrapers, a grid of crisscrossing wires in the sky, and, barely visible at the bottom of the

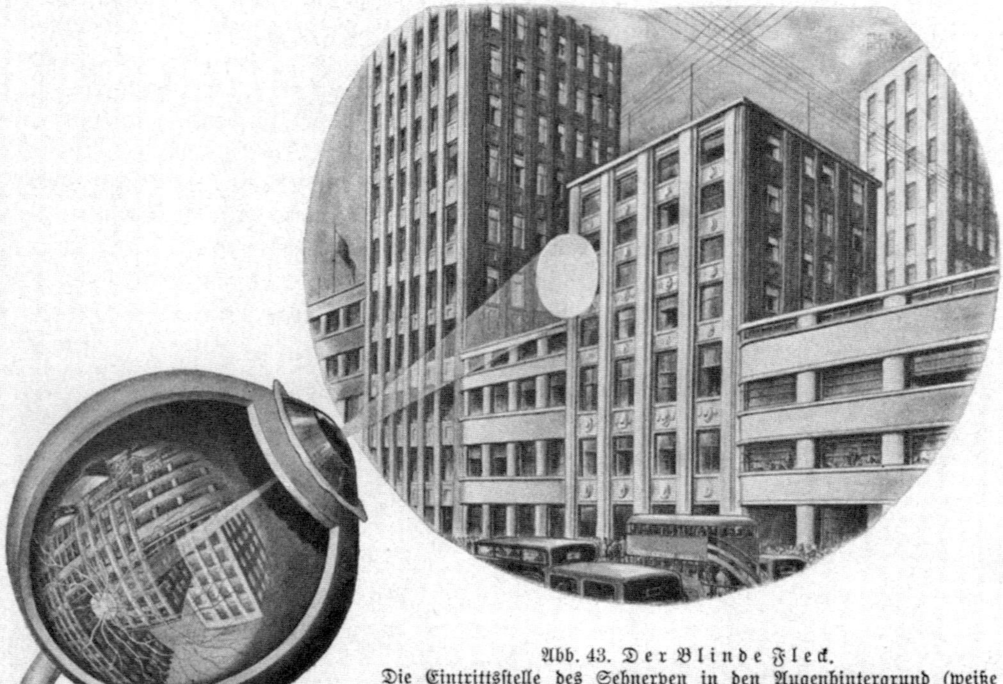

FIGURE 4.3. *"The Blind Spot,"* Das Leben des Menschen *(1931), 5:67. Artist: Alwin Freund-Beliani. National Library of Medicine. Copyright Kosmos Verlag, Stuttgart.*

view, automobiles (again a scene based on a source photograph of the iconic modern city, New York). The image's brilliant conceit is to turn the blind spot into a source of illumination, a spotlight or searchlight projected from the eye to the city. (Adding to the modernizing frisson of the picture: searchlights were then, like skyscrapers, a novel technology, first made familiar by use in the night sky battlefields of the Great War and then, after the war, in movie premieres over the night sky of the industrial city.) At the same time, the city is introjected into the eye, where the ruled lines and sharp boundaries are warped in a disorienting curvature, like M. C. Escher's post–World War II visual puzzle prints.

BODY FACTORIES AND CUTAWAYS

Le Corbusier, the celebrated architect of modernism, famously argued that a "home is a machine to live in."[9] Kahn's images argue that a body is machine to live in, and also that the image is a machine to live in. As we have seen, Kahn's body factories figure and naturalize the technology and social relations of the industrial era; they do service as a homunculus of the self in modernity. The attention they received from contemporary readers and reviewers attests to their rhetorical salience. They call out to the viewer: you are a compilation of technologies; here is an image for our times, a picture that synopsizes the industrial human condition. The body becomes a robot (and in some places Kahn does invoke that staple of imaginative fiction, the mechanical man).

"Der Mensch als Industriepalast" was Kahn's most famous body factory. But Kahn and his artists made many body factories, played many variations on the theme. The body factory came to Kahn and his artists from different sources and underwent considerable development over the course of Kahn's career. A peek at a small selection of Kahn's lesser-known body factories will help us think about the rhetorical effects and development of the genre.

"Human Respiration" is a transitional piece, an early response to the cutaway profiles in *Wunder in Uns.* Printed as a book illustration in *Das Leben des Menschen* and also as a small stand-alone monochrome poster, it may well have been a prototype or audition for "Der Mensch als Industriepalast." (If so, artist Georg Helbig didn't get the job.) The profile cutaways in *Wunder in Uns* only dealt with the digestive and nervous systems and neglected the respiratory system. In "Human Respiration," Kahn rectifies that omission, but he adds a twist: his "Human Respiration" profile cutaway is a man of steel. Except for a brick column spine (a metaphor brought over from *Wunder in Uns*), almost everything else in its imagined body is made of metal. There is a suspension bridge, central duct, and piping—lots of piping. Two inconspicuous little workers toil within; one feeds a furnace in the brain, and the other directs some salivary spray. The rest of the operation is automated. The plumbing, which stands for the lungs and airways, is busy, overcomplicated like a Dadaist machine. The valentine-shaped sheet-metal heart, the playful gusts of breath coming from the nose and mouth, and the charming toy men performing sweaty labor perhaps signal that Kahn and Helbig are trying for a bit of whimsy (a mood that the rigorously orchestrated "Der Mensch als Industriepalast" tamps down but does not entirely suppress). But the final result is a bit off: the steel structures inside the body look a bit silly, like a clanking robot.

The anthropomorphic profile cross section of "Human Respiration" and "Der Mensch als Industriepalast" is a form that gestures both toward and away from the three-dimensional body. "The Biology of the Smell of a Roast" takes a more estranging tactic. The figure is still legibly human, but it is more generatively modern in content (a head with laboratories and manufacturing machines) and stylistically modernist in design. Its schematic figure is a composition of geometrical lines, shapes, and lots of negative space within the figure, abstracted, almost ideogrammatic (that surrealist eye where a homunculus replaces the retina!). In displaying a human body opened up, the cutaway, like a dissection, dramatizes a claim: here is a view of what is normally hidden from view—not the body's anatomical structure, but its technological processes. In "The Biology of the Smell of a Roast," that revelation is neatly laid out as a flowchart that takes the form of an exhibition display: a grouping of window scenes, with display type and small captions, where posed mannequins and industrial machines demonstrate the subject for viewers.

The small sliver of scholarship devoted to Kahn's body factories concentrates on content and doesn't consider the formal aspects of the cross-sectional cutaway. However, the cutaway deserves consideration. When Kahn and his artists lifted it from *Wunder in Uns,* it had already been in use for centuries in technical illustration, schematics, and

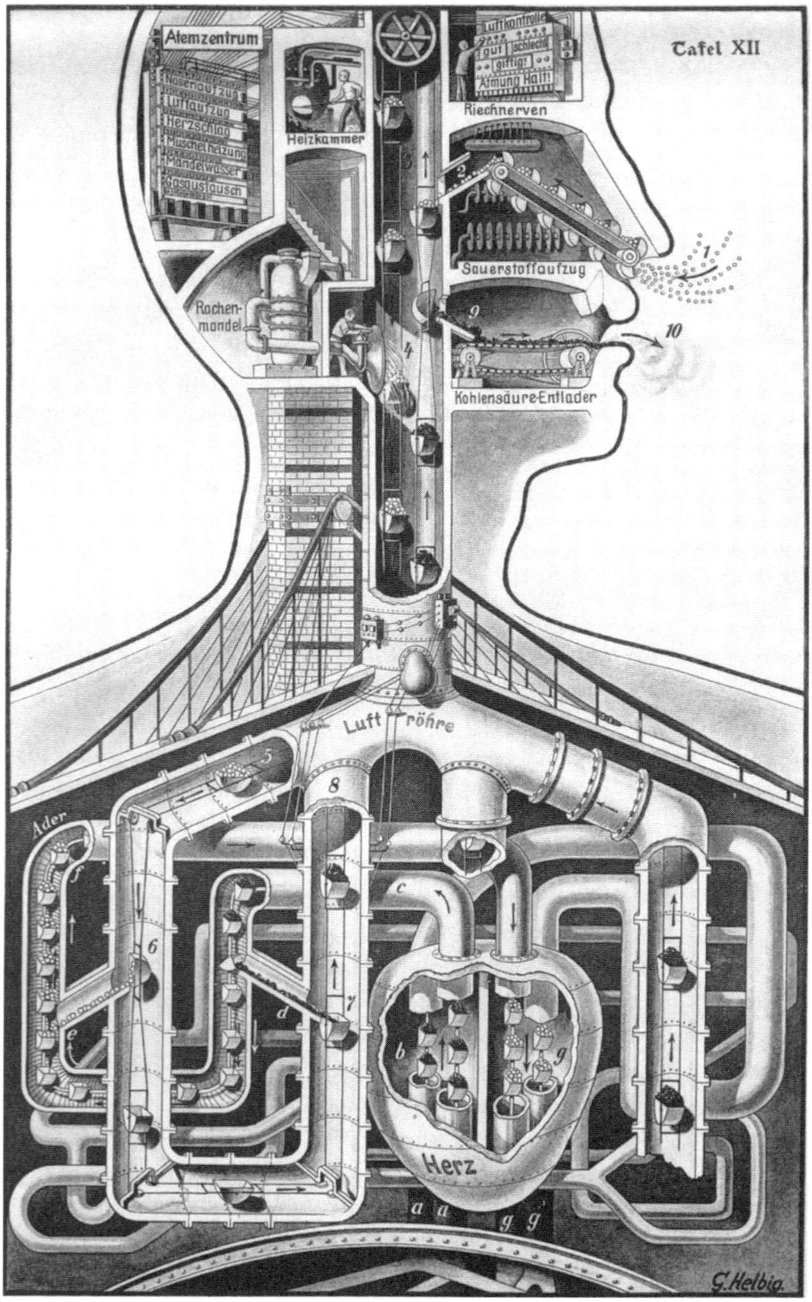

FIGURE 4.4. *"Human Respiration,"* Das Leben des Menschen *(1926), vol. 3, pl. 12; also sold as a 44 × 84 cm black-and-white poster. Artist: Georg Helbig. National Library of Medicine. Copyright Kosmos Verlag, Stuttgart.*

renderings in architecture, mechanics, set design, and engineering. In the middle decades of the twentieth century, the cutaway frequently appeared in exhibition displays, advertisements, engineering diagrams, architectural renderings, and illustrated books and magazines. No one has written its history, and I will only attempt a brief one here.[10] My goal is limited: to set out the iconographic sources of the cutaways that appear in Kahn.

Cutaway devices began to proliferate with the print revolution. Seventeenth-century Jesuit polymath Athanasius Kircher featured cutaways in his publications, most spectacularly a volcanic Mount Vesuvius. Cutaways also appear in the plates of the *Encyclopédie* (1751–72) of Diderot and d'Alembert—the cross section of a mining operation, the interior of a theater stripped of a portion of its outer wall, and, most memorably, the inside of an equestrian statue.[11]

These were imagined scenes, but the cutaway diagram had a dynamic relation to actual objects that were peeled, sawed in half, or sliced to reveal their innards. In the nineteenth century, there were anatomical cross sections (produced as specimens in actual human flesh, as well as

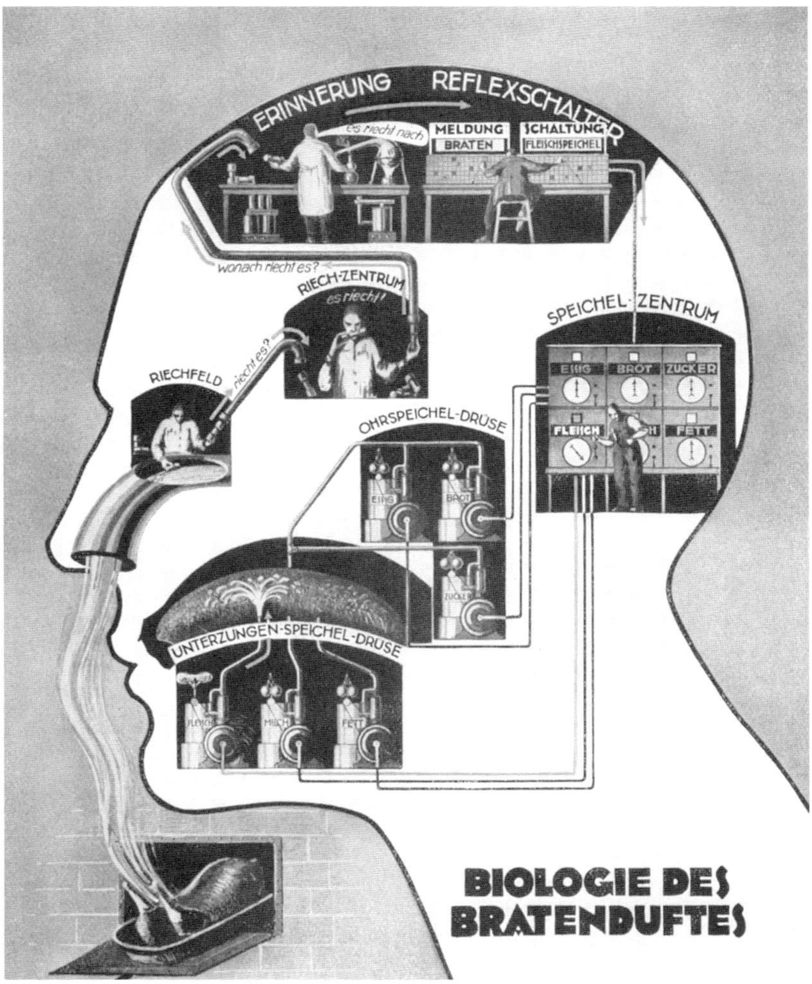

FIGURE 4.5. *"The Biology of the Smell of a Roast,"* Das Leben des Menschen *(1926), vol. 3, pl. 15. Three color. Artist uncredited. National Library of Medicine. Copyright Kosmos Verlag, Stuttgart.*

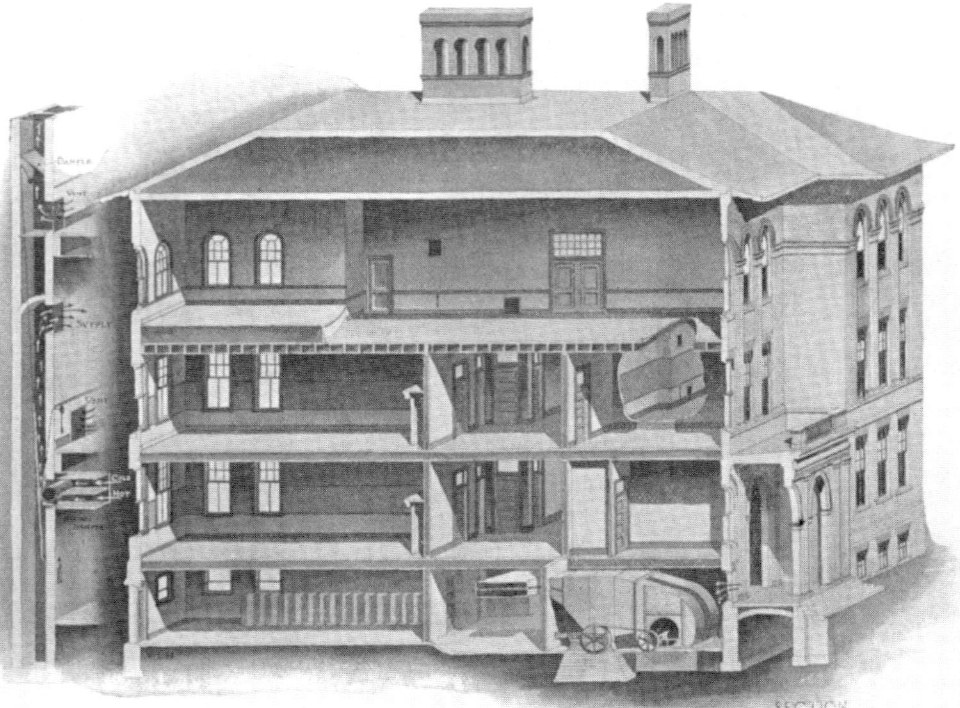

FIG. 84. AGASSIZ SCHOOL, BOSTON, MASS.

FIGURE 4.6. *B. F. Sturtevant,* Ventilation and Heating *(Boston: B. F. Sturtevant, 1896), 145. Courtesy Ken Swezey. A cutaway cross section shows the architecture of ventilation in the style of a dollhouse-type exhibition model. Arrows indicate the flow of air.*

models and two-dimensional illustrations) and special display versions of mechanical devices that had openings or doors, or that omitted a front or back wall, to show their inner workings. Perhaps the most common three-dimensional application of the cutaway was the dollhouse, which in the nineteenth and early twentieth centuries underwent technical development both through a dedicated cult of hobbyists and the emergence of the mass-manufactured toy industry. Another iconographic source was the theatrical stage, which opened its curtains on designed cutaways of furnished interior spaces. There were also scale models of stage sets. In dramatizing interiority, the theatrical cutaway externalized it, making the opaque private interior into a legible public space.

The cross-sectional display technique was obviously useful, and as extractive mining, architecture, industrial manufacture, and associated industries developed in scope and sophistication, so did cutaway illustrations. These in turn inspired new three-dimensional cutaway models and tableaux that appeared in large expositions, trade shows, museums, and other display venues. As the cutaway device became an ever more familiar part of the visual lexicon, illustrators and authors began exploiting it to make purely imaginary scenes for educational purposes, satirical commentary, or pure pleasure.

There was, of course, much overlap among all of these. By the late 1910s and 1920s, the cutaway could often be seen in the illustrated Sunday papers and monthly magazines. Perhaps its most prolific purveyor was *Popular Science Monthly.* Every issue featured cutaways that showed the interior of a device, factory, or work site where readers could see production processes in sequence, often staffed with busy little workers, a visual form that had some affinities to the sequential storytelling and image/text hybridity of its contemporary, the comic strip.[12]

On their terrain—the illustrated human body—Kahn and his artists developed the form and its rhetorical possibilities. They used internal arrows, numeration, and alphabetic coding to guide readers through a sequence of bodily compartments, left to right, top to bottom, a pictorial form that converges with writing. The cross section could be flat, like a floor plan, flowchart, or anatomical slice. It could also peel off the outer skin, rip open a facade, or neatly snip away some surface covering. It could reveal the innards in illusionistic perspective to convey depth and complexity: the interior as the space of dynamic action and interaction. Or it could flatten the three-dimensional object and turn it into a simplified, almost static, diagram. Kahn and his artists used all of these devices, singly and in combination, in relays or nested multiples.

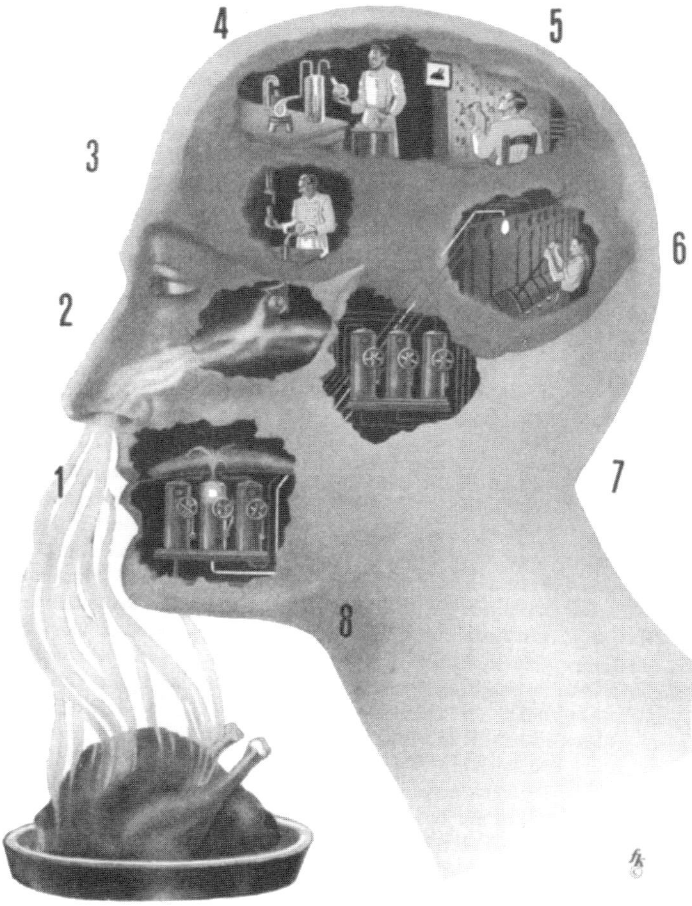

FIGURE 4.7. *"It Smells of Roast!"* Der Mensch Gesund und Krank *(1939), 1:323. Artist uncredited. National Library of Medicine. "The Biology of the Smell of a Roast" (cf. Figure 4.5) is here redone to be more dynamic and engaging.*

In *Der Mensch Gesund und Krank* (1939), many of the cross-sectional profiles are reduced to minimalist diagrams and anthropomorphic flowcharts. But things could also go in the opposite direction. "It Smells of Roast!," a reworking of "The Biology of the Smell of a Roast" by a different (also uncredited) artist, makes the figure friendlier, more human, by softly rendering the head as a drawing, not an outline, located in the space (the shading implies that it is lit by a light source), with a smooth surface of skin. Instead of ruled squares, the skin and skull are torn open in jagged pieces to reveal industrial scenes within, like the multiple vignettes of a film or photomontage. It's as though the 1920s-model body factory—the trope of the dollhouse populated by a labor force of tiny mannequins—was no longer sufficiently kinetic to convey the dynamism of the body in modernity.

DYNAMIC BODIES

Gesture plays no role in "Der Mensch als Industriepalast" and other body factories. For the most part, Kahn's industrial palaces are stationary fixtures. Motion takes place within the body, but the body itself moves not at all. The presiding metaphor is the body as building or container, with interior paths of conveyor belts, chutes, and pipes. After the pill-like molecules of air and food enter the perimeter through mouth and nose, they move along on tracks in regulated movement. The body factory is the outer perimeter of an unmoving human silhouette wrapped around a microindustrial order.

That was one way to show the body in modernity. But something more was wanted. If the physical universe, the industrial economy, the human psyche, politics, the evolution of biota, and everything else was beset by, or made up of, churning, incessant "dynamic forces" *(dynamische Kräfte),* then so was the human body.[13] And that needed to be dramatized for modern readers. A dynamic age needed dynamic bodies.

Kahn and his artists tried to evoke that feeling by showing bodies in motion, frozen in "the decisive moment" (photographer Henri Cartier-Bresson's still resonant phrase).[14] Contemporary newspaper photography, advertising, and cinema were profound influences. A blurb for Kahn's 1942 *First Aid Popular* boasted that the "illustrations are so arresting that the reader feels as though he were watching a film and forgets that he is engaged in the perusal of a scientific work."[15] Elsewhere, Kahn wrote that "modern biology" is "as interesting and exciting as the best news reel."[16]

But Kahn's illustrations evoked dynamic motion in ways that cinema could not. Powerful forces—depicted as bold lines, arrows, or rays—surged within ("The Soccer Player," Figure 4.8), or they streamed from outside onto and along body surfaces ("Sunbath!" Figure 3.4). At the same time, bodies and body parts themselves took on a dynamic appearance, as though shaped by machine processes and designed by industrial designers. In some illustrations, the body is almost transparent or has a reflective sheen. Skin and other body surfaces take on the look of shiny metal, molded glass, Bakelite, or plastic wrapping. Like the industrially produced manufactures that were wrapped in skins of shiny plastic, glass, or metal, Kahn and his artists offered industrially produced bodies as objects of desire.

In "The Physiology of the Knockout," a powerful force—a knockout punch—conveys energy from the punch of a disembodied gloved hand to a point of impact on the body's surface and then to points interior. The illustration, which appeared in a 1929 volume of *Das Leben des Menschen,* shows two ways that a boxer can be knocked out. For centuries, the disembodied hand was used as a graphic pointing device. It directed the reader's attention to "look here!" But in "The Physiology

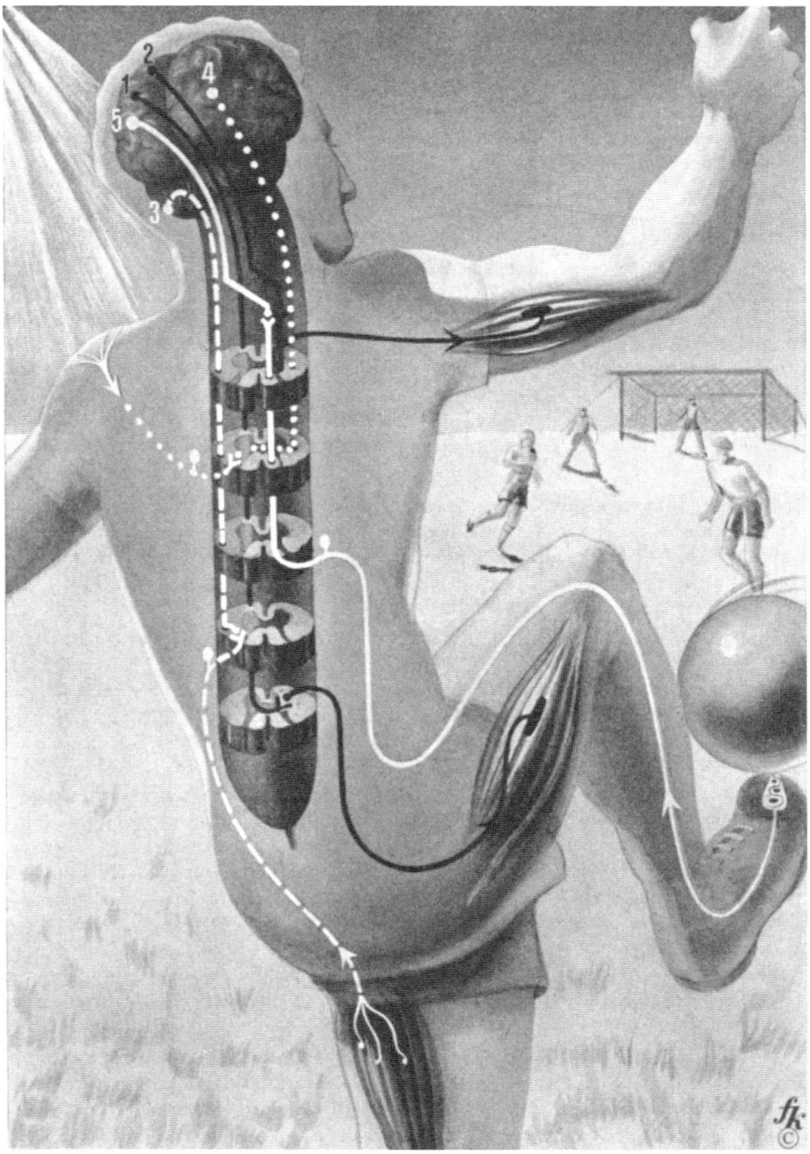

FIGURE 4.8. "*The Five Main Lines of the Spinal Cord . . . and Their Functions, Shown in a Soccer Player,*" Der Mensch Gesund und Krank *(1939), 2:159. National Library of Medicine.*

of the Knockout," the disembodied gloved hand has a "doing" function: it instigates action.

Kahn was perhaps dissatisfied with the illustration; the boxer is too delicate to evoke the aggression and violence of modern boxing. A second version, retitled simply "Knockout!," appears in the 1939 *Der Mensch Gesund und Krank*. The differences are telling. The boxer is beefed up and the scene made more dramatic. The number of figures is doubled; they are placed close together and overlapped to suggest movement in time. Another punch is added, suggesting a flurry of blows (which again materialize out of thin air, unconnected to any opponent). The new version now (misleadingly) appears to show a sequence from initial blow to unconsciousness on the canvas, although the caption explains that each figure is the recipient of a particular kind of knockout punch. Influenced by motion picture kinesis, "Knockout!" intensifies the reader's experience. An American reviewer, thrilled by the dynamism of Kahn's visuals, and perhaps with "Knockout!" in mind, writes, "Kahn pulls no punches. One amazing fact succeeds another, and vivid analogies strike blow upon blow."[17]

"The Soccer Player" and the "Knockout!" boxer heroically confront the reader in angled motion; they dominate the foreground. Unlike Kahn's body factories or silhouette figures, they present life in modernity as a vertiginously sensory experience. They are made to look and move a lot like the figures shown in contemporary movies, newspapers, magazines, and posters—figures that Kahn's readers were well accustomed to enjoying.

The figures do double duty. They model dynamic exterior motion, but they also seem to host within the interior of the body evocative

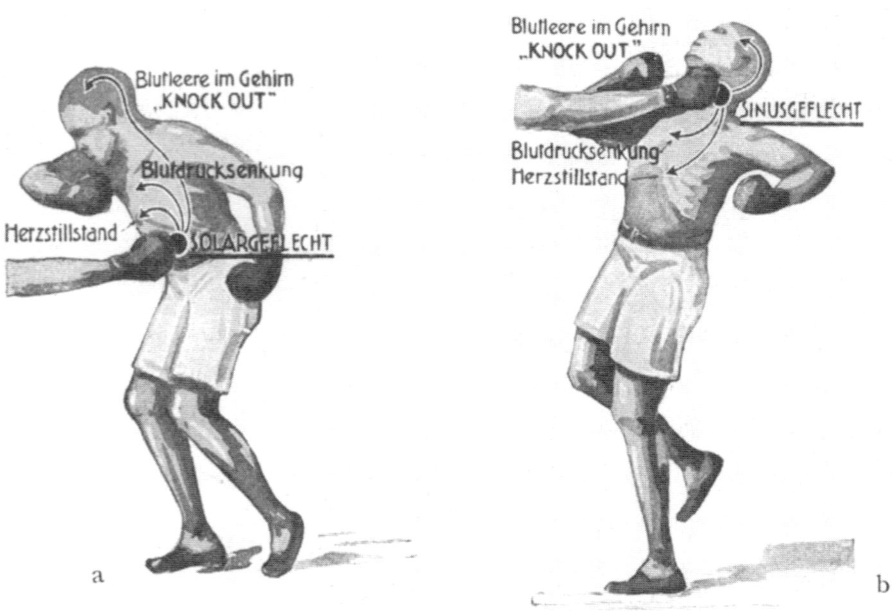

FIGURE 4.9. *"The Physiology of the Knockout,"* Das Leben des Menschen *(1929), 4:119. Artist uncredited. National Library of Medicine. Copyright Kosmos Verlag, Stuttgart.*

signs of force. In the "Knockout" of 1929, arcing arrows travel inward from the site of impact; in the "Knockout!" of 1939, little swarm of jagged lines spread between spine and liver; in "Soccer Player" and "Sunbath!" (Figure 3.4), there is a complex mix of arrows, dotted, dashed, and wavy lines and arcs.

Straight, circular, jagged, elliptical—Kahn's radiant lines have an iconographic history that goes way back. They may be seen in coins, mosaics, and the marginalia of old manuscripts. They run rampant in early modern treatises on geometry, astronomy, optics, physics, medicine, and engineering as well as in hermeticism, natural magic, and alchemy—an intoxicated domain of "occult qualities" and hidden forces.[18] And in religious art, they work as an iconography of spiritual power—radiant coronas, haloes, spheres, and auras.

There are other sources. From the print revolution, the engraver's burin contributes dashed, dotted, and wavy lines in series, parallel, and crosshatched thickets, all of which soon appear in freehand pen-and-ink drawing. The typographer contributes more dots and dashes, straight rules, and repurposed punctuation marks. By the nineteenth century, there is a well-stocked iconographic storehouse of lightning bolts, stars, exclamation points, glows, ripples, auras, beads, bubbles, and droplets—marks that appear in all sorts of illustrated popular medical and scientific texts, especially in anything to do with electricity, magnetism, or mesmerism. The icons also show up in the satirical drawings of Grandville, Wilhelm Busch, and other cartoon fantasists. By the 1910s and 1920s, they are increasingly recruited to serve as a visual

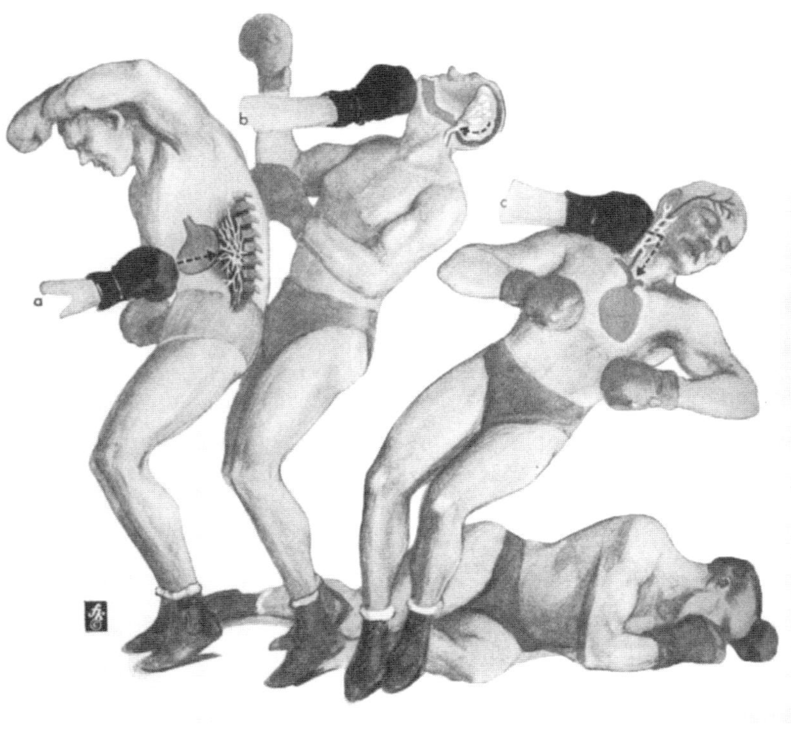

FIGURE 4.10. *"Knockout!"* Der Mensch Gesund und Krank *(1939), 1:212. Artist uncredited. National Library of Medicine.*

vocabulary of subjective sensation, impulsive action, and emotion—the brilliantly legible marks that signify emphatic pain, astonishment, fear, disorientation, anger, violence, noise, and commotion in comic strips and animated cartoons like *Felix the Cat*, *The Katzenjammer Kids*, *Buck Rogers in the 25th Century*, and E. C. Segar's *Thimble Theatre*.[19]

RADIANT MODERNITY

What especially connected these symbols to modernity was Wilhelm Röntgen's 1895 discovery of the X-ray, Marconi's work on radio waves, and the Curies' work on radioactivity. These hidden forces, recently discovered or invented, could make the interior of the body visible, glow and burn at a distance, and convey electrical messages through the air. Widely publicized, they were regarded as amazing wonders of the modern world. The public was gripped by a mania for rays and waves: X-rays, ultraviolet rays, radium rays, infrared rays, cosmic rays, gamma rays, brain waves, sound waves, and radio waves—as well as a cultural imaginary of exotic, miraculous, and deadly forces soon to be discovered or invented. Around the turn of the century, these forces start to receive representation in illustrated fiction, movies, and advertisements (especially for light bulbs and ultraviolet sunlamps). They show up in the works of Kahn, who was also charmed by radiant rays, beams, and waves, and eager to cater to his readers' fascination with the same.

Kahn took almost as much interest in physics as he did in medicine. He allotted considerable space in his publications for physics and physical forces as the setting for embodied life. *Der Mensch Gesund und Krank* opens not with the human body but with a discussion of astronomy and atomic particles, then goes on, section by section, to X-rays, cosmic rays, sunspot radiation "swarms," ultraviolet rays "of living substances," "hypothetical earth rays," "radioactive rocks," "action currents of the heart," heat, sunlight, gamma rays, alternating electrical current, and even "death rays."[20]

Science fiction was then emerging as a popular genre. Illustrators depicted rays, beams, and waves as sources of playful or not-so-playful disruption. They could annul gravity, make things larger or smaller, destroy buildings, even kill. But they could also protect, heal, rejuvenate, beautify, and confer strength—all sorts of miraculous things. Either way, the miraculous force served as an open-ended metaphor for the dynamism of industrial modernity. If the illustrated ray, beam, or wave made visible the hidden physical forces that surged in, through, and on the surfaces of our bodies, it perhaps also stood for the invisible social, political, and economic forces that touched everyone but were hard to narrate or see. For Kahn, rays were not cause for anxiety but for enchantment: that we live in a world of rays, he rhapsodized, was "one of the greatest discoveries of our times." Everyday life was like the world we "dream about": "radiant"![21]

Consider "The Nervous System Is a 'Relay System,'" a double diagram with double waves. The cropped spreading concentric circles signify the emission of invisible radiant energy. They also denote bodily sensation, heat, the content of a radio transmission, and maybe an electrical tingle. The scene is the nighttime metropolis (perhaps again New York). The shadow figure is paired with an elevator building, which metaphorically corresponds to the human body. The dark space of the body has an interior illuminated display, a simple electrical/neurological circuit. In parallel, the elevator building, also a darkened space but with lit windows, has a penthouse homunculus—the building's brain—who is the terminus of the relay. The telos in both cases is conscious apprehension.

An earlier version (Figure 1.2) fills in the urban setting with automobile traffic, street lamps, lit windows, pedestrians, and a disorienting

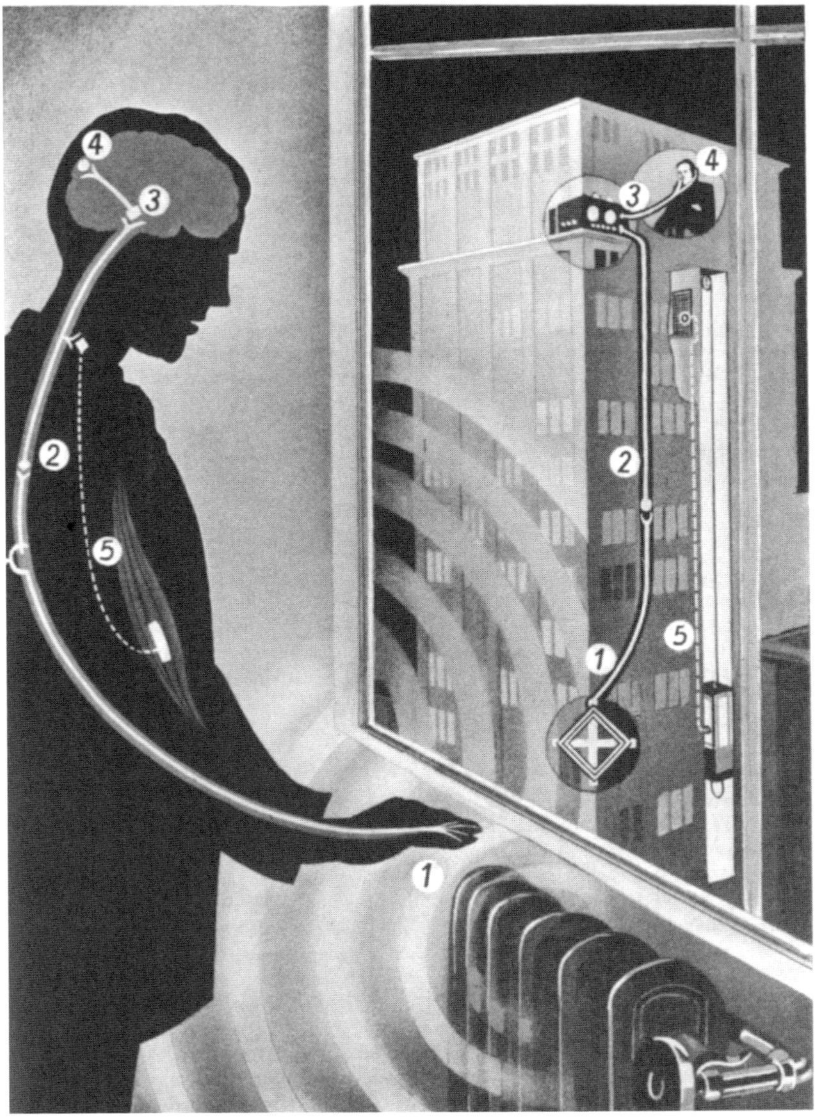

FIGURE 4.11. *"The Nervous System Is a 'Relay System,'"* Der Mensch Gesund und Krank *(1939), 2:143. Artist uncredited. National Library of Medicine.*

image within the image: the shadow body itself is figured as an image on an illuminated billboard on the side of a building, as a message transmitting for public reception. The caption explains that the body, like the building, is the housing of a radio transmission system. The brain is the bodily equivalent of the man in the penthouse who listens to the radio; the nerve ending at the skin's surface, which receives the sensory input, is a radio antenna. All of this doubling stands for the reader's relation to the illustration (and to the entire medium of printed images). More than that, the illustration argues that sensation is a chain of mediations produced by outside stimuli carried by technological devices, both inside and outside the body. This makes things complicated; it goes beyond doubling. An ever-growing network of teleradio connections (like the ever-growing network of representational correspondences) crosses the boundary between self and society. The human body becomes a device of devices, one linked to an external web of devices in the industrial city. The distinction between self and technological environment collapses. Electromagnetism happens both inside and outside us.

Both versions of the illustration link radiant night light to electricity. In the 1920s and 1930s, cities all over the world were being electrified and lit up. New York was the model, with the lights of Broadway, illuminated skyscrapers, and the never-ending flow of headlamped traffic. In Germany, the enthusiasm for the lighting and light design of the night city culminated in the 1928 Berlin im Licht festival, a week

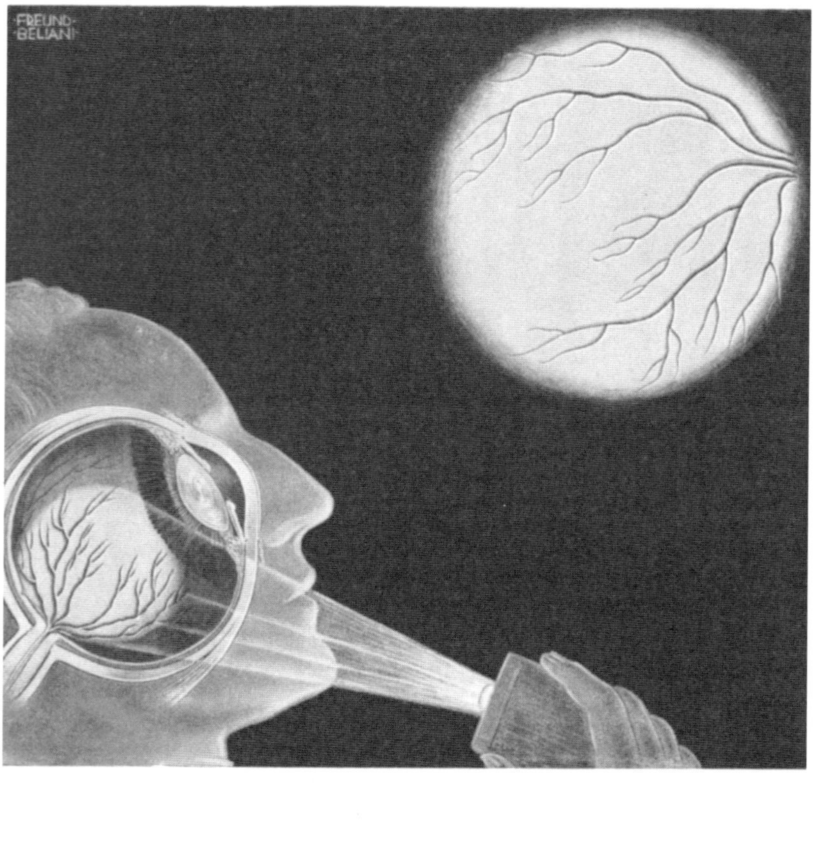

FIGURE 4.12. *"The Purkinje Vein Figure,"* Das Leben des Menschen *(1931), 5:61. Artist: Alwin Freund-Beliani. National Library of Medicine. Light projected into the interior of the bizarre, disproportionately large eye is remediated as a moody external projection that demonstrates the Purkinje veins. Copyright Kosmos Verlag, Stuttgart.*

of dazzling evening light displays. But this was only a small part of the passion for new electric light technologies. The motion picture and motion picture industry, founded on techniques of electric light projection, had a devoted mass following. Neon lighting for signs was new and spreading rapidly. High-intensity arc lamp searchlights could be seen sweeping the night sky. The mystique of gaslight was transferred to the night world of electric illumination, *modernité noir.*

The nervous system illustrations don't only refer to visible or palpable radiance. They also conjure up the new and miraculous technology of beams and waves that had only recently come to pervade everyday life with disembodied speech and music (as well as unaccountable static, squeals, and hums): wireless radio, which was then in the early days of broadcasting. (Radio, of course, was not entirely invisible or impalpable; the units came with vacuum tubes that lit up and got hot, and the apparatus of the transmitter, antenna, radio set, plugs, and wires could potentially give shocks to the unwary.)

Consider another dark image, "The Purkinje Vein Figure." In profile and in cutaway, a man shines an electric flashlight into his own (disproportionately large) eye to demonstrate the pattern of veins at the back of the eyeball, making a strange light show that traverses from interior to exterior. The technique, invented in the early nineteenth century by Czech physiologist Jan Purkinje, had become by the 1920s a routine part of the ritual of the modern eye examination.[22] The reader is presented with the vein pattern both in the cutaway of the man's face and in the mysterious vein planet hanging in the night sky. Looking up, the gracefully lit, almost sculptural figure (brilliantly rendered by artist Alwin Freund-Beliani) stands in for the self-surveying and enlightened reader. Enlightenment is revealed in sublime darkness. Humanity, enlightened, emits light. Kahn comments, "No one who perceives . . . this effect for the first time can conceal his delight in the appearance of this enchanting little vein tree." Here again, Kahn and his artist feature an invention that was, to them, a device of industrial-era wonder: the focus-beam battery-powered electric flashlight (so familiar to us) only became widespread in Europe in the 1910s and 1920s; it still retained the power to serve as a signifier of the modern.

Kahn and his readers anticipated that more technological wonders were on the way. In "The Doctor of the Future," he presents a doctor in his office monitoring a patient who is on a steamship (in midocean) on the other side of the planet, via radio waves that transmit medical information in both directions. The wireless wave transmissions correspond to displays of graphed heartbeat and brain waves. The setting is the modern world of ocean liners, elevator buildings, big glass windows, modernist furniture, and microphones, as well as something new, television, which was widely expected to be the next transformative media technology.[23]

Industrial technologies of radiation paralleled, perhaps channeled, the protean forces of natural radiation. In "The Biological Effects of

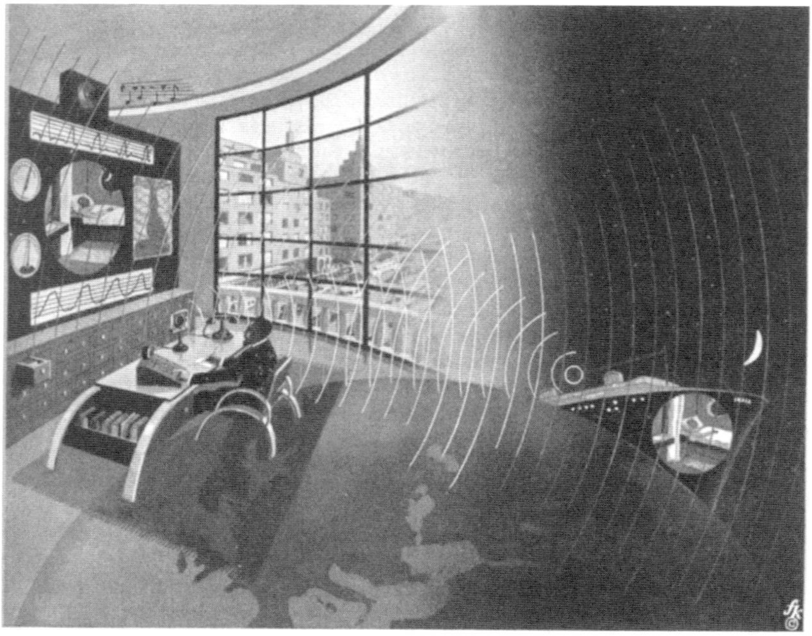

FIGURE 4.13. *"The Doctor of the Future,"* Der Mensch Gesund und Krank *(1939), 1:199. Artist uncredited. National Library of Medicine.*

Ultraviolet Light," sunlight, with its spectrum of visible, infrared, and ultraviolet light, is treated as a natural technology and represented with the same iconographic vocabulary used for radio and other devices of industrial wonder. Like those devices, it occurs on the inside and outside the human body, and travels in both directions. The science of rays is not just about the rays that are cast upon us but also about those we generate. We, who live in radiant nature and amid radiant technologies, are radiant beings: "The interior body radiates" and generates healing "growth rays." This identification of rays with life makes for an odd technoenvironmentalist perspective, obscuring the fact that many forms of natural and man-made radiation are deadly. For Kahn, the radiant outer environment fuels and sustains our radiant internal environment—and that intoxicates. Kahn exults, "The brain, blood, muscle and gland, every cell in our body is a microscopic transmitting station—Man does not [only] live in a radiant world, but is himself a brilliant being! This is the great, perhaps the greatest, discovery of modern science."[24]

The focus on Kahn's body factories tempts us to identify him with the nineteenth-century mechanist tradition of Helmholtz and du Bois-Reymond, scientists who sought to demystify the world through an antivitalist biophysics of mechanics, objective measurements, and mathematical equations. But if Kahn was a mechanist, he was an enchanted one. In his illustrations, the body pulsates with radiant energy. His modernity was not Max Weber's disenchantment (or the Bauhaus's abstractionist functionalist purity, though some of his artists

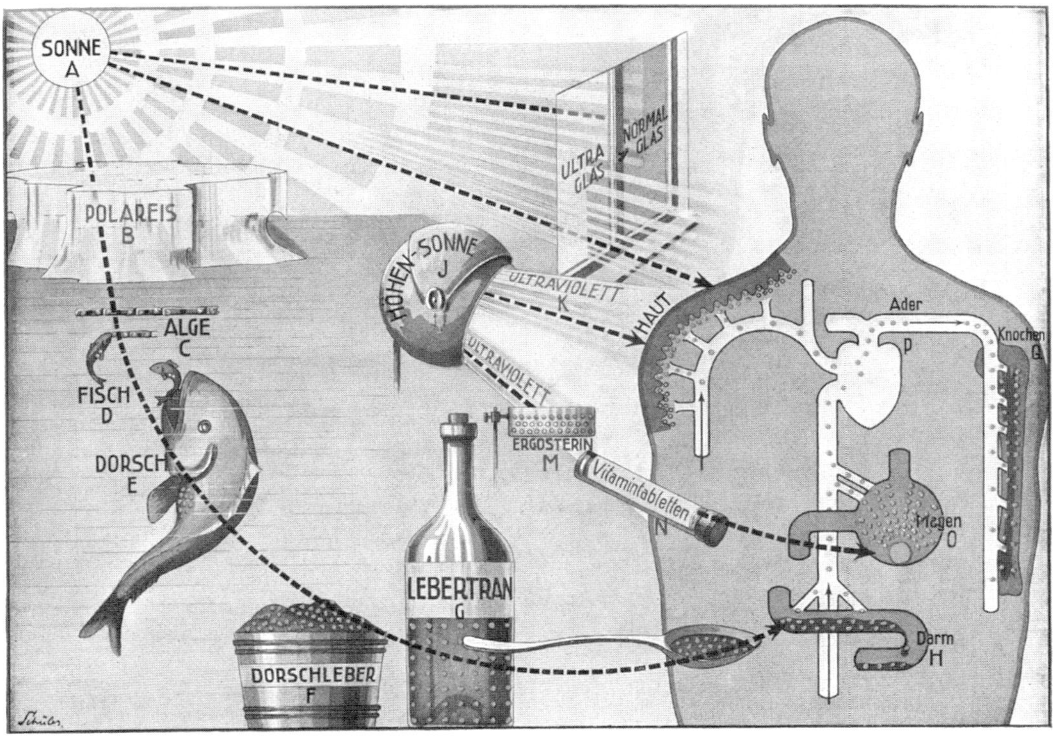

FIGURE 4.14. *"The Biological Effects of Ultraviolet Light,"* Das Leben des Menschen *(1929), vol. 4, pl. 27. Two color. Artist: Fritz Schüler. National Library of Medicine.*

were aligned with that). Rather, his modernity was charged with the grand synthesis of German Romanticism. His world was made up of primal powers. He explained these as physical forces, but they were also a quasi-Nietzschean mystical life force. Kahn and his artists thus sometimes seem to be reaching for joyful rapture, a never-attainable representational swoon.

SYNOPTIC MODERNITY

Kahn concludes his 1939 meditation on rays with "Man in the Radiant World," an illustration that brings together different types of rays around the figure of a man, with rays also radiating outward from his heart.[25] The illustration is a clumsy synopsis, a simultaneous view of elements. Unlike Kahn's principal pictorial mode, which is to visually explain how things work, visual synopsis is a genre that sums up something by showing the things that constitute it. The term "synopsis" usually refers to a brief description of contents, a short summary. But visual synopsis is usually selective—a medley or a highlight reel, not a full table of contents, a graphical overview that depends partly on what looks good and what fits together.

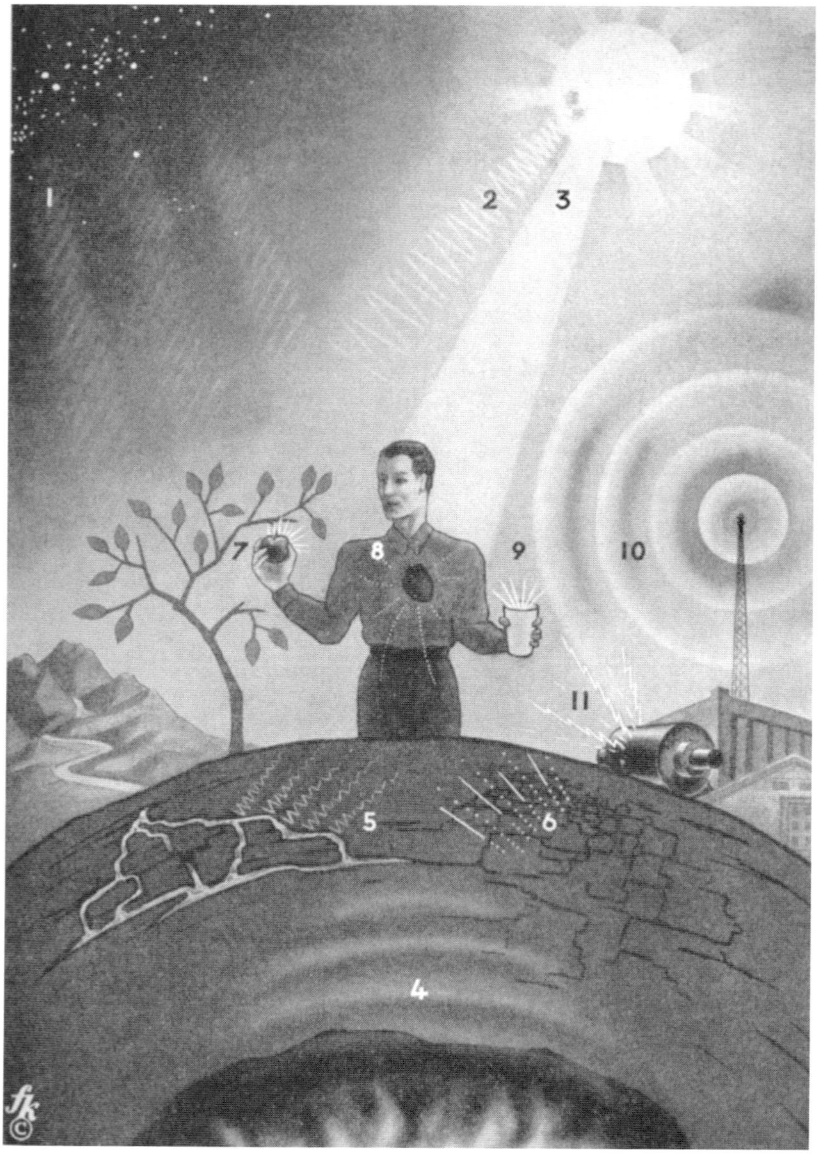

FIGURE 4.15. *A visual synopsis. "Man in the Radiant World,"* Der Mensch Gesund und Krank *(1939), 1:17. Artist uncredited. National Library of Medicine.*

The form, once again, did not originate with Kahn. For centuries, title page and frontispiece illustrations had presented a gloss of a book's contents in imaginative displays that combined typographical elements with allegorical and mythological figures, fantasy architecture, and icons that alluded to what was treated within. Such images were cast in the form of entrances of buildings, with attendant caryatids, bas-reliefs, and freestanding statues, thereby suggesting that readers would be entering into a larger order of things: the realm of civilization, reason, art, beauty.

By the 1920s and 1930s, visual synopsis rarely followed those conventions. It was a staple of commercial illustration, used on dust jackets and film posters, and in magazine illustrations and advertisements. Synopses were not necessarily modernist, but in juxtaposing and choreographing disparate elements, they provided juicy opportunities for commercial artists with modernist inclinations. Kahn's artists used synopsis to mimetically evoke the jumbled swirl of modern experiences—modernity's proliferating disorienting technologies, spaces, objects, textures, and sensations (Plate 5). An assignment to make a synoptic illustration was a chance to play with the picture, fool around, and tease the viewer.

Take, for example, Roman Rechn's dust jacket for volume 3 of *Das Leben des Menschen,* which brings together some of the book's themes—the digestive system and the physiology and chemistry of nutrition—as an asymmetrical modernist mise-en-scène. In the foreground, a boy looks directly at the reader (Plate 6). His coffee cup is a surrogate smile, while behind him a riotous field of monstrous tongue-like (phallic!) intestinal cilia rises up. Behind the boy's left shoulder, a floating manufactured disk (a chemical diagram of food) falls apart, scattering debris made up of its modular elements. In the book, the human body is made to splice together incompatible objects of different manufacture, origin, scale, texture, angle, and velocity. So does the dust jacket picture, which combines photography, mechanical draftsmanship, and freehand drawing. It almost calls out to the reader, with the direct address of advertising: Enjoy this.

Rechn, a committed modernist, was Kahn's most brilliant visual synopsizer. In "The Iris-Key," the assignment is to make the iris into a synopsis. An eye, stocked with a mishmash of detached body parts, represents the iridological method, which is to read the iris as a diagnostic map of the entire body. Changes that appear in specific locations in the iris reveal the presence of disease in distant parts of the body.[26] But there's more: the eye, and the body fragments collected within it, appear to be spinning, like a roulette wheel. It's as though Rechn designed the image of the eye to stand as the "period eye" of modernity, as if he used his assignment as an opportunity to conflate our view of the eye with the eye's view of us. We are asked to see the body world as a disorienting jumble of proliferating, recombinant body parts rotating around a central axis: the eye.[27] This somehow synopsizes our experience of embodied life in the present moment.

KAHN'S WORLD: AN ICONOGRAPHY OF THE UNIVERSAL BODY

"The Iris-Key" does one other thing: it equips the iris with circular lines of latitude and longitude, a grid that suggests a kind of correspondence between eye, body, and world in modernity. Over the course of the 1920s and 1930s, images that supported a universalist view of humanity gained rhetorical salience as fascism and the politics of exceptionalism grew in power and ferocity. Hitler raged against the Jews, a transnational race, but also against "democratic-pacifistic internationalized people" who, he claimed, "contaminated" the nation.[28] The National Socialists invoked the unique moral/racial superiority of the German

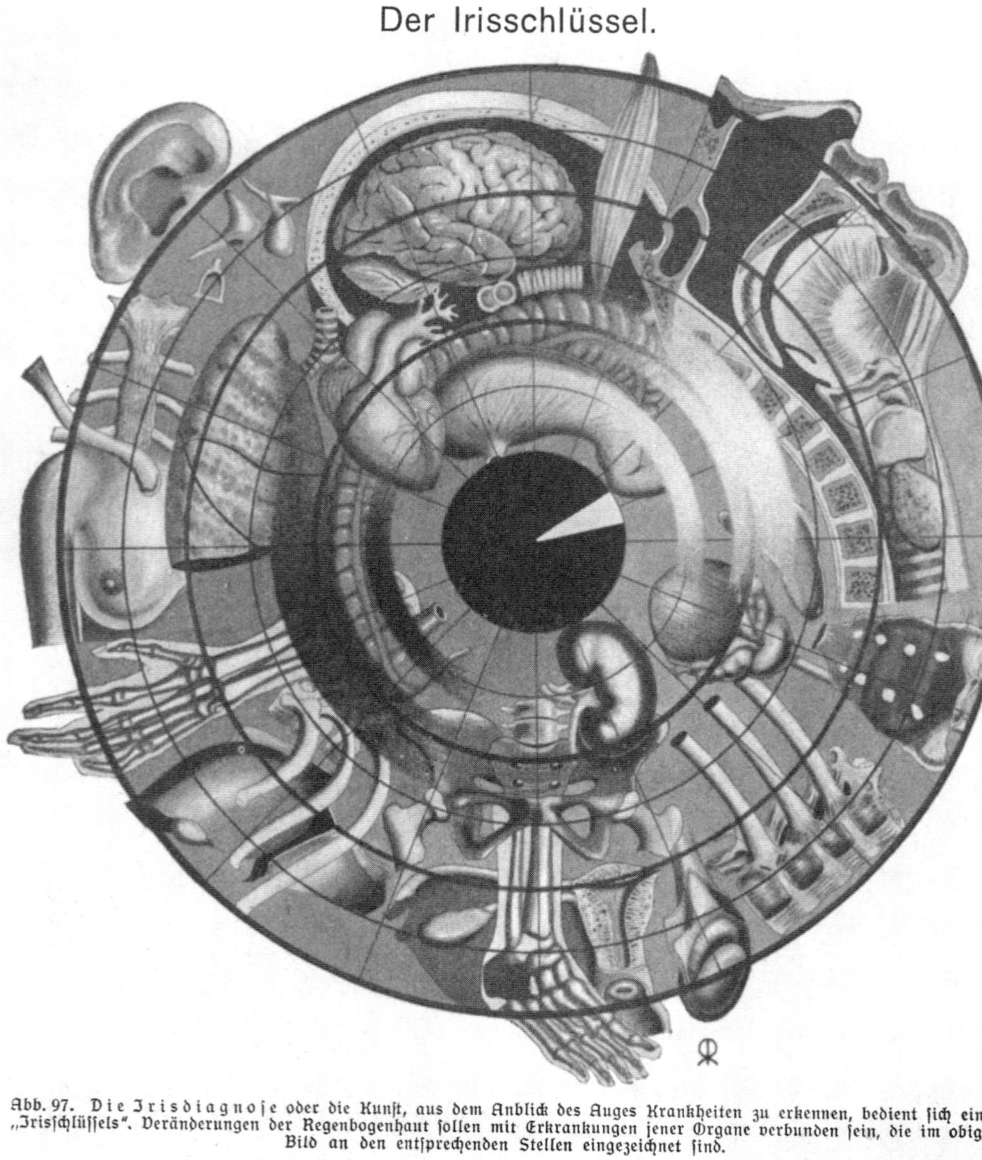

FIGURE 4.16. *"The Iris-Key,"* Das Leben des Menschen *(1929), 4:129. Artist: Roman Rechn. National Library of Medicine. Copyright Kosmos Verlag, Stuttgart.*

Volk and *Reich* to justify their program of permanent war, persecution, torture, and murder. As an antidote, the universalist opposition offered in one register modern science, and in another modern art. Both claimed to discover and express universal truths that transcend and dissolve national and ethnic borders, differences, and prejudices. The "votaries of science," Waldemar Kaempffert argued in a 1942 essay on "Science, the Machine and Democracy," "constitute an international brotherhood." Science, in tandem with "machine methods" and "mass manufacturing," is "a force in achieving true internationalism,"

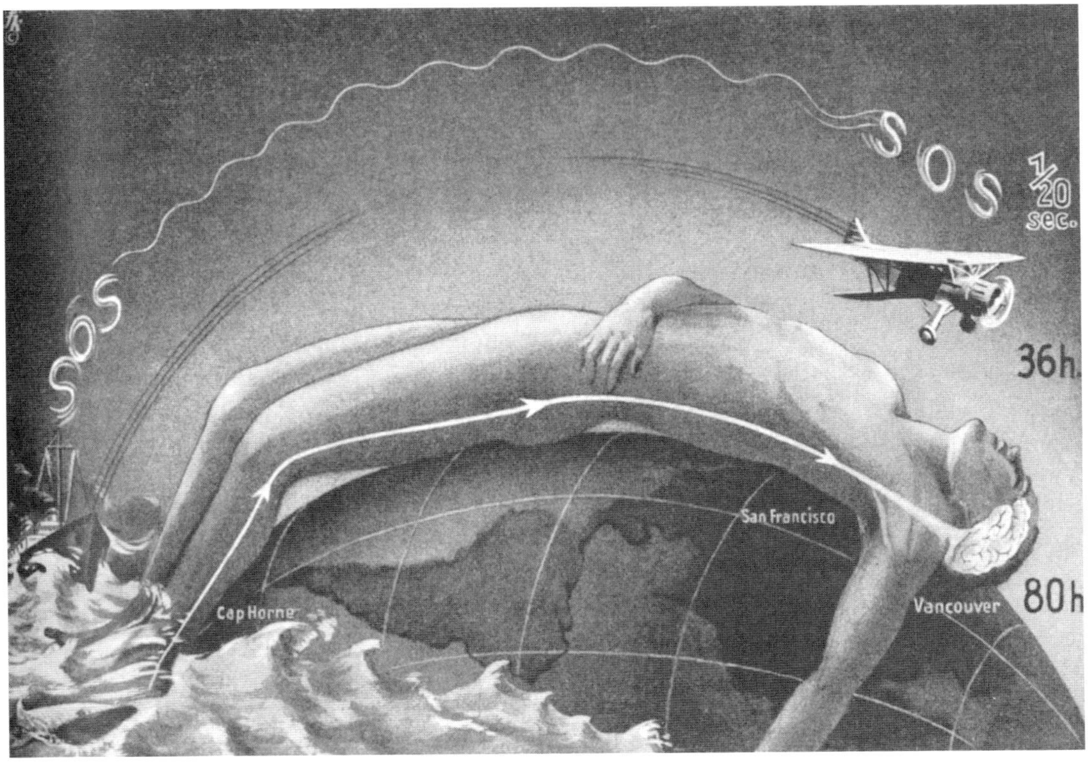

FIGURE 4.17. *"The Speed of Thought—Surpassed by Technology!"* Der Mensch Gesund und Krank *(1939), 2:137. Artist uncredited. National Library of Medicine.*

a "subjugation of self and country . . . for the good of the [human] race."[29] Science, a global project that belongs to all humanity, works to affirm the human as our primary identity, against racist or nationalist conceptions that promote one group at the expense of all the rest. This one world–ist, one race–ist humanist scientific internationalism gained in intensity with the start of World War II, the formation of a global antifascist alliance, and the mobilization of scientists and artists in the great struggle.

Abstracted representation of the human figure—strategically emptied of specificity (often in some neoclassicist idiom that signified communion with a transcendent international high culture)—contributed to this universalist gestalt. As we have seen, Kahn deployed many such figures. But universalism had an even more explicit iconography: stylized representation of the planet. The earth could be figured as a shaded, illusionistic globe floating in space or as a flattened diagrammatic circle (as, for example, in the United Nations logo, which was designed in 1945). In the 1930s, depictions of the globe in political cartoons, travel posters, logos, and children's books, along with the installation of globes and world maps in classrooms and bourgeois households, were a routine feature of everyday life. This was the cultural milieu that philosopher Martin Heidegger was responding to when he marked the present moment as "the age of the world picture" and proclaimed, "That the world becomes picture . . . is what distinguishes the essence of the modern age."[30]

Not surprisingly, then, Kahn and his artists piled on by showing bodies and body parts in conjunction with the planet earth. In "The Speed of Thought—Surpassed by Technology!" a naked man, signifying the body, lies recumbent on the planet, encircled by wireless radio signals and an orbiting airplane, his corporeal measurements and processes scaled up to global proportions. It's an impossible scene, an imaginary "aperspectival" perspective (perhaps borrowed from the Universal Pictures logo, which also featured a biplane circling the globe). According to historian Lorraine Daston, "Aperspectival objectivity became a scientific value when science came to consist in large part of communications that crossed boundaries, nationality, training and skill."[31] "The Speed of Thought" is premised on the universalist belief that humans the world over share identical bodily processes and features, and that this physiological commons corresponds to a techno-scientific commons that (amazingly) exceeds the human organisms that originated it. Global humanity is metaphorically composed of, and globally connected via, technologies of transportation (the steamship and airplane), telecommunications (wireless telegraphy), mapmaking (the globe itself), and naming (the geographical place-names that correspond to the head, heart, and feet of the figure).

Modernity is global; the global perspective is modern. "The Speed of Thought" ratchets up this mutually constituted universalist ethos to almost apocalyptic intensity by the Blakean styling of the figure: a universal man, prostrate, naked, lacking any cultural or ethnic markers. For us, of course, the figure is marked at birth. We, the readers of the twenty-first century, have come to regard all universalisms as in some way parochial, as artifacts of place and time. That was not the case in the late 1930s. For Kahn, place and time were linked to exile and dispossession. In that moment, placelessness must have seemed like both a moral necessity and the human condition.

All of that in a little picture lesson on the transmission of sensation: "The speed of thought—has been surpassed by technology! A nerve current travels at a rate of 250 kilometres per hour. A person reaching from Cape Horn to Alaska would first feel a shark's bite after eighty hours, and it would take just as long for the volitional impulse to retract the foot to pass from the brain to the toes."[32] The visual comparison, designed to give readers a proportionate sense of the mechanics and timescale of human sensation, revels in the disproportions of metaphorical global scale and the instantaneity of radio communication.

In "Man Walking Is a Pendulum Machine," disproportionality is an estranging tactic. The illustration and accompanying text suggest that the globe is like a large ball and that walking on it is a circus trick. The picture is meant to illustrate a lesson about how we manage walking

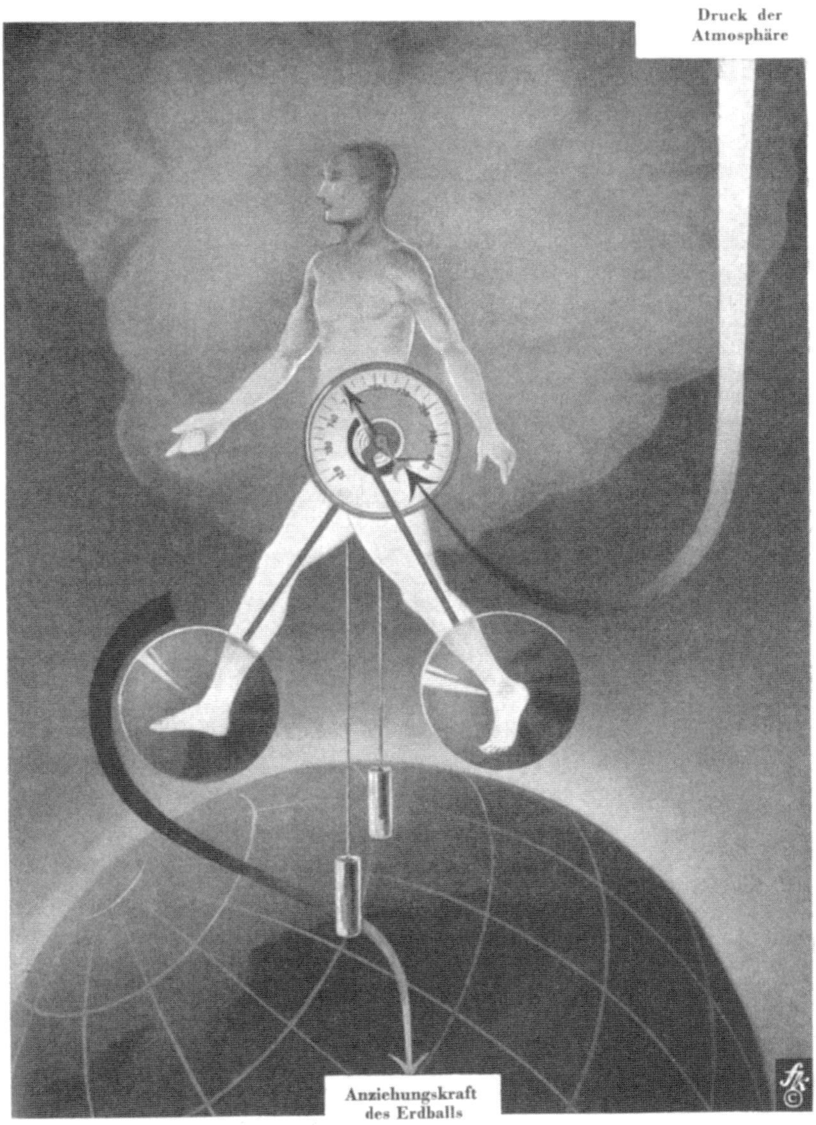

FIGURE 4.18. *"Man Walking Is a Pendulum Machine,"* Der Mensch Gesund und Krank *(1939), 2:175. Artist uncredited. National Library of Medicine.*

using our skeletal structure (the "precariously constructed bone tower known as Man") to work against gravity with the aid of atmospheric pressure (represented by the moody cloud behind the walker).[33] Here again the illustration says much more than the text. It marries an abstracting neoclassicist modernism—another expressionless naked Man—to the technology of gauges and precisely balanced pendulums. Together, figure and tilted globe model universal modernity.

Kahn's text describes walking as a circus trick that requires human beings to mobilize air pressure as they walk on "thigh stilts" to balance in gravity. However, the figure of the pendulum man doesn't take up the metaphor; instead, it appears to march and at the same time dangle, puppetlike, with feet in the air, not on the ground—things not easily reconciled. Above the earth, a gray, portentous cloud looms behind the eerily lit figure. Evocative and beautiful, the picture does its explanatory work: it teaches a lesson on the mechanics of walking. But also suggests an unnarrated backdrop: clouds of fascism and war—the dark side of modernity—threaten to overwhelm the global universal. Man strides (purposefully?) into the utopian future, or into the apocalypse.

IMAGE HYBRIDS: PHOTO MANIPULATIONS AND JUXTAPOSITIONS

Kahn's illustrations mix together different kinds of objects—organic body parts and industrial technologies—and different kinds of pictorial elements—figures, numbers, clouds, city scenes, directional arrows, globes, silhouettes, wave patterns—to teach science lessons that evoke a sense of enchantment in modernity. They also sometimes mix together different types of images, produced in different ways and different eras, such as engraving, lithography, drawing, and painting.

For Kahn and his artists, this was mostly standard operating procedure. Trained by everyday experience, readers were not surprised to encounter a mishmash of visual elements. But in one genre—photomontages and painted photographs—the mash-up directly invited readers to attend to the combining of disparate elements in a single image as a distinctively modern way of making pictures.

Media theorists Jay David Bolter and Richard Grusin call the photomontage, and other kinds of re-presentation, "remediation": something that occurs whenever a new medium emerges. By incorporating the techniques and content of earlier media, new media acquires a preexistent vocabulary and syntax of signification, and stakes a claim to cultural significance by paying homage to, rivaling, and refashioning the productions of predecessor media. Bolter and Grusin focus on digital media, but they rightly insist that remediation does not only belong to the media-voracious digital age. Earlier media also remediated predecessors: photography remediated the forms and genres of painting and drawing; film remediated the forms and genres (and many of the scripts) of the theatrical stage; television remediated film, vaudeville, and radio; and so on.[34] However, remediation is never only a linear progression. The remediated past restructures our view of the present even as current productions restructure our view of the past. There is also horizontal and reciprocal remediation: media contemporaries remediate each other—for example, smartphones remediate personal computers, which remediate smartphones.

Hungry for images to populate their pages and pictures, Kahn and his artists were great remediators. They made halftone versions of engravings and lithographs taken from previous centuries, used alone or in combination with new material. And they used halftone photographs and line art from contemporary publications and exhibitions.

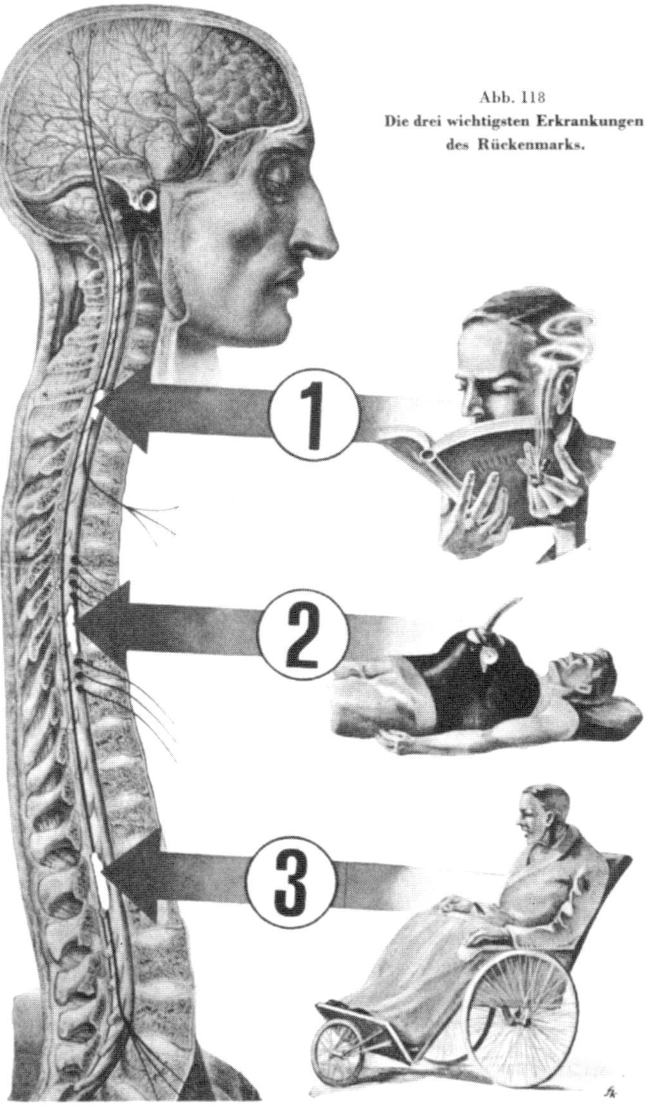

FIGURE 4.19. *"The Three Most Important Illnesses of the Spine,"* Der Mensch Gesund und Krank *(1939), 2:161. Artist uncredited. National Library of Medicine. The spinal figure, taken from a mid-nineteenth-century chromolithograph by Bourgery and Jacob, is here equipped with diagrammatic elements and painted vignettes.*

As we have seen, Kahn even remediated his own material. His artists painted over and drew over; they excised figures and backgrounds. They combined and recombined, making all sorts of changes, minor and major.

In all of this hypermediated remediation, the photograph had special qualities and uses. Almost from its inception in the first half of the nineteenth century, photography was acclaimed as an emblematic technology of modernity, a historic achievement that transformed everyday life. The photograph was something more than persuasive rendering. It was a work of "mechanical objectivity" made by light on treated paper or glass; it was not dependent on human skill, imagination, or bias but rather was a transcription of a particular thing in a particular place at a particular time.[35] As such, it had a higher epistemological status—a "reality effect."[36] Full of extraneous information and details that had no ostensible function, the photograph had the appearance of a disinterested view. It seemed not to make truth claims; it just appeared to be true. It had an evidentiary value that no eyewitness account, descriptive drawing, or painting could compete with. This in part explains why the photograph, once it had become cheaply reproducible in the form of halftones, so thoroughly supplanted engraving as the dominant pictorial mode of the illustrated newspaper.

Even if intended to make some particular point by showing some particular thing, the photograph continually surprises by showing other things, trivial or important, that no one has noticed: a dimple, a crease in a shirt, the bend of a wire, a person strolling by. The photograph undertakes an eccentric survey of surfaces, textures, foreground, and background; it freezes a fixed but somehow open-ended view in an image. It is powerfully convincing. Photographic realism became the new standard for what art historian Martin Kemp calls "the rhetoric of reality."[37]

As such, it had profound cognitive effects. Vision came to be reconceptualized as a camera taking continuous pictures over time. Every view was a photograph; experience was a motion picture. Siegfried Kracauer in 1927 wrote, "In the illustrated newspapers, the world is turned into a photographable moment and the photographed present is completely eternalized."[38] Those contradictions make the photograph amenable to surrealism and psychoanalysis; photography seems to live in the borderlands between yesterday and today, consciousness and unconsciousness. It captures and makes evident all of the visual noise that we normally tune out or forget.

Views can, of course, be staged, and so can photographs. From the very beginning, photographers took delight in using their epistemologically privileged technology to show things that were obviously contrived or impossible. People painted on photographic plates and prints to improve on reality or to goof on it (or to cover an artifact of light or a blemish of the photographic process). They also sought to make photographs more effective by painting out the background or other material deemed irrelevant to the purpose at hand, be it for educational purposes or merely to enhance the viewer's pleasure. They took delight in making photographs do unphotographic things—to have photographic patches of reality freely consort with the impure world of paint and line drawing, or consort with other photographs made in some obviously different place or time. People loved to fool around with the size, shape, and setting of photographic images within the frame of the picture or on the page. A photograph could be restaged to do things that previously only purely imaginative media could attempt: political satire, time travel, fabulism, metaphor, pure nonsense.[39]

After World War I, Dadaists, constructivists, surrealists, and other modernists developed photomanipulation and montage into a recognizable genre. They made pictorial mash-ups that juxtaposed disparate photographic, painted, and design elements to make visually arresting, often brilliant images.[40] Their photomontages playfully rehearsed and reshuffled the prolific jumble of accidental juxtapositions that were the stuff of print culture and everyday life. By mixing the flotsam and jetsam of mass-media illustration and typography, bits of scientific diagrams, reproductions of fine art, and anything else that came to hand,

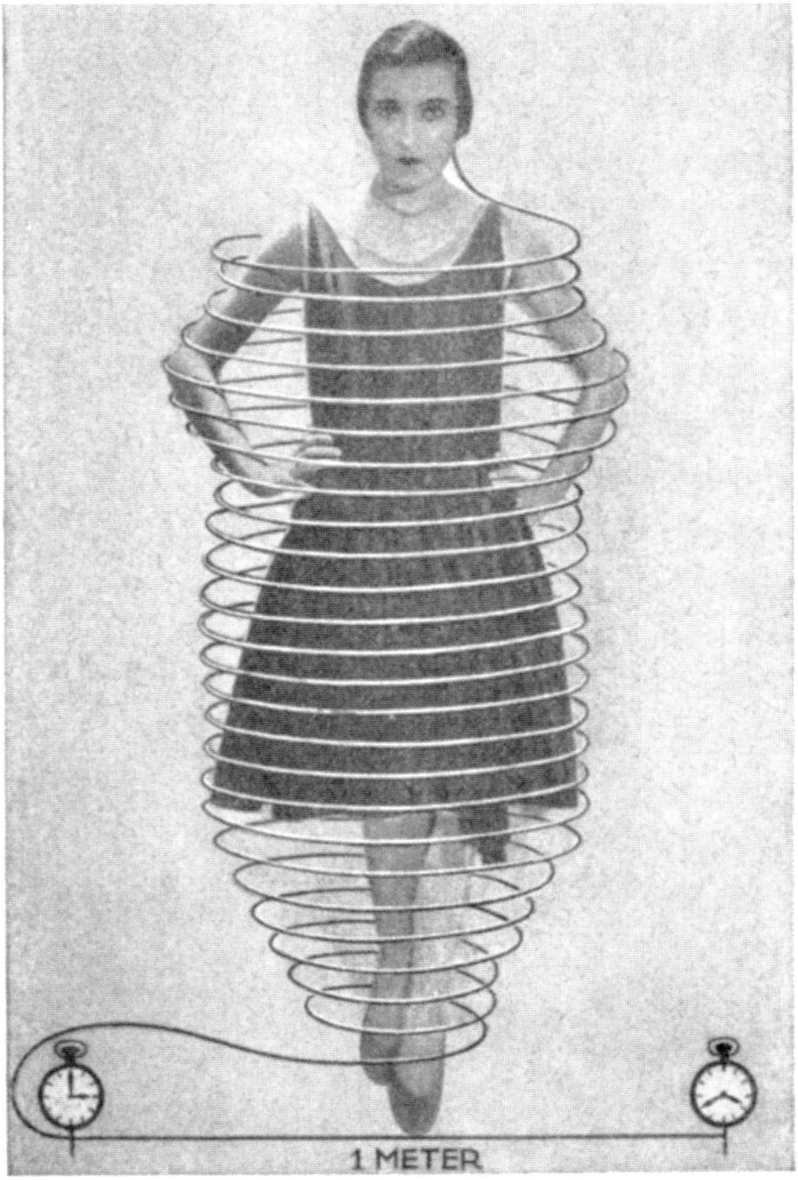

FIGURE 4.20. *"The Daily Growth of Hair,"* Das Leben des Menschen *(1929), 4:236. National Library of Medicine. "The human body produces daily 30 meters of hair. . . . Every 40 minutes hair grows one meter." In this playful photo manipulation, hands on hips, a young woman stares provocatively at the reader. A spiral (of hair) cages her or envelops her like a force field, or suggests that she is spinning like a top. Copyright Kosmos Verlag, Stuttgart.*

they made eye-catching dystopian/utopian critiques of capitalism, fascism, and utilitarian reason that turned the world upside down and sideways.

Under the influence of that prankish disorderly aesthetic, Kahn and his artists also took to photomontage. Some of their pieces take a rough, almost Dadaist approach, using found images, or they mimic the cheeky look of contemporary advertising (Figure 4.20). A few others (Figures 3.1 and 4.21) look as if John Heartfield, the genius left-wing political photomontagist, had been recruited to make seamlessly brilliant science lessons. Nothing in Kahn's written text takes notice of the fact that "Human Cell Radiation" looks like a stigmatic icon of Christian mysticism. The image is spooky, sexy, perfect—a fever

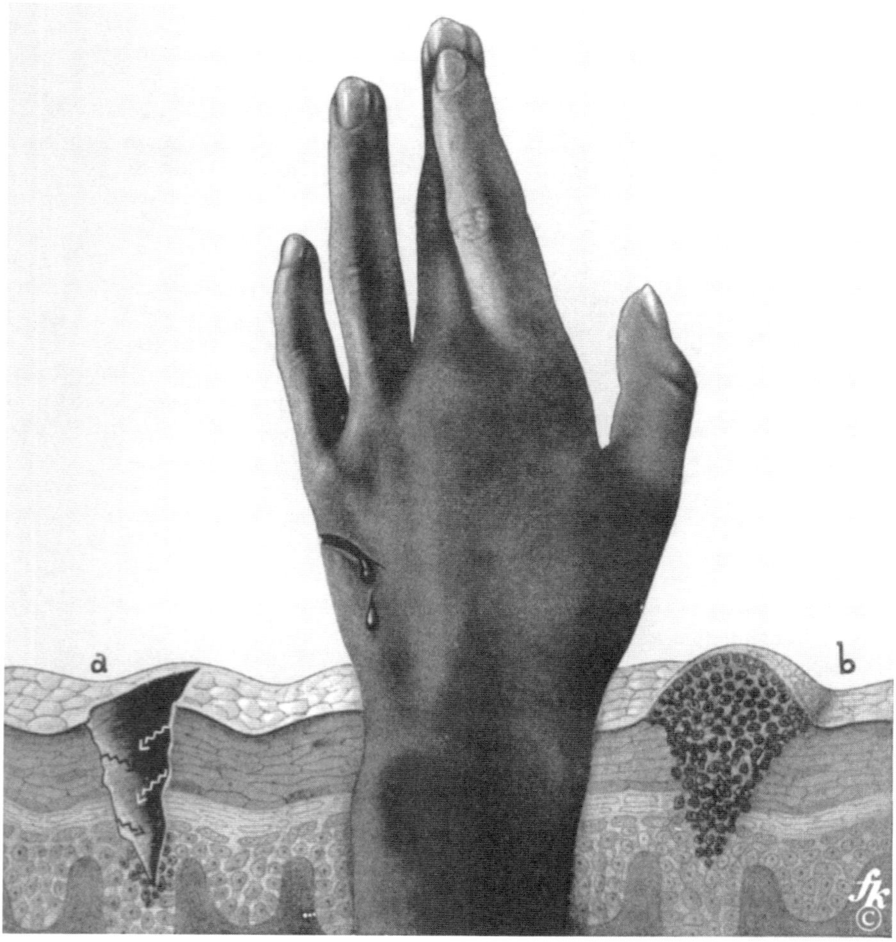

FIGURE 4.21. *"Human Cell Radiation,"* Der Mensch Gesund und Krank *(1939), 1:31. Artist uncredited. National Library of Medicine. A bleeding hand grows out of the soil of the skin, illustrating the physiology of the body's repair mechanism. The skin has a seductive velvety sheen; the index and middle fingers cross, seeming to caress each other. "The wound on the hand heals, because the damaged cells send growth rays into the area."*

dream; it is too evocative to serve as a lowly prop for Kahn's discussion of how the body repairs small wounds.

We don't know the identity of Kahn's brilliant photomontagist (or if there was only one)—none of Kahn's photomontages is credited—but one gets the sense that in "Human Cell Radiation," the artist was aspiring to something greater than the assignment. Rudi Feld, a poster artist and set designer who collaborated with Fritz Lang and other film directors, remarked on the special resemblance between photomontage and modern subjectivity: "Before our inner eye, there appears . . . an inner photomontage. . . . The public rushing by can only be chained to us if we succeed in projecting the . . . material onto the outside world."[41] In Feld's poetics, remediation occurs in the transit from psyche to psychotechnical production and back again. In modernity, consciousness itself is a photomontage. The photomontagist restages that and captivates the swarming public; he enchains them to the image, and by extension its maker.

THE SCHEMATIC BODY

Kahn and his artists were inventive in applying the principles of the photomontage to the cause of visual explanation, and in a few instances, they produced some extraordinary pictures. But they were not innovators, and their photomontages don't shine quite as brightly as those of El Lissitzky, Aleksander Rodchenko, John Heartfield, László Moholy-Nagy, or Hannah Höch. When it comes to the scientific diagram, however, Kahn and his artists shine intensely. The human body is transformed into a purified engineering schematic composed of rhyming pictographic elements and symbols. Detail is suppressed. Emancipated from corporeality, the body enters the domain of pure design; it undergoes a functionalist, abstractionist—very modernist—beauty makeover.

While a modernist diagram like "The Two Divisions of the Nervous System" may look familiar to us, illustrations of this type were fresh when first introduced to Kahn's readers. In the mid-1920s, modernist engineering and design diagrams were beginning to appear in advertising and exhibitions. Around the same time, schematic representations of the human body and organs start to appear in Kahn's publications as well as in the "Red Vienna" exhibitions of Otto Neurath, and only a bit later in the exhibitions of GeSoLei and the Deutsches Hygiene-Museum and in the animated diagrams of filmmaker Walter Ruttmann.[42] In these images, the dictum of "form follows function" was turned on its head: function could be made legible by the artful arrangement and presentation of forms, carefully balanced between positive and negative space, light and dark and gray (Kahn's diagrams were mostly monochromatic), precisely rendered and labeled using the tools and templates of the drafting table.

Kahn used a number of different artists for his body diagrams, but his crucial artist-collaborator was Otmar Trester, about whom we know almost nothing except through his work in *Das Leben des Menschen*. "Trester" (as he signed himself) was a gifted mechanical artist, pen-and-ink draftsman, and letterer. He especially favored a Futura-like sans-serif lettering system (and occasionally Sütterlin script)—both recently invented—to give his diagrams an internally and serially consistent modernist look that met the standards of high-end graphic design.[43]

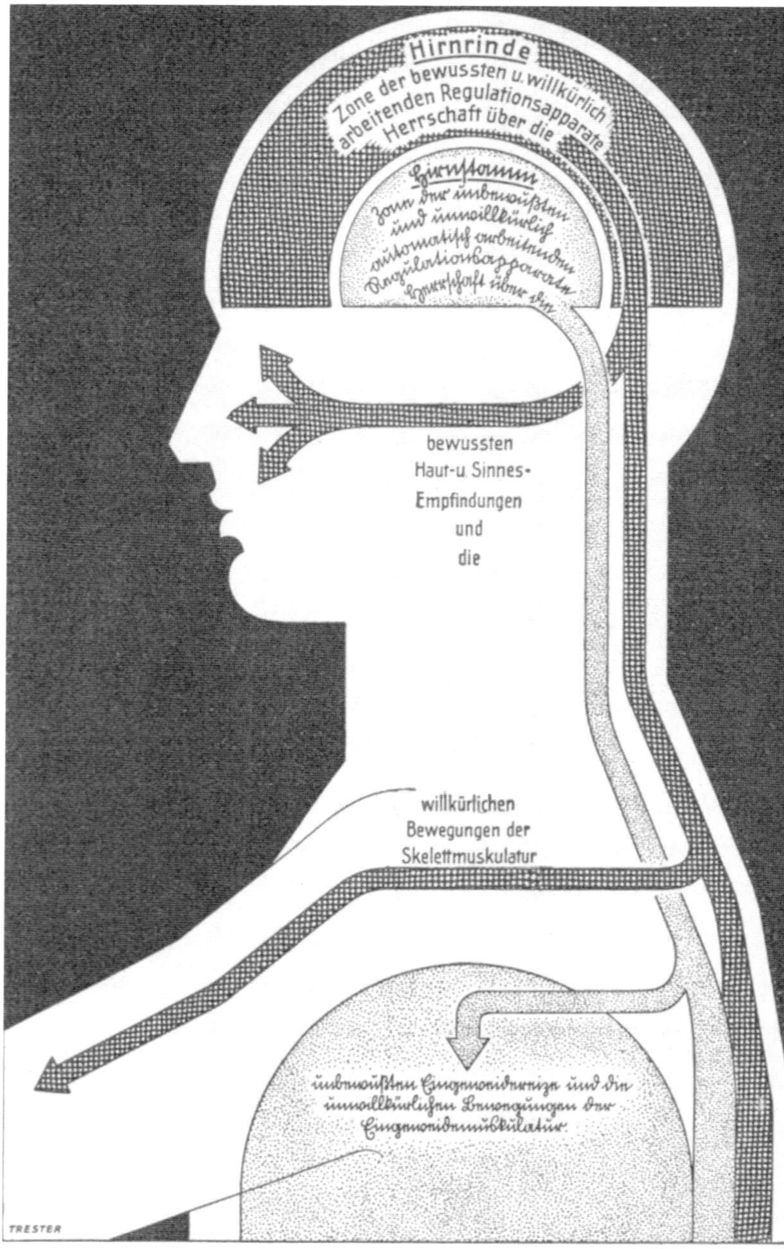

FIGURE 4.22. *"The Two Divisions of the Nervous System" (animal/voluntary and vegetative/involuntary),* Das Leben des Menschen *(1929), 4:99. Artist: Trester. National Library of Medicine. Copyright Kosmos Verlag, Stuttgart.*

In "The Two Divisions of the Nervous System," Trester's graceful pen reduces the human to a bare ruled outline. *Hirnrinde* (the cortex) stands out like a logo or brand name, while the words below correspond to advertising copy. Arrows emerge from the brain, conveying nervous force outward and downward. The universal human and dynamic physiological process converge, make a purified work of art and science in one minimalist profile cross section.

Commenting on the diagrammatic plates of the *Encyclopédie,* Roland Barthes tells us they are "both a didactic work, based . . . on a severe demand for objectivity (for 'reality'), and a poetic work in which the real is constantly overcome by *some other thing*. . . . By purely graphic means, which never resort to the noble alibi of *art*, Encyclopedic drawing explodes the exact world it takes as its subject."[44] In Trester's schematic diagrams, modern art, with its radically reductionist compositional principles, austerely sensuous lines, and precision-machined sensibility, overcomes the real. However, no alibi is in evidence, and "explosion" is the wrong metaphor. Trester's body schematics reduce the body meditatively, without resorting to explosion. "The real . . . constantly overcome by *some other thing*" is a pleasure principle that is gracefully harnessed to the obligation to inform. At the same time, Trester's body schematics perform a wish: that the natural design principles of organic adaptationist evolutionary theory will harmonize with the abstractionist functionalist design principles of the Bauhaus and De Stijl.

Strangely, this way of doing things opens up possibilities for visual rhyming and punning that can go beyond the functional requirements of visual explanation and lead away from pure abstraction. In "The Food Journey through the Intestines," the uncredited artist uses the faces of clocks to indicate that the processing of food through the digestive tract occurs over two hours, measured in half-hour increments. At the same time, the clocks seem to rotate like capstans (alternating between clockwise and counterclockwise), whipping the food (and the intestines themselves) along the path. That sleight-of-hand device mimetically conveys a sense of rolling motion. It also suggests the passage of film through a motion picture projector or a manufactured object along an assembly line.

While Trester and other artists were developing techniques of showing and evoking the flow of physiological processes for Kahn, in America, Frank and Lillian Gilbreth were theorizing and instrumentalizing the flowchart. Self-described "efficiency experts" who began their careers as followers of F. W. Taylor, the Gilbreths introduced the "flow process chart" (now usually just called a flowchart) at a 1921 meeting of the American Society of Mechanical Engineers.[45] There they demonstrated a rigorously structured method for notating the steps of an industrial process. The Gilbreths wanted to make flowcharts that were universally legible, with ready-made components for easy assembly—a method of mass manufacture for diagram makers. Kahn's flowcharts were designed to visually explain process, how

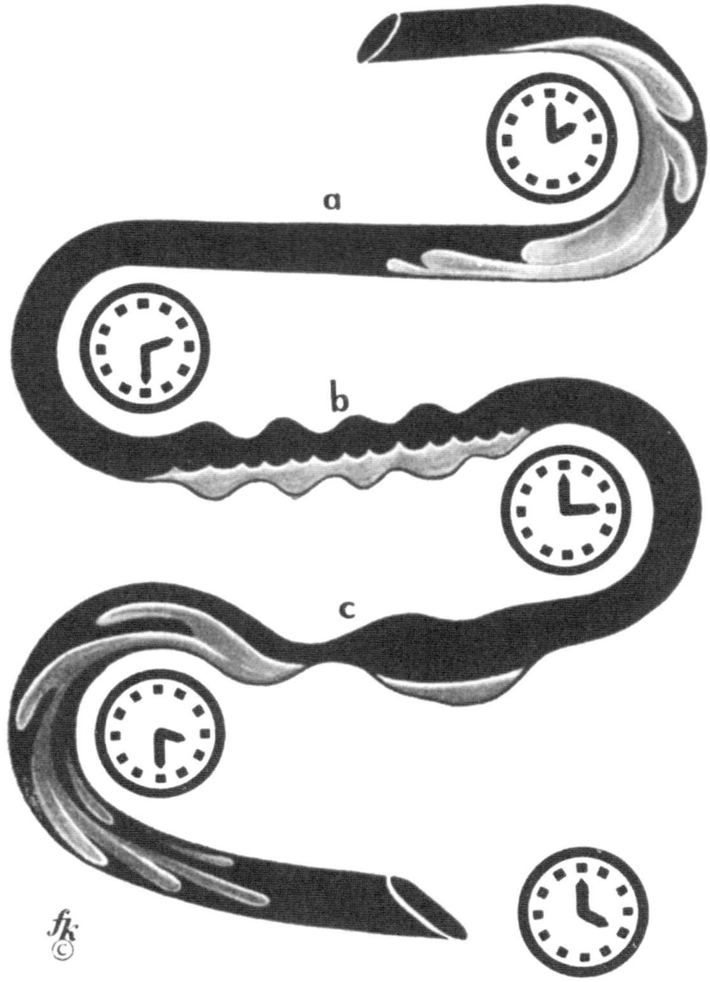

FIGURE 4.23. *"The Food Journey through the Intestines,"* Der Mensch Gesund und Krank *(1939), 1:354. Artist uncredited. National Library of Medicine.*

things work, but the Gilbreths' flowcharts were designed to be a design tool for the processes of production, selling, finance, emergency response—anything, even design itself. Their great innovation was the design of the design of process: to create a formally limited, standardized visual vocabulary that could show multiple pathways instigated by different decisions, steps, and contingencies. They stocked their iconography with geometrical forms, pictograms, letters, arrows, and mathematical symbols that corresponded to a kind of engineering symbolic logic—and called on professional societies all over the world to adopt their system. This would be facilitated by the mass manufacture of sets of plastic templates of the symbols for use in technical drawing. It was an immensely productive approach that bore fruit in the 1940s and in the postwar era in industrial engineering and systems analysis.

Kahn's artists were not so systematic. They dreamed up bodies and body processes out of many of the same visual elements that the Gilbreths used, but they were tasked with making charts that were beautiful, playful, and mimetic as well as neat and practical. Unlike the

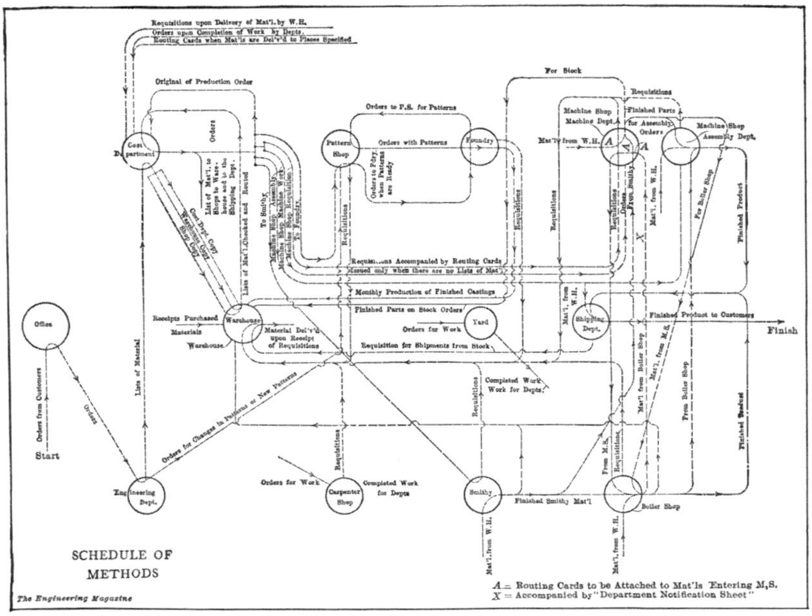

FIGURE 4.24. *"Routing Diagram for Materials and for Printed Forms in a Manufacturing Plant,"* *Willard C. Brinton,* Graphic Methods for Presenting Facts *(New York: Engineering Magazine, 1914), 18. National Library of Medicine.*

Gilbreths, they wanted lay readers to imaginatively visualize and feel processes, not just notate them.

Did Kahn and Trester ever come into direct contact with the Gilbreth system? We can't say, but we do know that, well before the 1920s, the practice of charting flow was already occurring in a variety of domains. Willard C. Brinton's classic 1914 manual, *Graphic Methods for Presenting Facts,* is full of "routing diagrams" and "process charts."[46] In plumbing, ventilation, chemistry, electronics, engineering, industrial manufacture, and all sorts of professions and trades, people were developing lexicons of lines and symbols to make diagrams that show pathways of process. Avant la lettre, flowcharting was in the air.

It was a modern thing to do, but not modernist. In the manuals of Brinton, Mary and Evart Routzahn, and Walter Dill Scott, a critical commentary was taking shape about what constituted a good diagram and an effective graphic.[47] Their books discussed what was legible, what was tidy, what was logical, and what was effective with viewers and readers and technicians. But they did not discuss what was beautiful, and they did not discuss what would remake the world. That was where the work of Kahn and Trester was salient: they harnessed modernist aesthetics to the task of making the schematic diagram—a type of object that was often considered to be a tedious bore—into a compelling visual experience for the nonspecialist public. In this, they were not just trying to do things that were efficient and rational and instrumentally useful; they were also contributing to a larger project: the building of a collective modernist *Gesamtkunstwerk.*

One wonders about the implications of their approach for the production of scientific knowledge. What impact, if any, did Kahn's

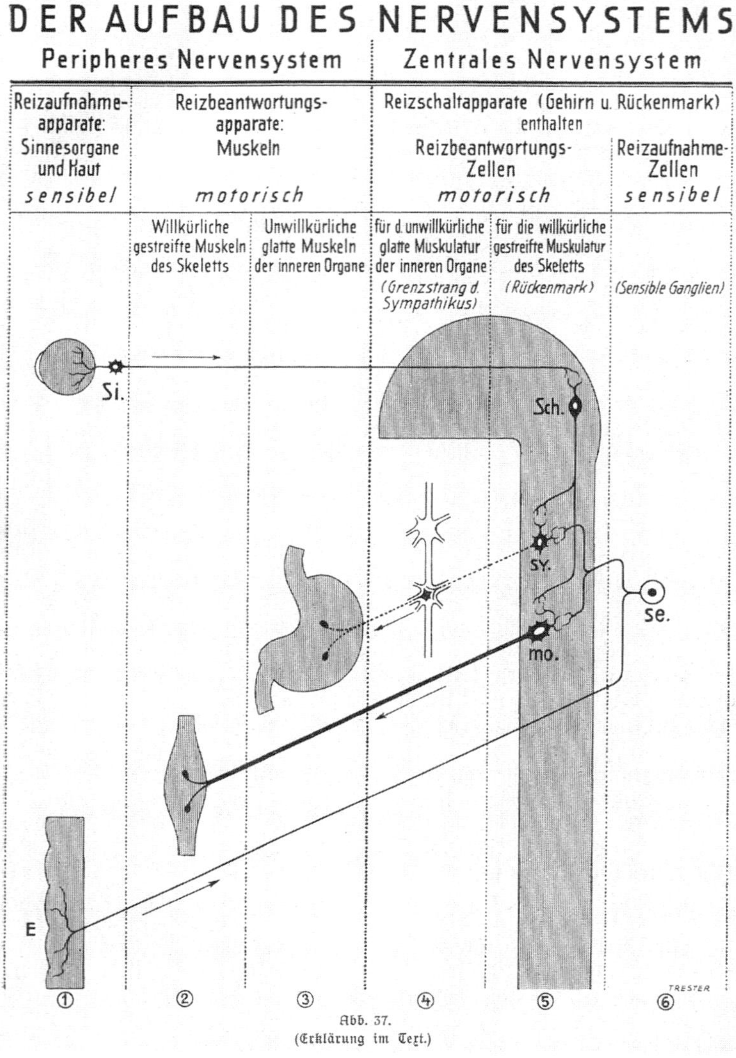

FIGURE 4.25. *"The Construction of the Nervous System,"* Das Leben des Menschen *(1929), 4:39. Artist: Trester. National Library of Medicine. Copyright Kosmos Verlag, Stuttgart.*

diagrams have on the development of technical schematic diagramming? A whole generation of scientists and engineers grew to maturity looking at Kahn and the many cultural productions that his work influenced. If the modernist body schematic was a kind of modeling, did its models provide those young readers with a visual vocabulary that cognitively shaped or fostered the development of new theoretical models? Did beautiful flowcharts contribute to new tacit standards and conventions of diagramming and help to promote new domains (and disciplines) of visualization? There are no certain answers to these questions. What is clear is that the modernist schematic diagram reconceptualized organic life in ways that had scant reference to anything that was observable with the naked eye or microscope. And that perhaps heralded the biomathematical universalist, cybernetic future that Norbert Wiener and colleagues would soon start to bring about.[48]

The compositional elements of Trester's "Construction of the Nervous System" only barely suggest the shapes of actual organic structures. Its most extraordinary innovation is its tabular columns, which remake the flowchart into a taxonomic table. The chart notates a sequence of events over time while also organizing its elements in space. The nervous system is divided into two categories, peripheral and central, which is then subdivided into subcategories, then divided again. That supports a larger concept: that bodily processes of development, structure, function, pathology, evolution, and even behavior and design can be usefully organized in patterns that are made manifest not just by the compilation of statistical data or the suggestion of organic structure but by abstract geometrical diagramming.

Again, there is an excess that goes beyond any reasonable pedagogic program—a compulsion to be modern in every register. In this new world, within text and picture, everything was exchangeable and transposable. Categories were graded by level so that equivalents could bear useful comparison. Tables were mobilized in pictures, and pictures were mobilized in tables. A picture *is* a table. A table *is* a picture. Pictograms stand for and with words, and both can be mixed with tables and diagrams. This might make for some confusion. If reductionist abstractionism and radical legibility were guiding principles, then abstractionism, carried to an extreme, could turn a flowchart into something glyphic, complex, and hard to read—a convoluted, almost baroque performance of the modern. Kahn's readers may well have been mystified, but they were also impressed.

THE DRAMATIZED STATISTIC

With the publication of Belgian mathematician-sociologist Adolphe Quetelet's pioneering 1835 essay on "social physics," which invented the modern discipline of social statistics, reformers all over the Western world began to advocate the collection, analysis, and presentation of data.[49] Statistics acquired epistemological authority and rhetorical power. It seemed almost self-evident that statistics could produce new knowledge and break old impasses, that the discipline could resolve questions empirically and scientifically, with inventories, comparisons, correlations, and predictions. Armed with what were termed "social facts," reformers believed they could persuade the public to submit to medical examinations and vaccinations, adopt hygienic practices, comply with doctors' orders, support health legislation, and so on. Data collection and analysis were powerful; they could tell us things we never knew or even suspected about society and the physical world. Thus did statistical presentation proceed in tandem with the formation of bureaus of health and labor, the adoption of laws and regulations, investment in municipal and state laboratories, and the development of sanitary infrastructure.

But what was the most effective way to present statistical data? In their initial enthusiasm, many reformers assumed that the mere presentation of graphs and tables would do the trick. However, it soon became evident that people paid little attention to barren recitations of raw data and percentages arranged in tables. Artless line graphs, pie charts, and bar charts just didn't do the trick. Progressive reformers Evart and Mary Routzahn argued that statistical diagrams, while "much used in exhibits intended for popular audiences," represent, for many people, "a disagreeable form of mental effort." It was hard to get the public to look at statistical data, let alone understand it. The only solution, then, was to make statistical displays appear "less technical . . . less dull" by inventing "variations in which the circles or bars or hills have been replaced by successions of concrete items, such as dollar signs, pictures of human figures, animals, or other appropriate objects."[50] The statistical chart had to entertain as well as instruct.

Thus, well before Kahn entered the scene—and well before the terms "pictorial statistics," "infographics," or "data visualization" were ever invented—the public was presented with printed material and displays that translated numbers into configurations of drawn, painted, and printed scenes, symbols, and figures. Statistical knowledge, often paired with snappy or provocative copy, was dressed up in cultural experiences that readers could identify with.

In some ways, Kahn's dramatizations of statistical data were not markedly different from what others were doing. But Kahn and his artists got better at it over time; they became increasingly ingenious and occasionally brilliant. To see

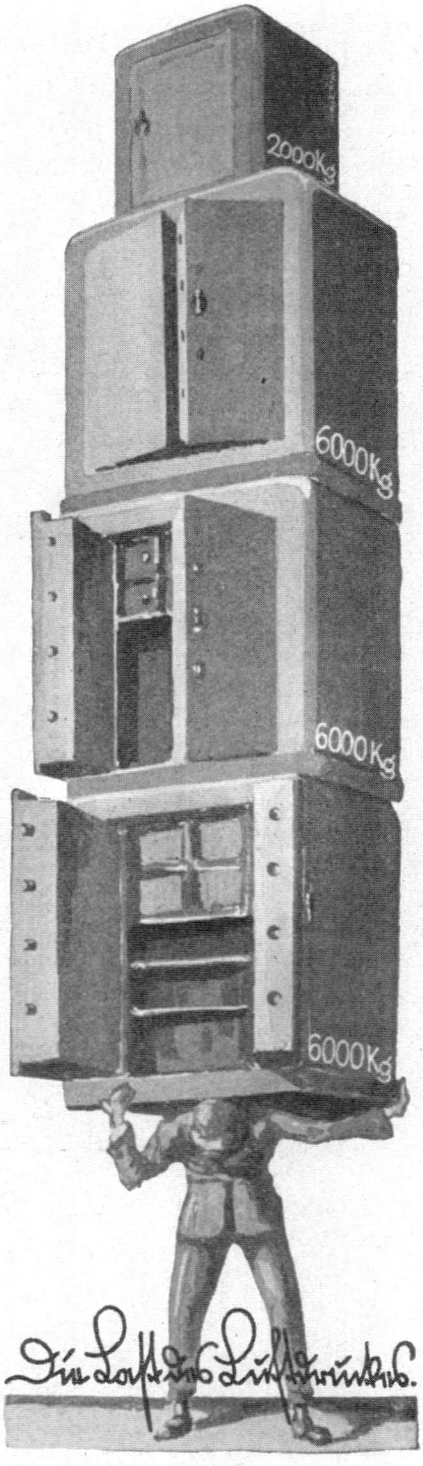

Abb. 41. 20 000 kg auf unseren Schultern! Die 300 km hohe Luftsäule lastet mit diesem Gewicht auf uns, wir verspüren den Druck aber nicht, weil er uns nicht nur von oben, sondern auch von unten und innen trifft und sich folglich ausgleicht.

FIGURE 4.26. *"20,000 Kilograms [of Atmospheric Pressure] on Our Shoulders!"* Das Leben des Menschen *(1926), 3:55. Artist uncredited. National Library of Medicine. A man in a business suit struggles to balance a stack of heavy steel safes, a circus trick. Is it also a comment on the burdens of modernity? (Life is a balancing act.) Copyright Kosmos Verlag, Stuttgart.*

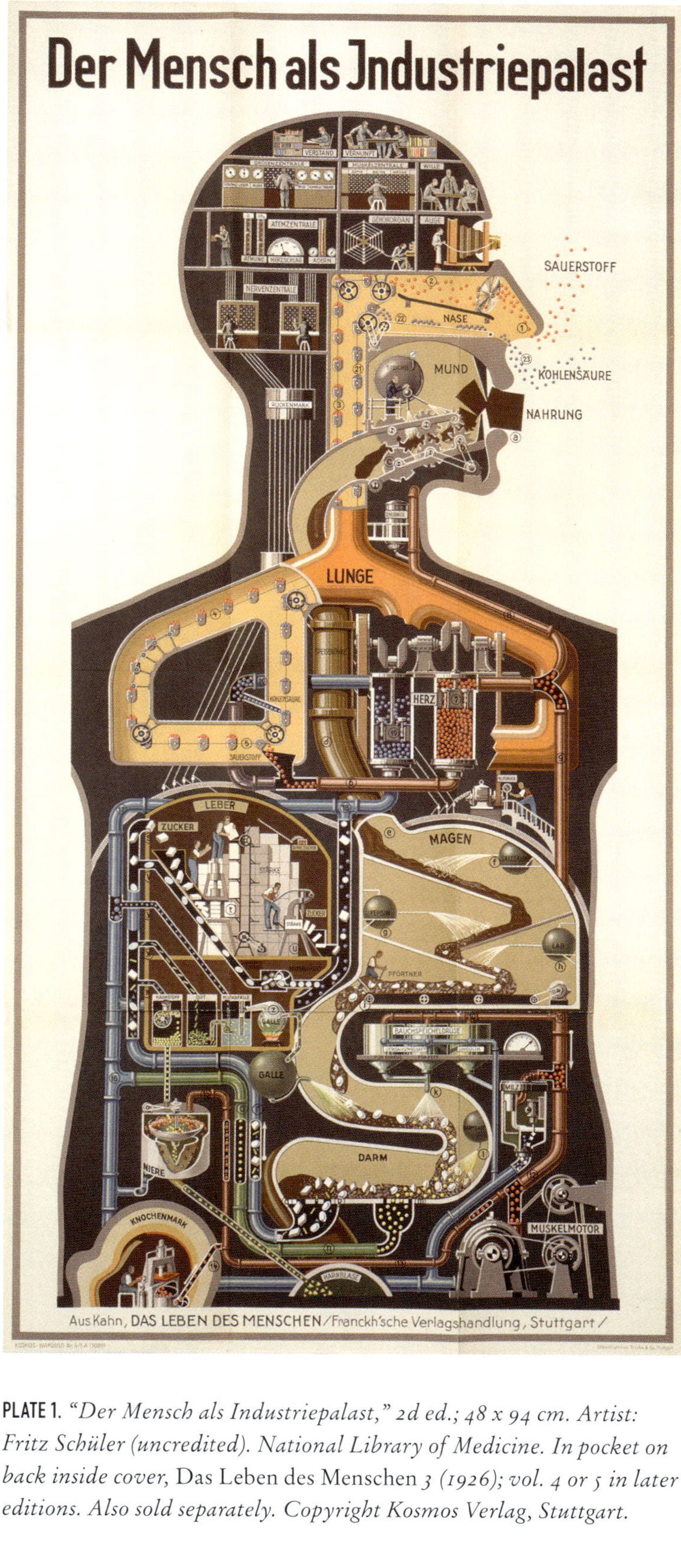

PLATE 1. *"Der Mensch als Industriepalast," 2d ed.; 48 x 94 cm. Artist: Fritz Schüler (uncredited). National Library of Medicine. In pocket on back inside cover,* Das Leben des Menschen *3 (1926); vol. 4 or 5 in later editions. Also sold separately. Copyright Kosmos Verlag, Stuttgart.*

Wunder in uns   Tafel XIII

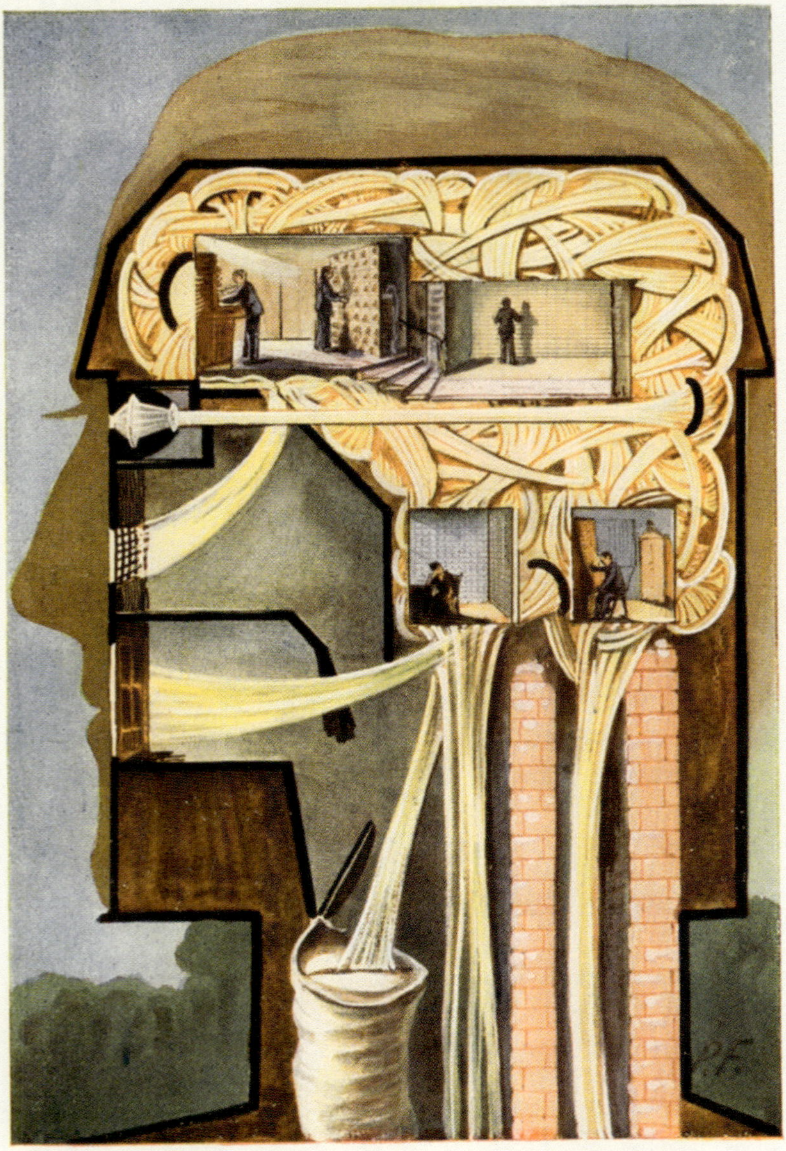

Abb. 26. Ein Versuch, die Einrichtung unseres Gehirns anschaulich darzustellen.

(Erläuterung auf der Rückseite.)

PLATE 2. *The model for "Der Mensch als Industriepalast": "An attempt to graphically depict the set-up of our brain" (from "A Look into Headquarters,"* Pictured Knowledge *(1917), colorized and redrawn by Paul Flanerky), in* Wunder in Uns *(2d ed., 1923), plate 13. National Library of Medicine.*

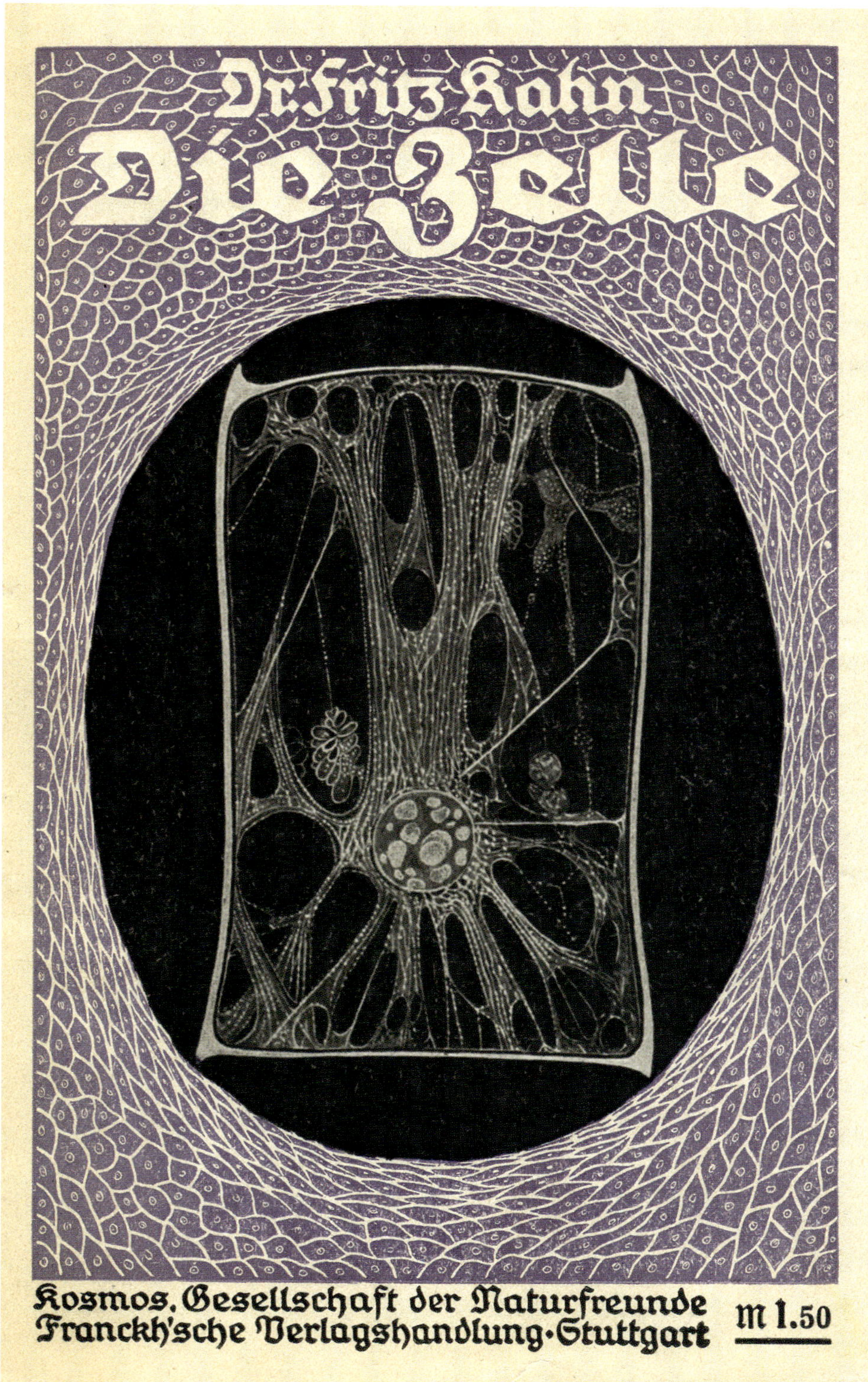

PLATE 3. *Cover illustration,* Die Zelle [The Cell] *(1919). Two-color. Artist: Georg Helbig. National Library of Medicine. The interior and outline of the cell suggest Art Nouveau's stylized organic lines and shapes: biology updated to a contemporary aesthetic. Copyright Kosmos Verlag, Stuttgart.*

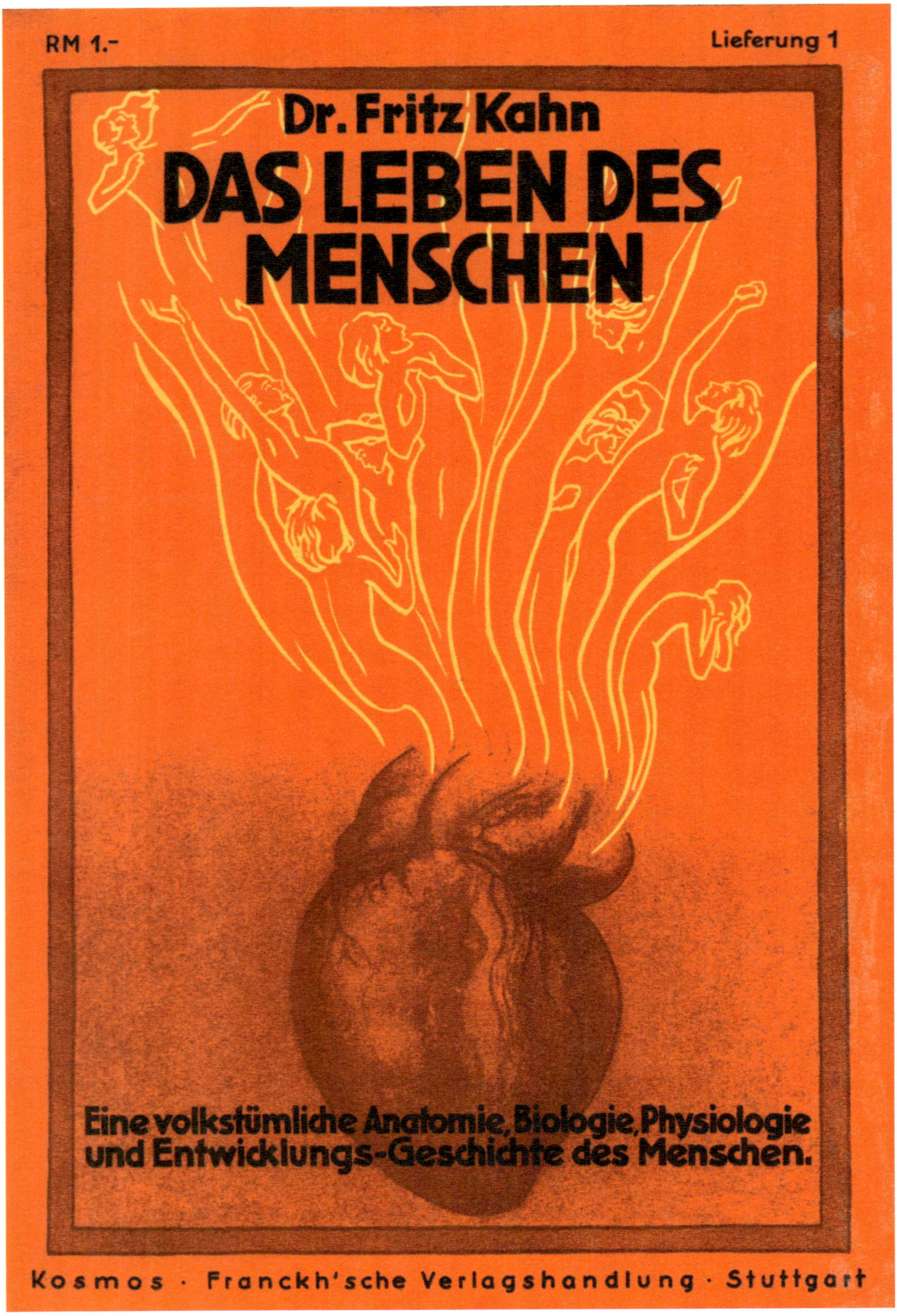

PLATE 4. *An eye-catching modernist paper cover for an installment of* Das Leben des Menschen *(circa 1930). Three-color. Artist uncredited. Leo Baeck Institute, New York. Copyright Kosmos Verlag, Stuttgart.*

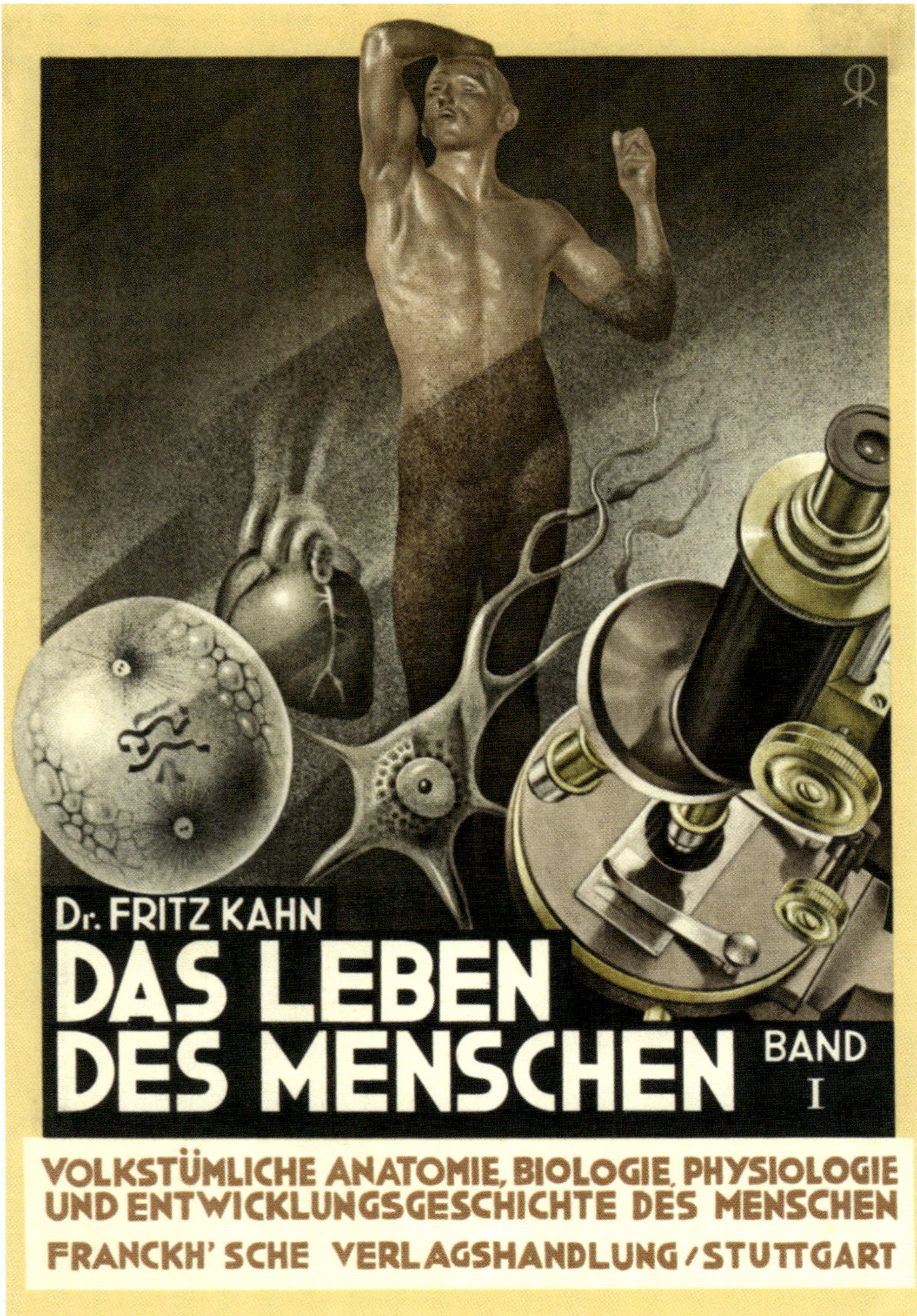

PLATE 5. *Dust jacket,* Das Leben des Menschen *(2d ed., 1926). Artist: Roman Rechn. Leo Baeck Institute, New York. Three-color. The figure of "Man" at center is taken from a photograph of Rodin's controversial 1876 statue* The Age of Bronze *(critics alleged that Rodin had cheated by making a life cast of his model). Is Rechn using the statue's contested realism to playfully reference his own copying practice, the incorporation of photographs into hand-drawn illustration? Copyright Kosmos Verlag, Stuttgart.*

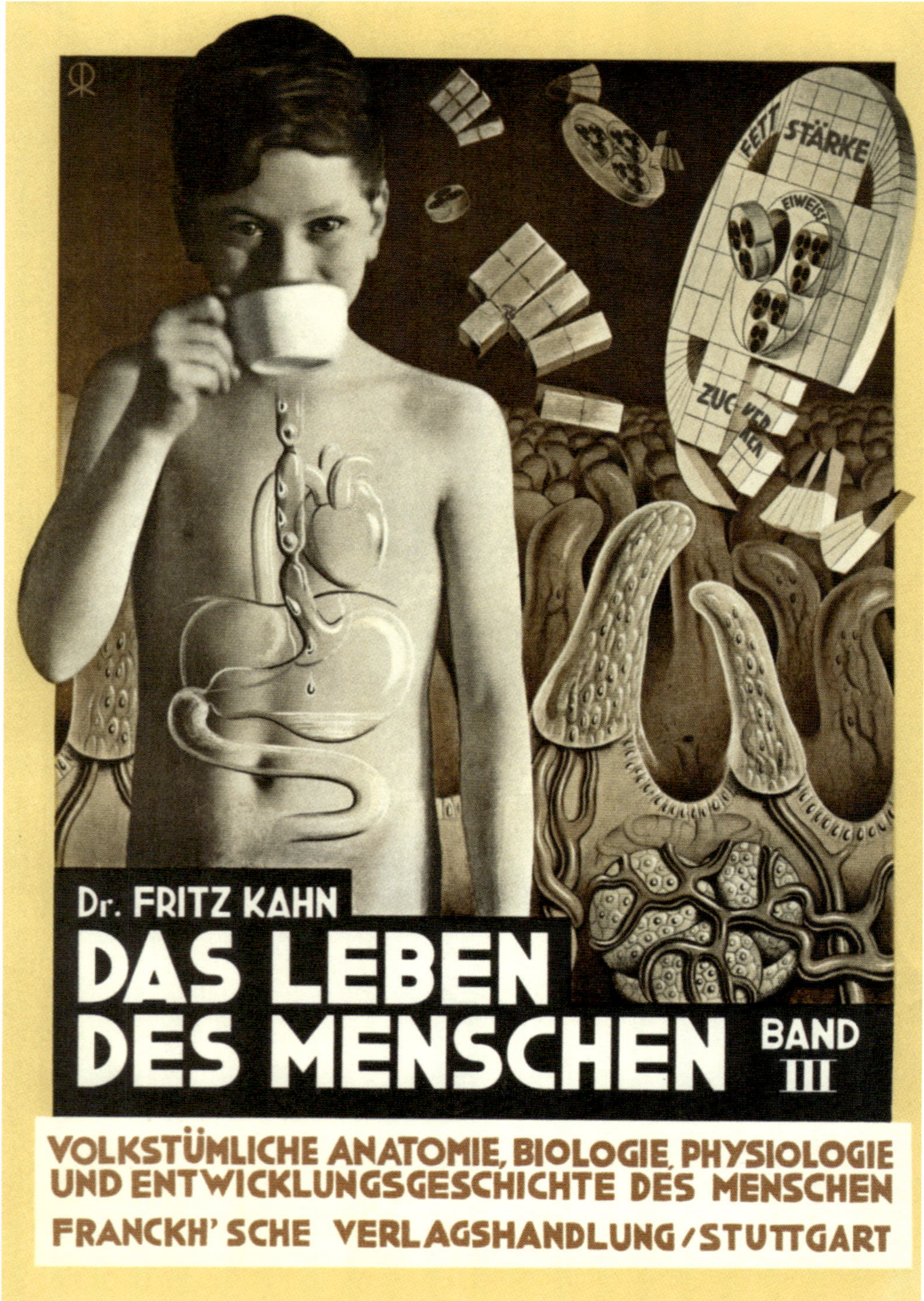

PLATE 6. *Dust jacket,* Das Leben des Menschen *3 (2d ed., 1926). Artist: Roman Rechn. Leo Baeck Institute, New York. Three-color. A naked modern boy drinks from a cup. The background elements represent digestion as mechanical/industrial debris floating in outer space and as the work of organic creatures in a monstrous phallic inner landscape. The boy's hidden smile, and the playfully elegant anatomical diagram superimposed on his body, suggests that there is pleasure in all of this. Copyright Kosmos Verlag, Stuttgart.*

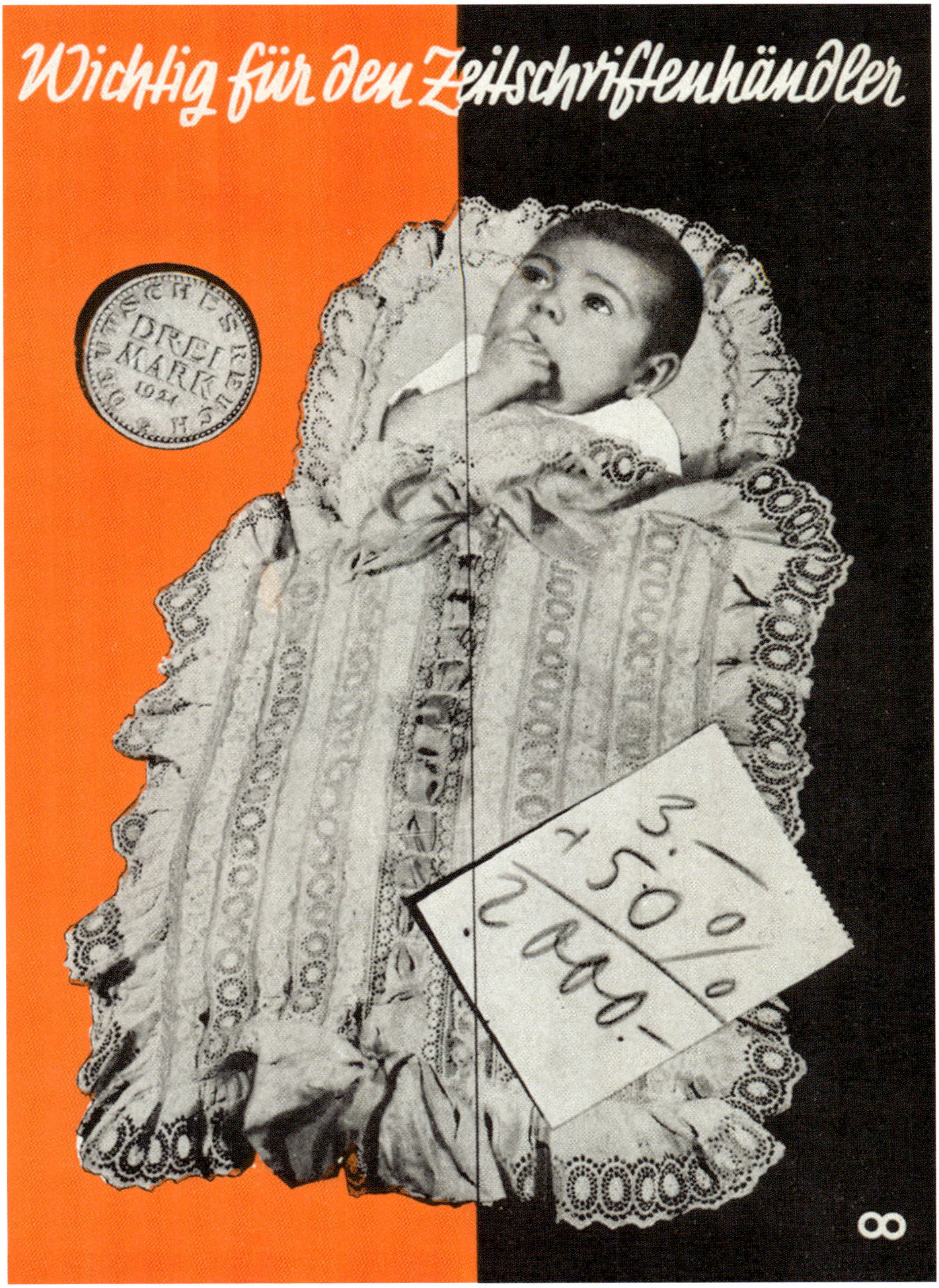

PLATE 7. *One of eight "sensational designs" from Werbekraft GmbH,* Gebrauchsgraphik *10 (1931): 11. Designers: "Plucer und [Arpad] Elfer." Clark Art Institute.*

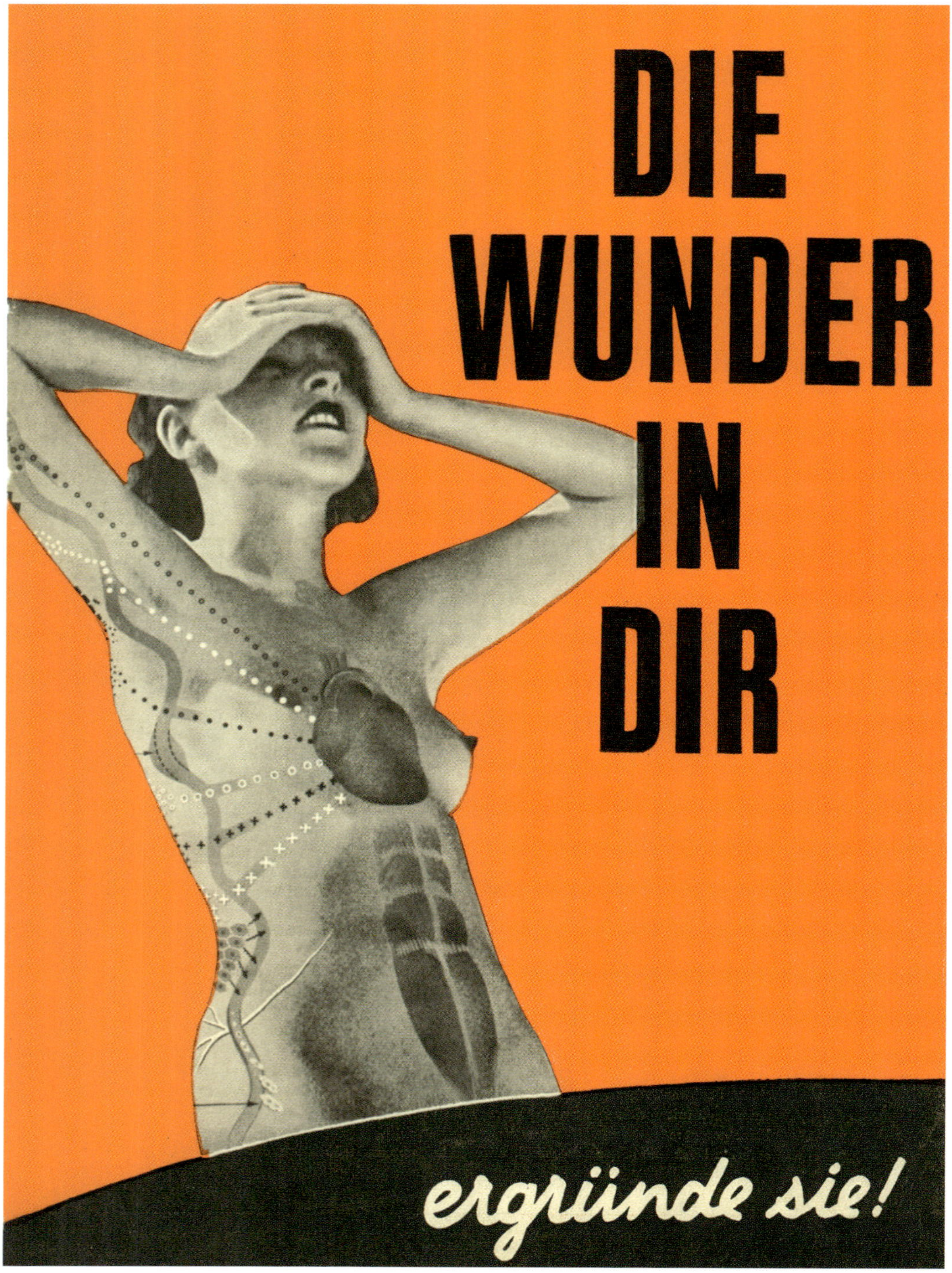

PLATE 8. *"The Wonder in You / get it!" Advertisement for* Der Mensch Gesund und Krank *(1939). Artist uncredited. Leo Baeck Institute, New York. The stylistic similarities between this and Plucer and Elfer designs featured in* Gebrauchsgraphik *10 (1931) suggest that Plucer and Elfer also created this advertisement. Other versions of the design (all uncredited) appear on the dust jacket of* Der Mensch Gesund und Krank *2 and page 253 as "Sunbath! The Twelve Now Well-Known Effects of Sunlight on the Skin" (see Figure 3.4).*

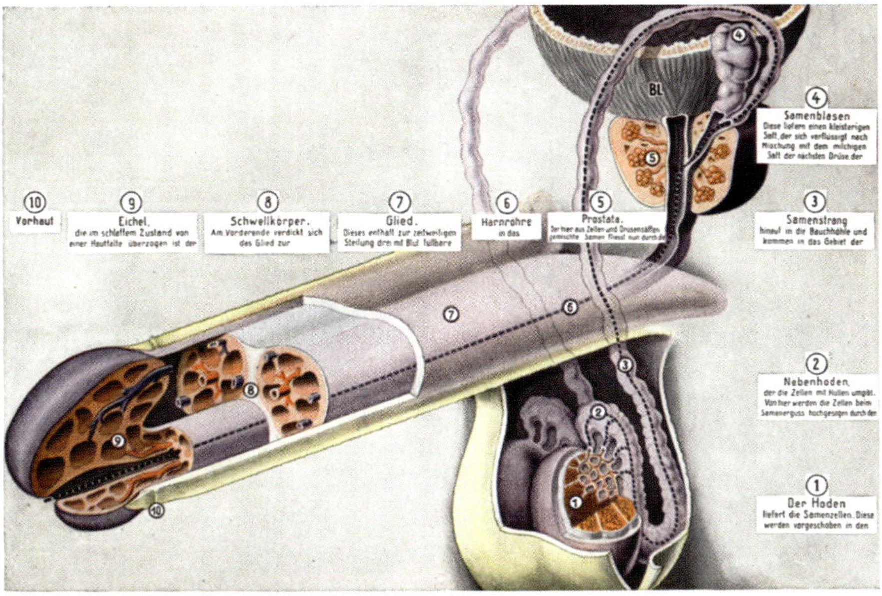

PLATE 9. *"The Male Sex Apparatus" ["Der männliche Geschlechtsapparat"],* Unser Geschlechtsleben *(Zürich: Albert Müller Verlag, 1937), plate 5. Artist uncredited. National Library of Medicine. The phallus looks a bit like a streamlined U-boat or a futuristic gun.* Unser Geschletchtsleben [Our Sex Life] *was Kahn's most translated and best-selling book.*

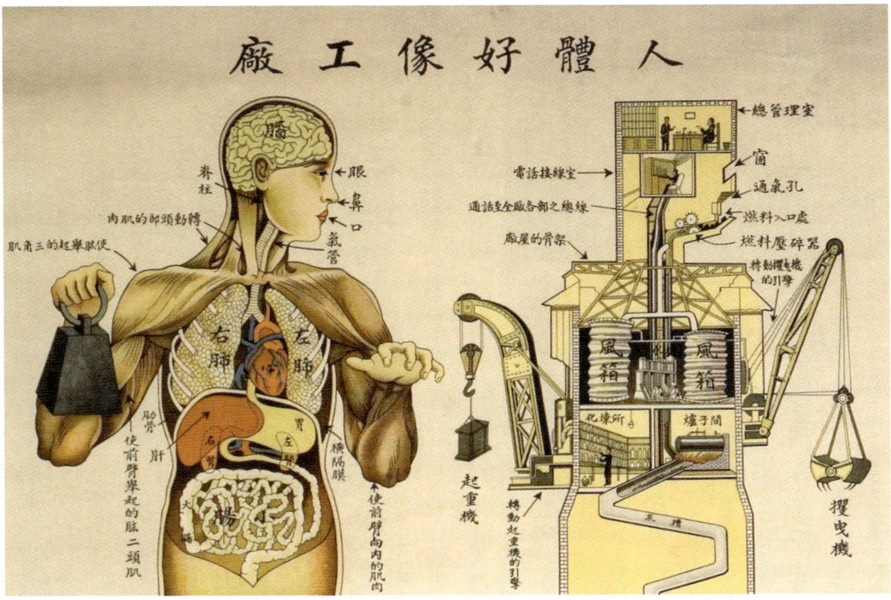

PLATE 10. *A Chinese adaptation of "Der Mensch": "The human body is like a factory," panel, children's health poster. Physiology and Hygiene Organization, Shanghai, 1933. Artist uncredited. National Library of Medicine.*

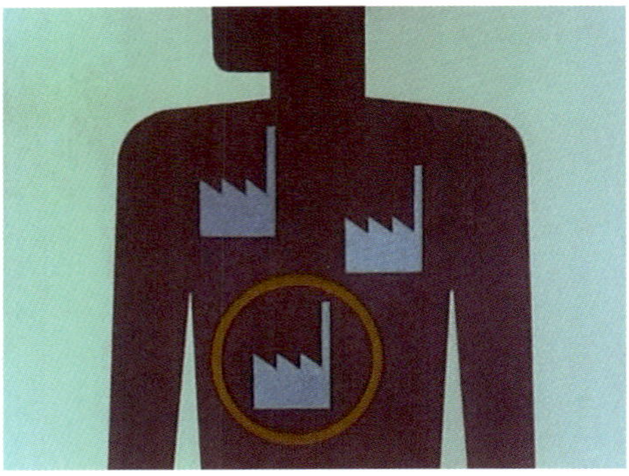

a

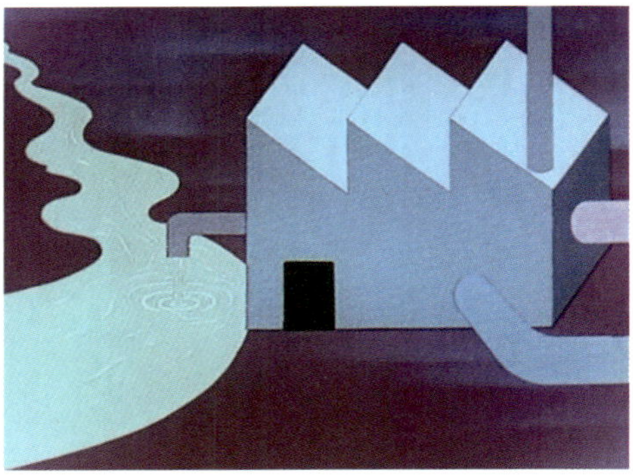

b

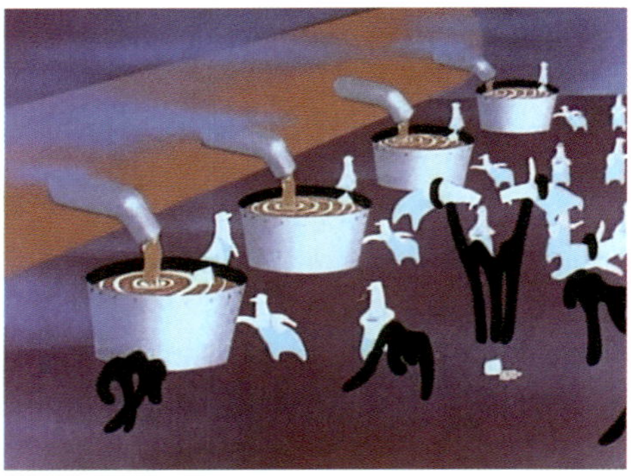

c

PLATE 11.
(a) The human body as factory complex. (b) A body-factory about to release effluent into the lymph stream. (c) Predatory black cancer cells disrupt industrial production. The Traitor Within *(American Cancer Society, John Sutherland Productions, 1946). National Library of Medicine. Footage printed by the permission of the American Cancer Society, Inc.*

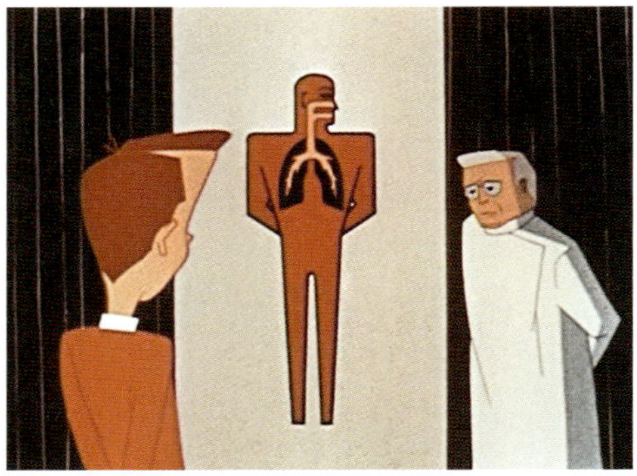

a

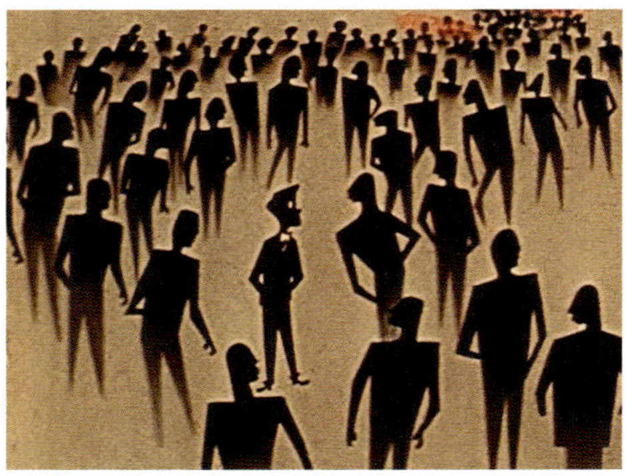

b

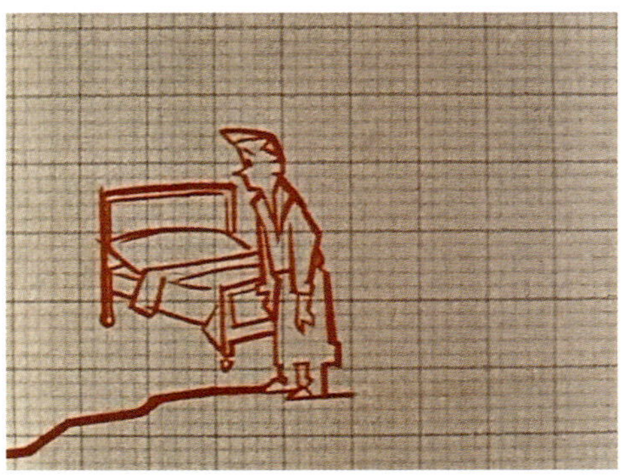

c

PLATE 12.
(a) Rodney and his doctor look at a diagrammatic anatomical wall chart of the respiratory system. (b) How the tuberculosis germ spreads in a human population in mass society. (c) Rodney starts to recover: a dramatized patient chart. Rodney *(National Tuberculosis Association, Film Graphics Inc., 1950). National Library of Medicine.*

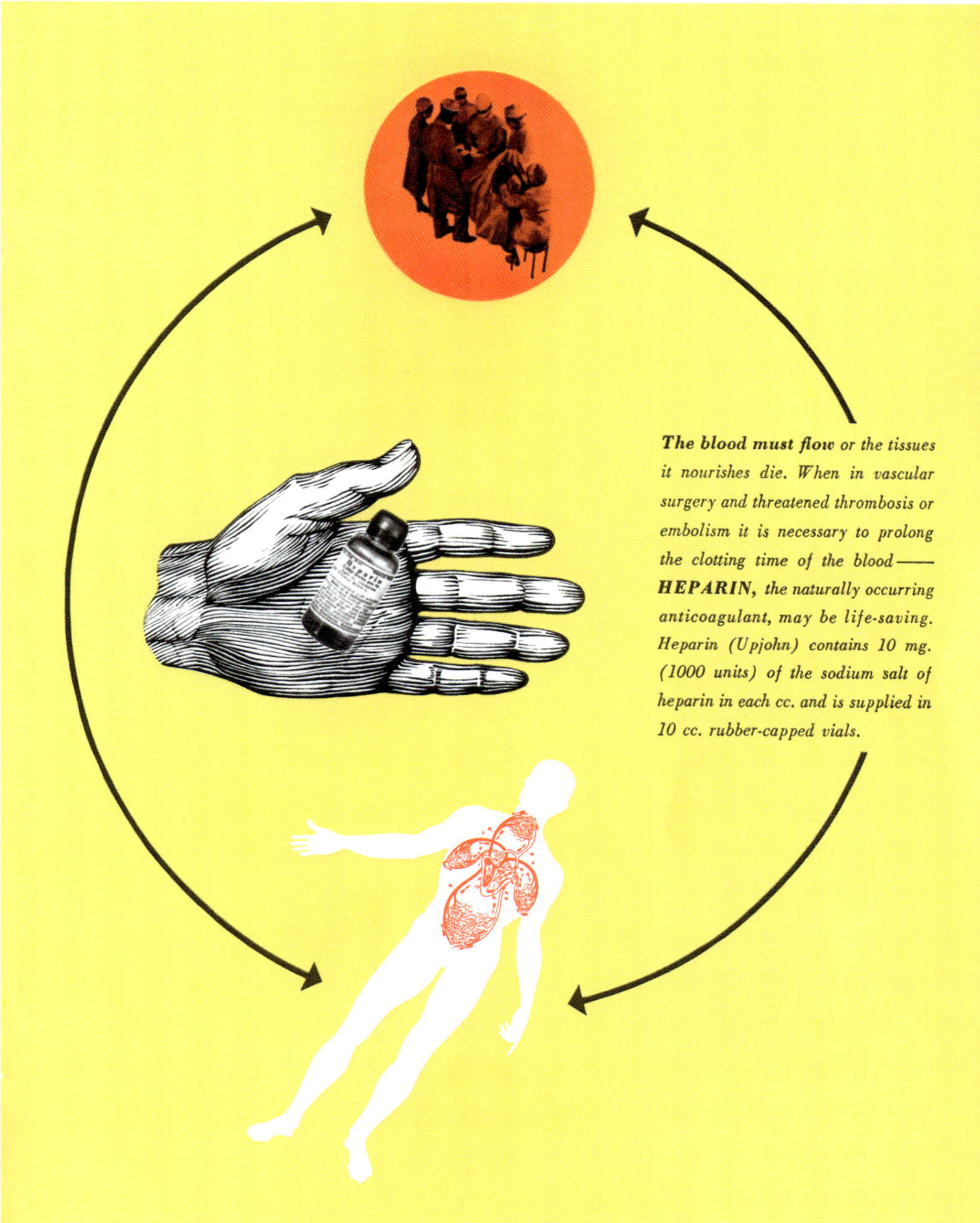

PLATE 13. *"Central control of blood pressure,"* Scope *4, no. 3 (1954): cover. Designer: Will Burtin. National Library of Medicine. Based on drawings and diagrams of the central nervous system from Jakob Christfried,* Atlas of the Central Nervous System *(Philadelphia: W. B. Saunders & Co., 1901).*

PLATE 14. *Advertisement for Heparin, a widely used injectable blood thinner,* Scope *1, no. 6 (1944): inside back cover. Designer: Lester Beall. National Library of Medicine.*

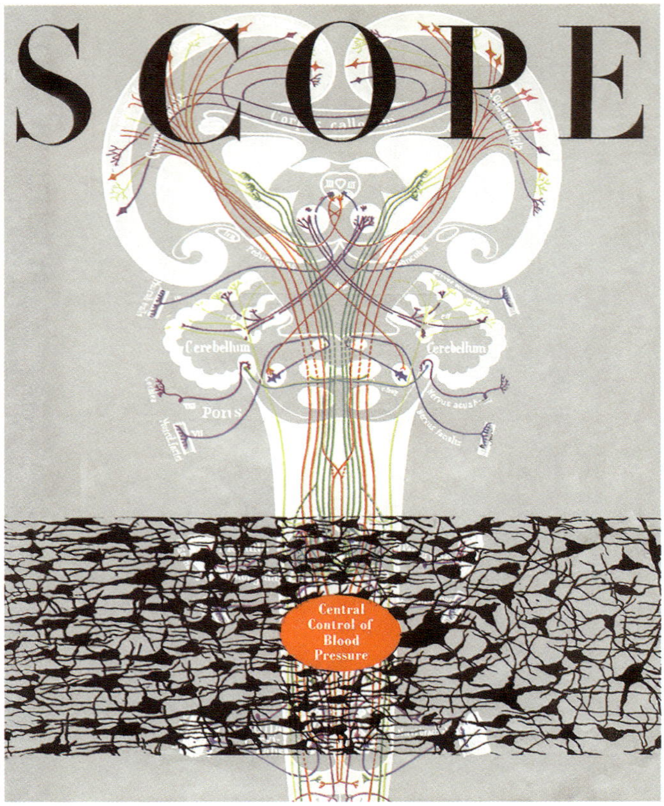

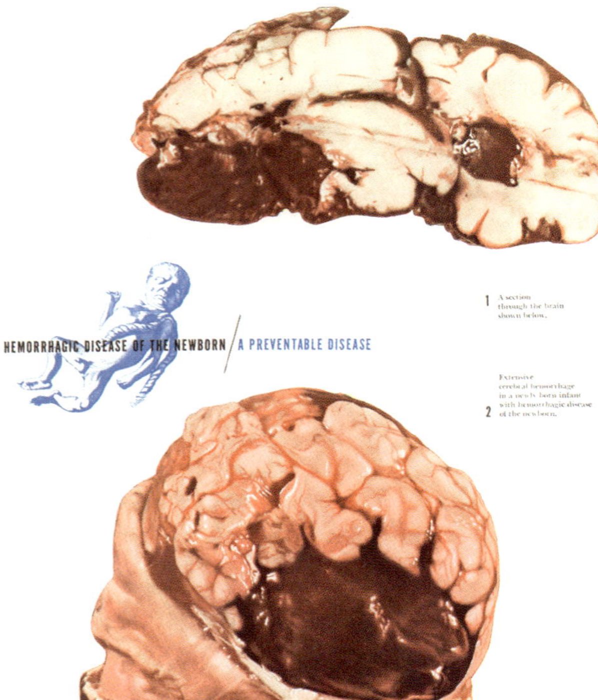

PLATE 15. *"Hemorrhagic disease of the newborn: A preventable disease,"* Scope *1, no. 2 (1942): 10 (foldout). Designer: Will Burtin. National Library of Medicine.*

the trend toward increasing dramatization, compare versions of "Air: Before Inhalation/After Exhalation" in successive editions of *Das Leben des Menschen*. In its original 1926 incarnation, the chemical conversion of air in the lungs is rendered as a dull pair of before-and-after pie charts. Three years later, Kahn put an artist to work on the figure, using the same data. In the revised version, a Janus-faced silhouette appears to simultaneously blow air bubbles of different chemical composition. Arrows in the lungs convey a sense of motion with the passage of gas. The lettering has been redone in a more modern sans-serif style. The addition of the silhouettes appeals to the reader's psychological tendency to seek out images of the self. The end result is nothing brilliant, only a pleasing symmetrical modernist composition—more complex, representational, and interesting than its predecessor. What had been an uninspiring graphic comparison was now something to look at.

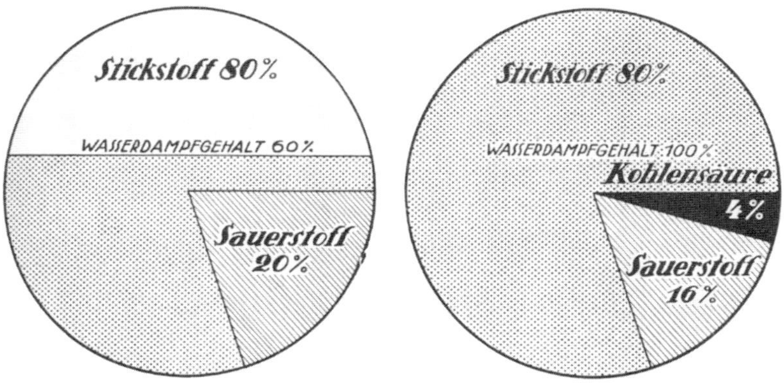

FIGURE 4.27. *"Air: Before Inhalation/After Exhalation,"* Das Leben des Menschen *(1926), 3:61. National Library of Medicine.*

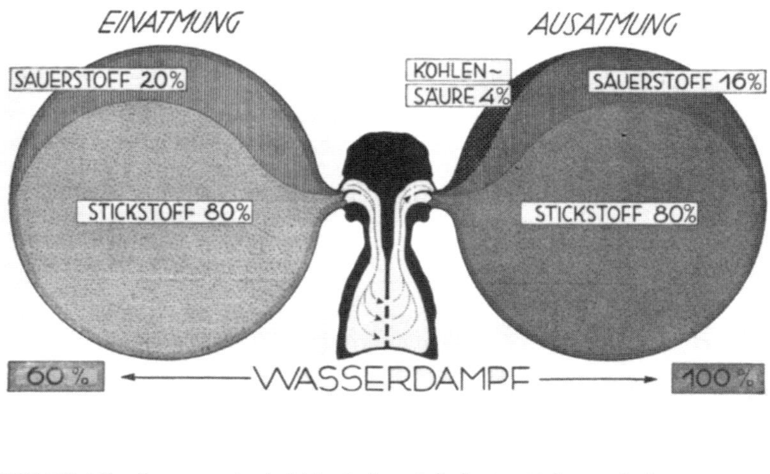

FIGURE 4.28. *The figure revised. "Air: Before Inhalation/After Exhalation,"* Das Leben des Menschen *(2nd ed., 1923), 3:61. Artist uncredited. National Library of Medicine. Copyright Kosmos Verlag, Stuttgart.*

Of course, there were many ways to be modern in the presentation of statistical data. Kahn and his artists often just fell in with a workaday commercial perspectival illustrationism (the style discussed in chapter 3) to depict modern industrial objects or scenes. "Expenditures of the German Population," for example, compares the sizes of built structures (not an obvious choice) to dramatize a statistical comparison: expenditures on bread versus alcohol versus public schools in the year 1909. The comparison supports the argument that the mass consumption of alcohol has unacceptable social costs and that it squanders money that could more usefully go to fund nutrition and education. The illustration simultaneously situates the quantified social problem and its solution in the modernity of the bread factory, the public school, and the industrially produced metal liquid storage tank—and in the modernity of the dramatized statistical illustration. The picture is essentially a beautifully executed version of the approach recommended by the Routzahns, well within the mainstream visual tradition of "social fact" reform, and an entirely generic piece of commercial art: modern in subject matter and style, but not particularly modernist in its aesthetic.

In later works, Kahn more often used statistical scenes to teach physiological lessons not intended to directly shape individual behavior or social policy. These images usually in some way place the body in

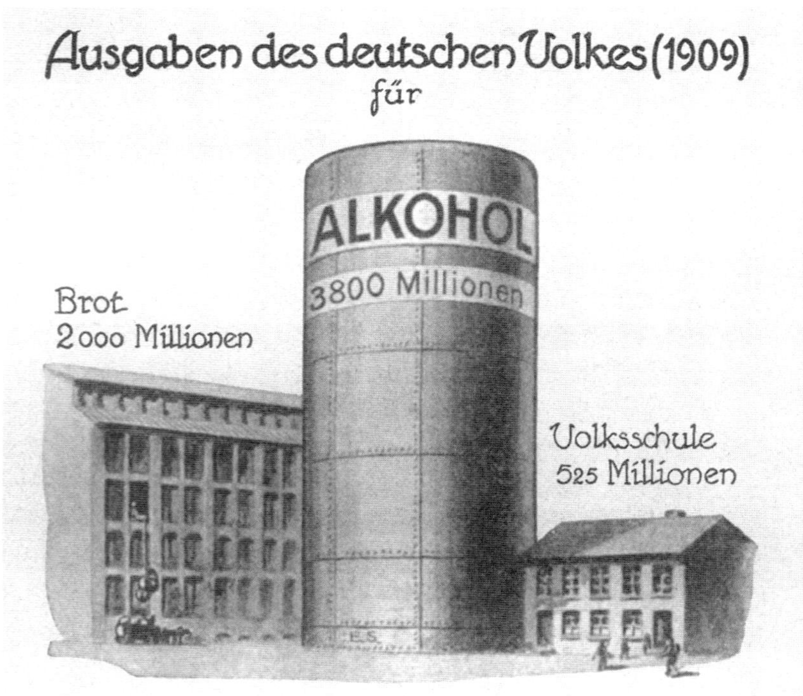

FIGURE 4.29. *"Expenditures of the German Population (1909) for . . . Bread, Alcohol, and Schooling,"* Das Leben des Menschen *(1926), 3:255. Artist: E.S. (Ewald Streletzki?). National Library of Medicine. Copyright Kosmos Verlag, Stuttgart.*

modernity and modernity in the body. The show of quantitative data bolsters the authority of the picture and its author, but it doesn't do much more than that. That is an effective pedagogical tactic in some illustrations, but "The Travels of Protein in the Body Is a Dice Game with Hundreds of Thousands of Combinations" looks more like a pretext to stage a surrealist entertainment for the reader's pleasure. Perhaps Kahn and his inspired artist (again uncredited) simply wanted to play a game.

PHYSIOLOGICAL ADVERTISING

In the 1920s and 1930s, as artwork became increasingly prominent in advertising, a good number of Kahn's illustrations copy the look of

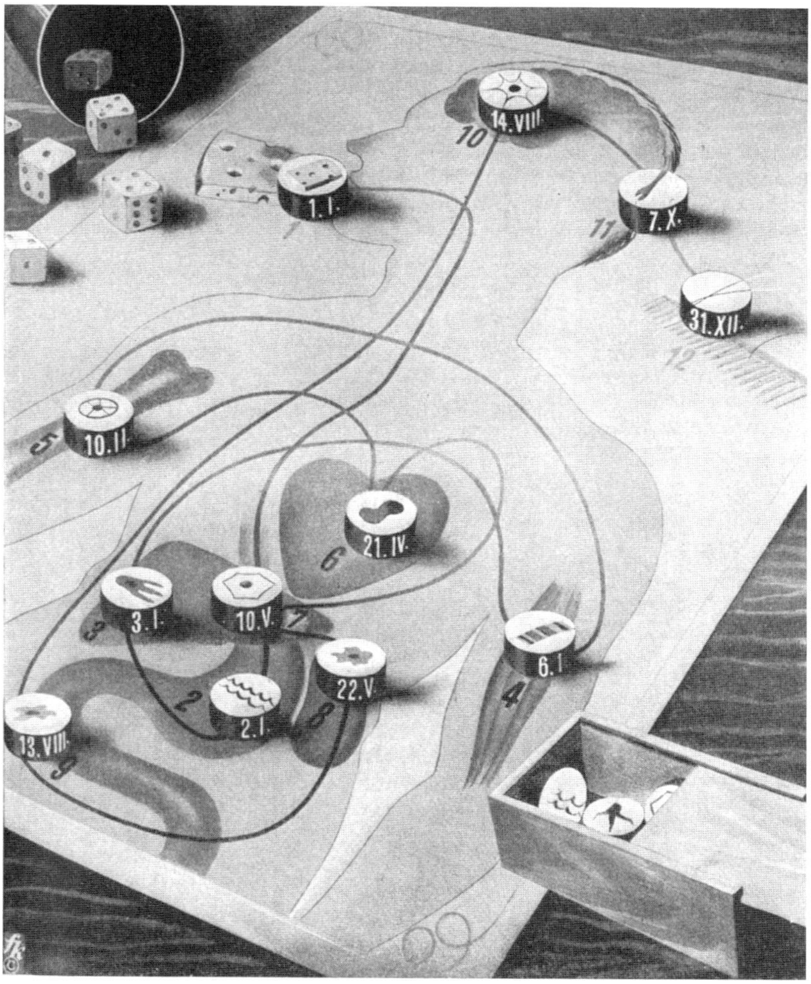

FIGURE 4.30. *"The Travels of Protein in the Body Is a Dice Game with Hundreds of Thousands of Combinations,"* Der Mensch Gesund und Krank *(1939), 2:15. Artist uncredited. National Library of Medicine. Digestion is compared to a board game. The probability of each body part receiving nourishment from a piece of cheese is represented by the throw of the dice.*

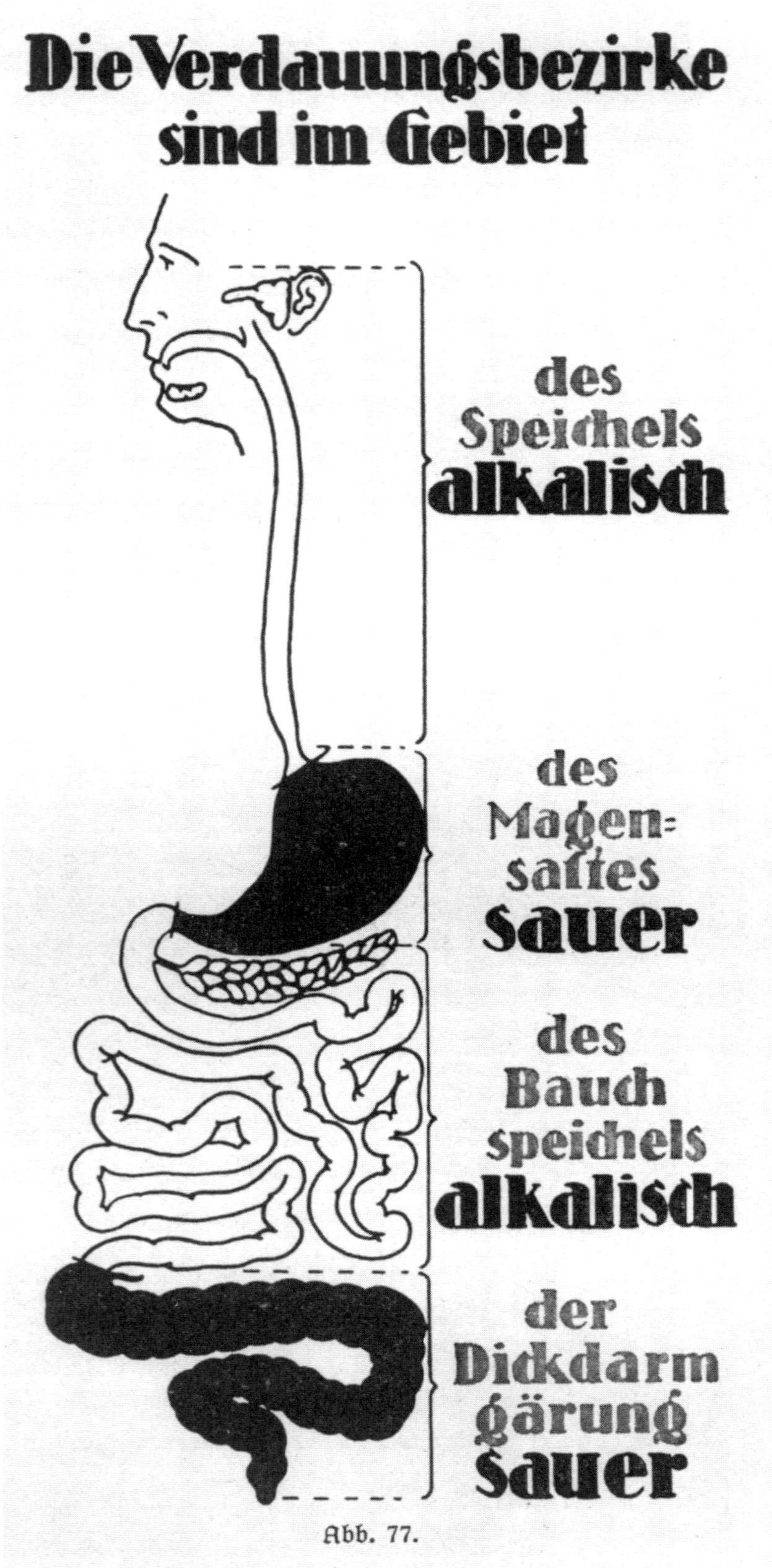

FIGURE 4.31. *A layout of boldface and lightface display type, in conjunction with a strikingly legible (and beautifully rendered) image, mimics the look of a contemporary advertisement. "The digestive zones in this region are: saliva—alkaline; gastric juices—acidic; pancreatic secretions—alkaline; colonic fermentation—acidic,"* Das Leben des Menschen *(1926), 3:111. Artist uncredited. National Library of Medicine. Copyright Kosmos Verlag, Stuttgart.*

contemporary advertisement, with display type, subheadings, arresting pictures, attractive models, and other features of commercial graphics. But in Kahn, the illustration does not aim to sell a product. Instead, it advertises the human body, its structure, and its functions.

That tactic, amply on view in *Das Leben des Menschen* (1922–31), is more highly developed in *Der Mensch Gesund und Krank* (1939), but only in some ways. Perhaps because it was published all at once, and not episodically, the latter work had a more modern and consistent designed look that did not permit headline-size display type within the frame of the illustration. That constraint ruled out the use of some formal elements. Even so, Kahn and his artists continued to eclectically borrow tropes and compositional tactics from advertising graphics—and mimicked their increasing sophistication.

"The Sensory Organs of the Head" uses the encircled face of a beautiful woman to present a lesson on cellular physiology and the senses. The setting is the home, the women's sphere. Haloed by a circular band around her head, the female figure resembles a Hollywood starlet. Within the spotlight, her head is tilted back and her lips parted slightly. The pose is ambivalent: Is she overwhelmed? Frightened? Sexually aroused? None of these is relevant to the lesson at hand, but all of them are relevant to the aims of the author and the artist, which is to get the reader to attend to the page. The illustration imitates the

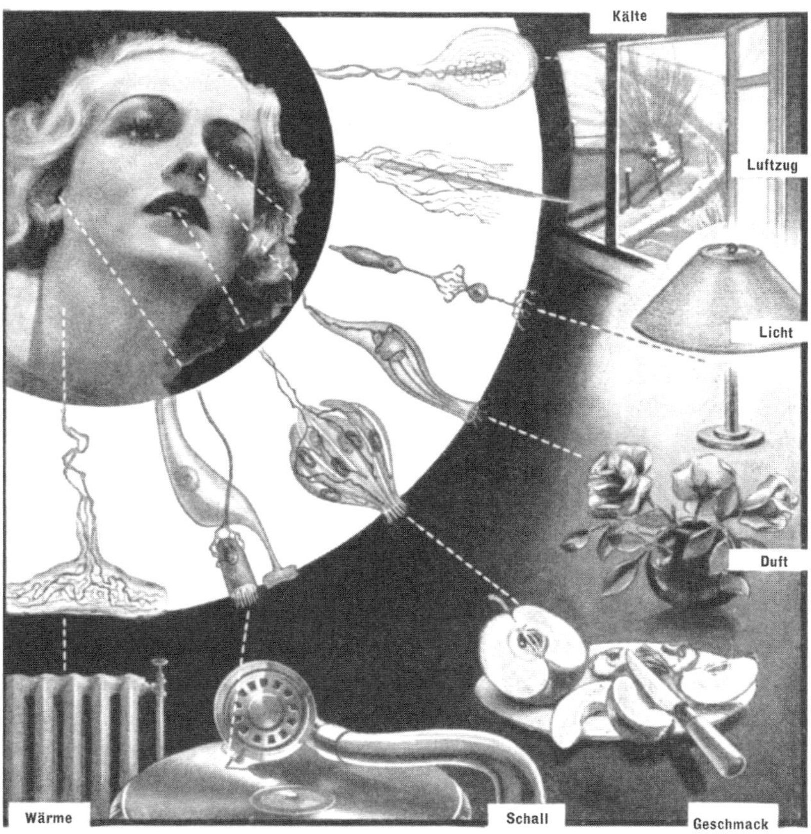

FIGURE 4.32. *"The Sensory Organs of the Head, Chiefly for Reception of Stimuli Arising at a Distance,"* Der Mensch Gesund und Krank *(1939), 1:265. Artist uncredited. National Library of Medicine.*

look of contemporary movie posters, glamor magazines, and cosmetics advertisements. The woman is bombarded by the proliferating sensual experiences of modernity. The specialized sensory receptor cells shoot out along radiating tracks of dashed lines that are launched from appliances, devices, commodities, and experiences of everyday life. Heat is represented by a steam radiator, sound by a phonograph, light by an electric lamp, cold by a draft coming through an open window. The cells, like futuristic creatures from outer space or surreal spermatozoa, try to penetrate the protective circle to gain access to the female subject and achieve the "reception of stimuli arising at a distance."

Everything about "The Sensory Organs of the Head" tells us that we are in the modern world: the image's aesthetic comes out of commercial illustration and advertising, not modernism. In "Four Ways to Deliver Drugs" and other illustrations, Kahn's artists copied the fashionable modernist look that was increasingly appearing in posters and magazines. There was also traffic in the opposite direction: "Der Mensch als Industriepalast" had commercial imitators. Pharmaceutical firms in particular adopted Kahn-style representation of the interior pathways of the respiratory and digestive systems and other body parts; they used Kahnian tactics, sometimes with a modernist aesthetic, to show process, motion, subjective sensation, and the effect of drugs within the human body.

THE CHEMICAL COSMOS

Kahn and his artists were especially challenged to show chemicals and chemical processes in the human body at the molecular level, where nothing is visible to the human eye. In nineteenth-century chemistry, molecular structures were mainly represented as strings of letters with subscript numbers, linked together by mathematical symbols in equations that notated chemical bonds and reactions. These equations became more complex and technical as chemistry became more sophisticated and specialized. As the field developed, there were sporadic efforts to visualize structure, most famously John Dalton's 1808 notation system, which featured symbols inside circles that could be arranged in patterns on the page, and August Wilhelm von Hofmann's 1865 stick-and-ball molecule models.[51] Chemists increasingly sought to describe chemical structures and reactions in three dimensions, but chemistry textbooks made surprisingly little use of illustrations and diagrams except to show how to set up and use laboratory apparatus and industrial equipment.

Chemistry is a notoriously technical subject. Kahn's text goes into it at great length and in great detail—surprising, considering his mass, nonspecialist readership. But he and his artists were hampered by the fact that the field of chemistry supplied no easily adaptable visual lexicon for the illustration of the chemistry of respiration, digestion, food substances, and cookery. That had to be invented.

They tried a variety of approaches. Chemical process might be rendered in two-dimensional schematics as a flowchart or in illusionistic perspective, in a decisive or representative moment, or as a sequence of moments. In the end, their solution was to develop their own pictographic system of organic chemistry.[52] Kahn's main artist-collaborator in this was Trester, but Roman Rechn, Fritz Schüler, and other artists also used the system. In "Der Mensch als Industriepalast," it can be seen the liver and stomach portions of the poster. It was a versatile lexicon of ideograms, well suited to the various talents of Kahn's artists.

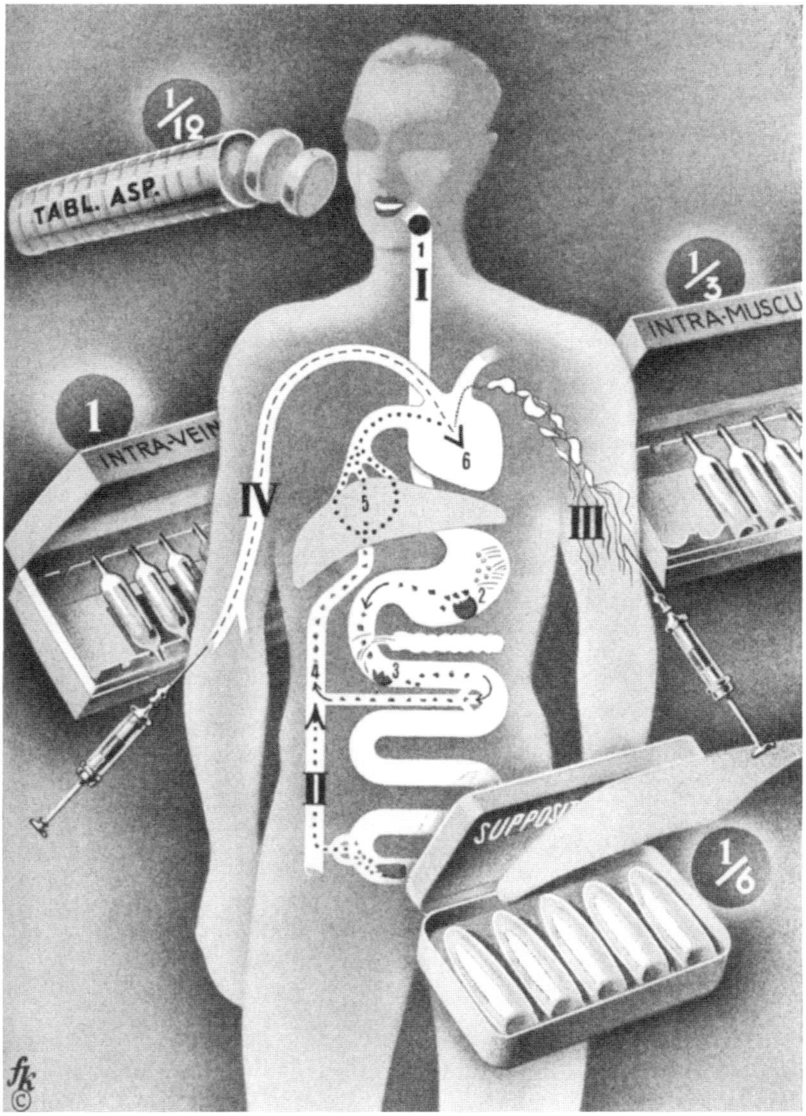

FIGURE 4.33. *"Four Ways to Deliver Drugs,"* Der Mensch Gesund und Krank *(1939), 1:343. Artist uncredited. National Library of Medicine. In its composition, shaded textures, and treatment of the figure (notably the lips), the illustration looks like a contemporary poster graphic.*

The components have the appearance of mass-manufactured standardized, stackable, interlocking modular puzzle pieces, as though hatched from the same industrial imaginary that gave birth to Lego (in 1949) and other modernist construction toys. The numerous minispheres are salt. Large black dots are amino acids. Three black dots in a circle make up a peptone (amino acid group). A circle with three peptones makes up albumose (peptone group). A circle with three albumose groups is a protein. The wedges are fat molecules, made up of one part glycerine (the tip) to three parts fatty acid. The rectangular bricks stand for different kinds of starch. Enzymes are represented as little keys that unlock the starch molecules and break them up.

Things mainly fall apart. Organic chemical food particles float in a dark void, like bits of scattered cosmic debris suspended not in fluid but in chemical inner space. The body is a *mikrokosmos* a modernist chemical universe, illuminated by some distant but invisible sun. The process of digestion is normally the subject of coarse humor, but Kahn and Trester endow it with majestic beauty, setting it entirely apart from the mess of embodied life as we experience it. (Not surprisingly, defecation and urination don't receive any representation.)

Again, none of this resembled anything that chemists were doing with visual modeling or notation. Historian of chemistry Lawrence Principe comments that nineteenth- and early twentieth-century chemistry "was about understanding actual molecular structures in a geometrical and architectural sense." Kahn's images "go in completely the opposite direction" by using an imaginary geometry to make an imaginary architecture. His depiction of the body as a chemical factory is worlds apart from contemporary chemistry, "more like seventeenth-century mechanical reductionism applied to chemistry (which was a total failure!)."[53]

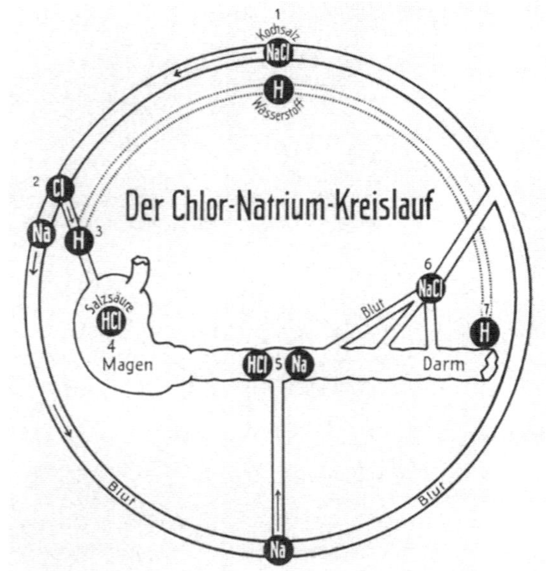

FIGURE 4.34. *A gracefully composed modernist schematic diagram that retains only the barest suggestion of figural representation in its depiction of stomach and intestines. "The Sodium Chloride Cycle,"* Das Leben des Menschen *(1926), 3:185. Artist uncredited, probably Trester. National Library of Medicine. Copyright Kosmos Verlag, Stuttgart.*

BODYSCAPES AND FANTASTIC VOYAGES

Kahn was especially fond of *mikrokosmos* as a visual and textual trope. The body as molecular universe was the highest degree of magnification and abstraction; the body factory worked in an entirely different register. But there was yet another microcosm that Kahn was delighted to present: the cellular and histological life of the body likened to an inner geography of caverns, rivers, pools, jungles, thickets, and reefs full of palpating tropistic organisms, grotesquely beautiful tubes, roots, and growths of all kinds.

In that mode, Arthur Schmitson (1857–early 1930s?) was Kahn's great collaborator. A well-respected technical medical illustrator—the only such artist in Kahn's stable—Schmitson had long experience in making precise, intensely colored, and sometimes disturbing pictures for books on forensic medicine, gynecological surgery, dermatology, surgical anatomy, and other demanding topics.[54] For most of his career, he worked in close collaboration with eminent professors and prestigious medical publishers, who only employed the most adept scientific illustrators. Kahn obviously valued Schmitson's résumé and relied on him for a variety of anatomical and histological assignments—work that called for an expert hand. But for volume 2 of *Das Leben des Menschen* (1924), he called on Schmitson for something else: two series of illustrations, almost books within the book, that stand radically apart in tone from all of the other images: "From the Life of a Wandering Cell" and "A Fairy-Tale Journey on the Bloodstream."[55] Readers are taken on a fantastic voyage inside the human body, which is depicted as land of enchantment and adventure populated by fairies and bloblike cell creatures. The two series are works of whimsy, more suited for a charming book of sugarcoated physiology lessons for

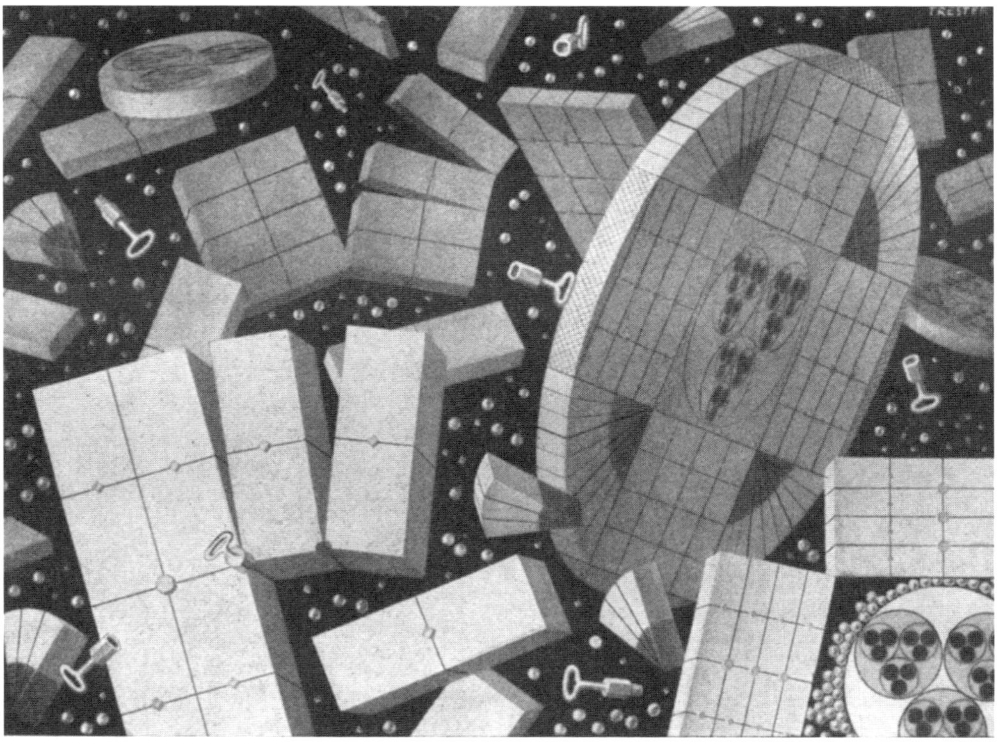

FIGURE 4.35. *"Digestion in the Mouth,"* Das Leben des Menschen *(1926), 3:101. Artist: Trester. National Library of Medicine. Chemical structure imagined in both two and three dimensions in the same diagram.*

children. (One suspects that they were originally created for just such a work and repurposed for inclusion in *Das Leben des Menschen.*)

The illustrated fantastic voyage was at the time a familiar genre of adult and children's fiction. Schmitson and Kahn were undoubtedly well acquainted with the subgenre of voyaging to the interior of the earth, which was famously figured as a primordial junglelike innerscape in Jules Verne's *Voyage au centre de la Terre* with illustrations by Édouard Riou (1864 and many editions thereafter).[56] In Kahn, as in Verne, the voyage inward is not just a journey in direction or scale but also in time—a journey back to the evolutionary past. The orientation of the universal human subject is reversed. Instead of a unified body shorn of detail and floating free of gravity in a global perspective or on the page of a modernist design, subjectivity wells up from warm, dark interior places populated by exotic creatures. The body is a sensuously filigreed corporeal *Umwelt*—the landscape generates flora and fauna. No sharp boundary marks the difference between foreground and background. The self dissolves; parts proliferate, branch off, float off. That indeterminacy is heightened by the fact that, unlike the 1966 film *Fantastic Voyage* (and unlike Verne's novel), Kahn doesn't feature any great heroes (or villains) journeying through the interior world.

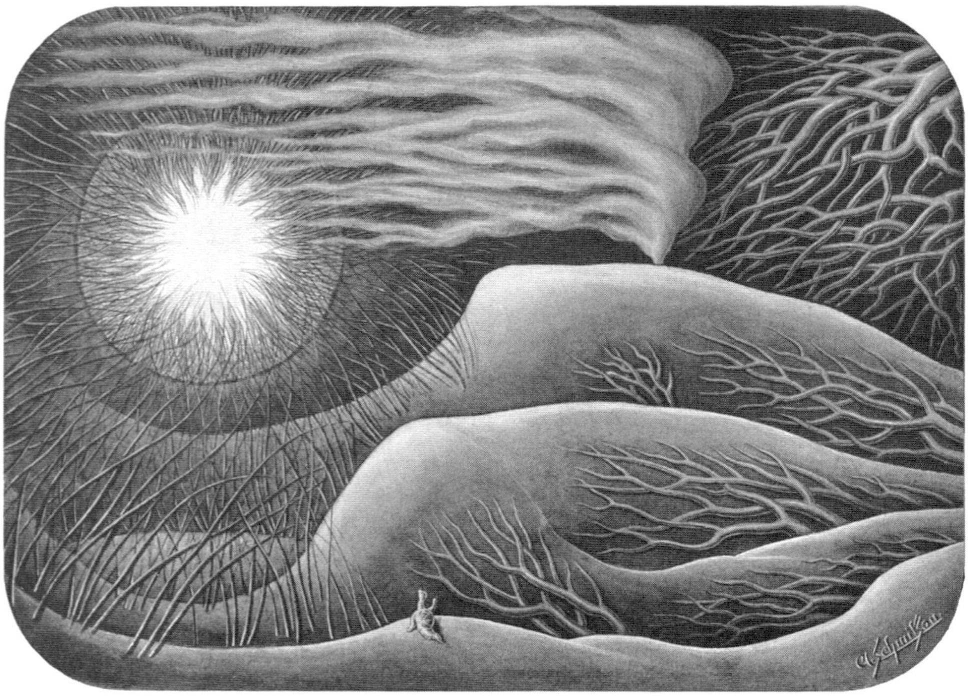

Aus dem Leben einer Wanderzelle IV
Auf den Muschelbergen des Naseninnern

FIGURE 4.36. *A nostrilscape. "From the Life of a Wandering Cell IV: On the Turbinate Bone Mountain of the Inner Nose,"* Das Leben des Menschen *(1924), vol. 2, pl. 33. Artist: Arthur Schmitson. National Library of Medicine. Copyright Kosmos Verlag, Stuttgart.*

Instead, he invites the reader to imaginatively identify with the scene of the voyage: the human body.

Aesthetically, the illustrations are unlike any others to be found in Kahn. They have no affinities to the Bauhaus-influenced stylings of Otmar Trester or Roman Rechn, or to the commercial realism of everyday magazine illustration, or even to the technical anatomical illustrations that Schmitson handled elsewhere. Instead, the bodyscapes have the look of the stylized organicism of Jugendstil, the late nineteenth-century art movement (a sibling to Art Nouveau) that reached its apogee in the decades before World War I. Born in 1857, Schmitson was perhaps the most senior artist to work for Kahn. That he would work in a Jugendstil-inflected style makes sense (even if nothing earlier in his career would lead us to expect such a move): Jugendstil was one of the modernisms of his generation.

It was a sensual, intoxicated aesthetic, a style that gave artists free rein to explore line, texture, and embellishment. By the 1920s, it had become a familiar sight in literary texts, advertisements, posters, erotica, ex libris plates, dust jackets, furnishings, architectural and interior ornamentation, and of course children's books. It could still delight audiences, but it no longer delivered the shock of the new. Even so, in

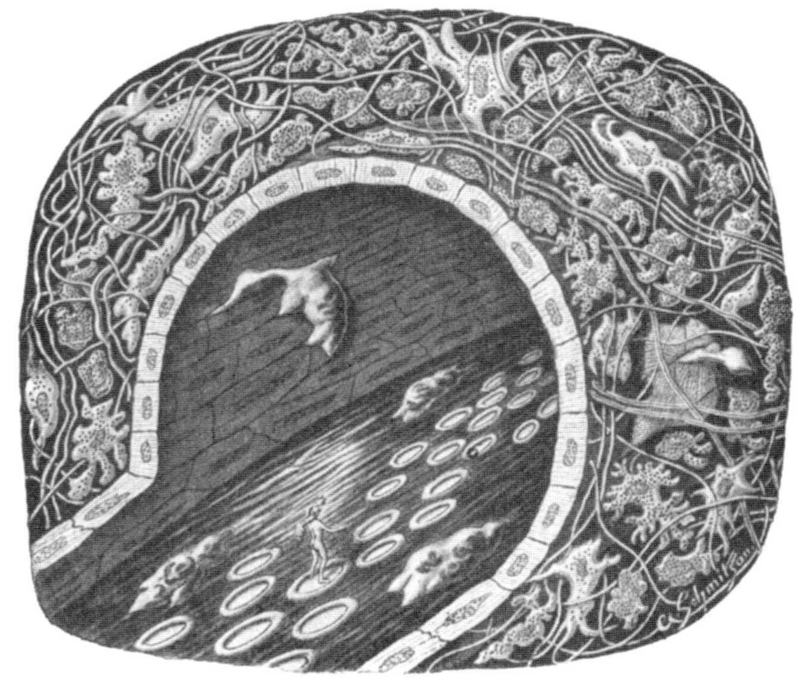

Abb. 203. Märchenreise auf dem Blutstrom II. Die Auswanderung einer Wanderzelle aus dem Blutstrom durch die Lücken des Aderrohrs.

FIGURE 4.37. *"A Fairy-Tale Journey on the Bloodstream 2. The Emigration of a Wandering Cell from the Bloodstream through the Gaps in the Vein Wall,"* Das Leben des Menschen *(1924), 2:291. Artist: Arthur Schmitson. National Library of Medicine. Copyright Kosmos Verlag, Stuttgart.*

the pages of a popular medical publication, it was a novelty. Jugendstil was especially amenable to imaginative decorative motifs that drew on, and sometimes wryly commented on, the subject matter of the books they designed and illustrated. Schmitson took full advantage. He made walls of cellular bricks and spooky thickets of nerves and veins; he wove together fanciful brocades of stylized organic structures taken from microbiology and histology. Only a few years earlier, Kahn had tried something similar in his collaboration with designer Georg Piek-Patrik. He attempted, with no great success, to incorporate microbial and histological decorative elements into fashion and home furnishing design. Piek-Patrik, an illustrator-designer of lesser talent, couldn't bring it off (see Figure 3.9). Schmitson, in contrast, brings it off beautifully.

After Schmitson, other artists also made bodyscapes for Kahn. In "The Nerve Cell" and "The Nerve Cord," the work of a gifted but uncredited illustrator, organic forms grow out of an abstractionist fantastic architecture. The human body becomes an impossibly beautiful

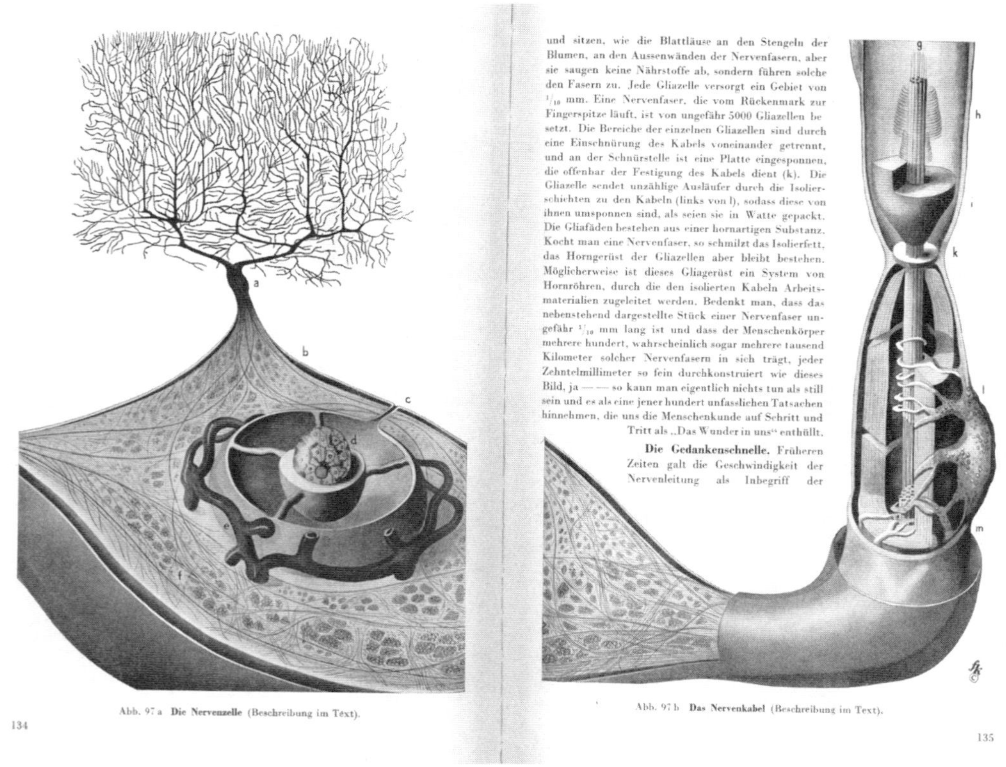

FIGURE 4.38. *"The Nerve Cell" and "The Nerve Cord,"* Der Mensch Gesund und Krank *(1939), 2:134–35. Artist uncredited. National Library of Medicine. The nerve cell and axon as science fiction/fantasy architecture, built on and connected to a sweeping cellular-patterned abstractionist landscape. At the rear grows a beautifully manicured bonsai-like nerve tree (based on the celebrated drawings of neuroanatomist Santiago Ramón y Cajal).*

fusion of living thing and manufactured construction. In these renderings, the body—shiny and translucent, with sensually smooth contours—appears as a streamlined modernist built environment (with no homuncular residents to clutter things up). The nervous system looks like the accomplishment of a visionary designer-architect or the outpost of an advanced civilization on a distant planet.

CORPOREAL ARCHITECTURE AND MECHANICS

Architecture has a long history in anatomical publication. Vesalius, like Galen and many authors thereafter, compared the skull, ribs, and bones to temples, palaces, dwellings, walls, and other structural elements as well as to simple machines.[57] In the 1830s, William A. Alcott based *The House I Live In; or The Human Body*, an enormously influential and popular illustrated anatomy book "for families and schools," on an extended architectural metaphor. Alcott introduced his account of the human body with illustrations of a series of houses, culminating with a picture of the frame of a wooden-frame house.[58] Like other mid-nineteenth-century popular anatomists, Alcott also featured illustrations of pulleys, levers, and other basic mechanical devices to explain the workings of muscles and joints. In the mid- and late nineteenth century, some anatomists (notably the Weber brothers, Eugene Meyer, and Julius Wolff) applied architectural and engineering principles to the study of musculoskeletal anatomy and motion. In their writings, they elaborately compared body parts to pumps, kitchens, and machines. A picture of an arch, lever, bridge, or temple might appear next to an anatomical illustration to reinforce the analogy made in the text. But metaphor did not appear within the frame of their anatomical illustrations or within the boundaries of their depictions of the human body.[59] That was a line that metaphor could not cross, until Kahn and his illustrators came along and made it look easy.

The architectural metaphor was ancient, canonical. Kahn referenced that in some illustrations by showing classical architecture and its devices: the arch, lintel, and column in the skull, vertebrae, leg bones, and so on. But, as in so many other areas, the imperative to be modern drove Kahn and his artists to up the ante. In *Das Leben des Menschen* and subsequent works, they created images that refer to contemporary architectural practices and styles—construction cranes, elevators, steel beams—to visualize musculoskeletal structures and principles. They also cooked up imaginary futuristic architectural forms.

In "The Inner Structure of Bones," a photomontage, Kahn and his artist take their image of the hip and heel bone (directly or via an intermediary source) from Julius Wolff's 1870 "On the Inner Architecture of Bones and Its Importance for Bone Growth."[60] That seminal article used an architectural comparison to propose a mathematical method of analysis of the structural mechanics of living organisms. According to Wolff, the same engineering principles that make iron truss bridges strong enough to support traffic (he shows a diagram of a bridge)—a series of opposing structural elements mathematically matched to combine to create tensile load-bearing strength—also make bones strong enough to support the weight of the human body.

Kahn did not merely reuse Wolff's images. He recruited objects that did not exist when Wolff published his article and added them to create a mixed-media montage of photographs and drawings. Wolff's diagrammatic line drawing of an 1870 truss bridge is transformed into a noirish modern steel structure (the base of the Eiffel Tower). The heel bone—glowing, within a modern shoe, as if materialized by an X-ray—gets its own updated simile: the crossed opposing spokes of an automobile wheel.

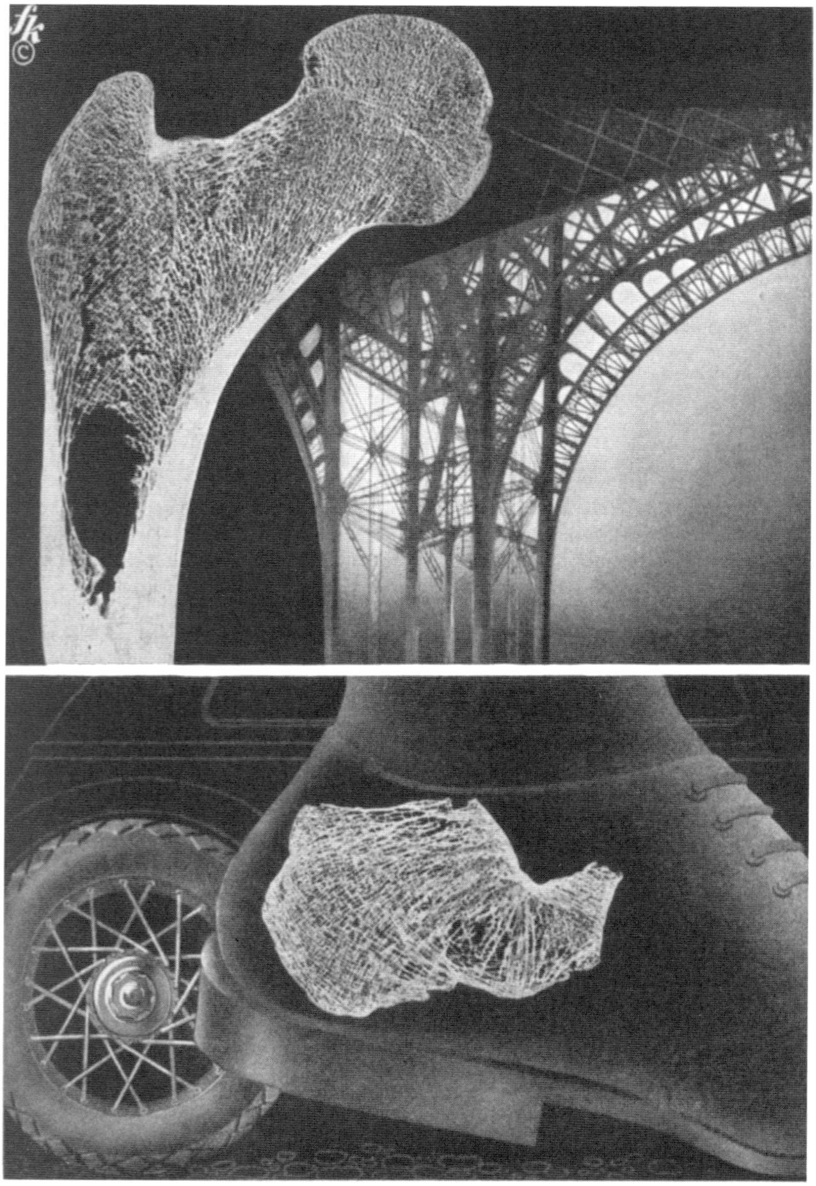

FIGURE 4.39. *"The Inner Structure of Bones,"* Der Mensch Gesund und Krank *(1939), 1:69. Artist uncredited. National Library of Medicine.*

PHOTOGRAPHIC BURLESQUE

The photographic image performs the modern.[61] This was especially true in the 1920s and 1930s, when halftone illustration in mass-circulation newspapers and magazines had become nearly ubiquitous, a fact of modern life that gave those publications an entirely different look than their nineteenth-century predecessors. Kahn knew all about the rhetorical effects of photography and worked them hard: his publications were full of photographic images. None of them originated in the studio of Kahn. They all came (mostly uncredited) from other print sources.

That wasn't difficult because the promiscuous use of photographic illustration was already standard operating procedure in the publishing industry (and, after the 1916 success of Waldemar Kaempffert's revived *Popular Science Monthly,* standard practice in popular science publishing). But the photograph was not by itself well suited to show processes or dramatize statistical relationships or imaginatively visualize metaphor. Besides, if everyone was using photographs, there was no distinction in it. To prosper, Kahn had to separate himself from the pack. Thus, drawn and painted images, eye-catching and often absurdist in their metaphorical mixture of things, became his trademark.

But photographs can also serve as eye-catching tokens of the absurdity of the real. In the pages of *Das Leben des Menschen,* readers are treated to a procession of specimens—images of pathology, microscopy, racial exotica, natural history, art, great moments in history, flashy scientific invention. Silhouetted to stand out on the page, the photos perform a halftone burlesque. Readers could ignore Kahn's words and go for a stroll through the pages, stopping here and there to take in a sideshow of real and amazing acts alongside the mind-bending illustrated art.

Mixed into the photo gallery were serious talk, lessons on healthy living and the prevention of disease, and an earnest narrative of the progress of civilization from antiquity to the modern present. This was a trope that Kahn loved to invoke in image and text: every chapter in *Das Leben des Menschen* begins with a quotation from some great author or philosopher. (When Kahn had to cut down on the amount of text in *Der Mensch Gesund und Krank,* he jettisoned this cultural, literary, and philosophical baggage.) Kahn marked his own filiation with the great chain of progress by presenting (poor quality) halftone reproductions of art masterpieces as well as reverential portraits of great men of medicine and science, as well as cultural figures such as Beethoven. But lest there be too much reverence, readers could pore over the much more numerous illustrations of regress—savages whose customs it was to remake their bodies into grotesque shapes, contortionists and other circus performers, people with deformities such as goiter, dwarfism, acromegaly, or microcephaly—all of the carnival tricks that nature plays on humanity and that humanity plays on itself.

The ultimate contrast and complement to all of this monstrosity was feminine beauty. Kahn thus presents two illustrations on facing pages. One shows an exotic figure: a nearly naked man with a radically

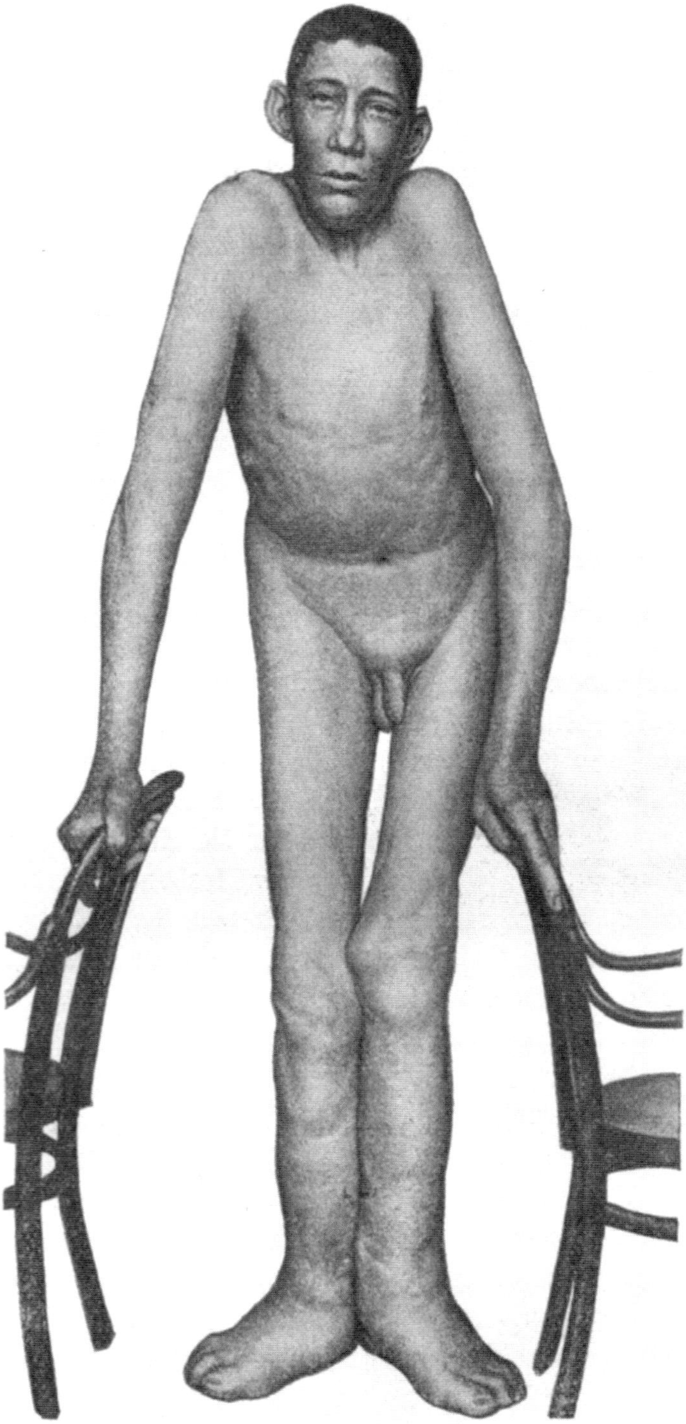

Abb. 232. Riesenwuchs infolge Überfunktion des Hirnanhang-Vorderlappens. Vermehrte Absonderung von Vorderlappensaft bewirkt vor allem Wucherung der Knochenenden, weswegen man dieses Krankheitsbild als den Spitzenwuchs (Akromegalie) bezeichnet. Man beachte das ungeheuerliche Wachstum von Nase, Kinn, Hinterkopf, Beinen und Fingern, z. B. den Zeigefinger der linken Hand. (Nach Sw. Vincent.)

FIGURE 4.40. *"Gigantism Due to Hyperactivity of the Pituitary Anterior Lobe,"* Das Leben des Menschen *(1926), 3:150. National Library of Medicine. Copyright Kosmos Verlag, Stuttgart.*

cinched waist, a "Mekeo Negro with Drawstring Waist Belt, the African Counterpart of the European Corset." (The caption is mistaken; the Mekeo people live on the island of New Guinea, not Africa.) The other, titled "The Influence of Lacing on the Female Body," shows side-by-side photographs of two sculptures, one fairly modern and one not, which just happen to show off two attractive naked women. The critique of corsetry was a staple of the popular health books of the nineteenth century. Kahn, here again almost reflexively modern, updates the old chestnut, making a new version for a new age, but the whole effort seems contrived, an excuse to show naked ladies and racial exotica. His tut-tutting critique of tight lacing, as well as the status of the sculptures as works of art, covertly gives readers the license to take a bit of tame erotic pleasure in the nudes. But the attack on corsetry seems out of place. By the 1920s, tight lacing was increasingly seen as old-fashioned and no longer much of an issue. In any case, Alexandre Falguière's 1896 statue (of Cléo de Mérode) could easily be taken as an example of a joyfully free-flowing and sexually emancipated modern woman (although not a 1920s flapper) rather than a woman deformed by corseting. Next to her, the Venus de Medici, with her concern for modesty, looks stiff and uncomfortably chaste. The appearance of the dark, nearly naked Mekeo tribesman next to two marble-white ladies turns the lesson on corsetry into a complete farce.

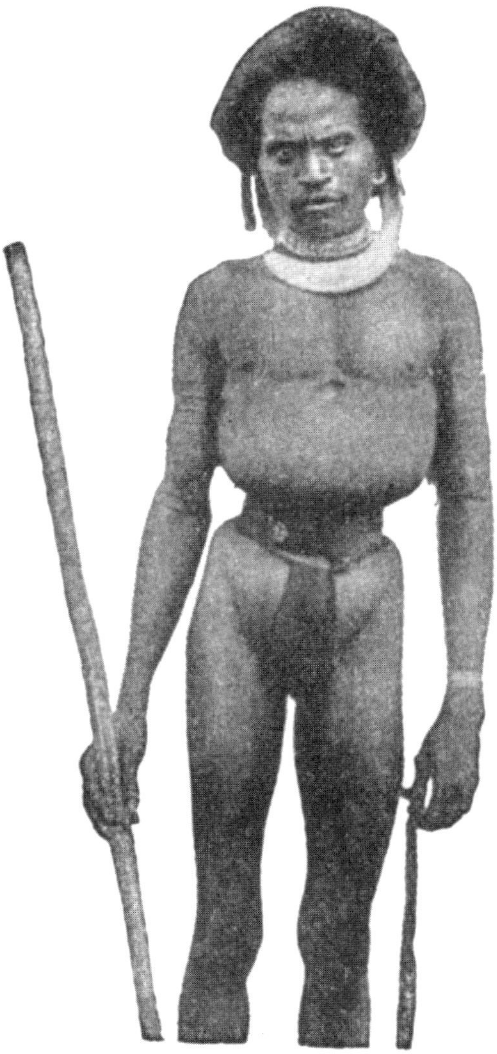

Abb. 68. Mekeoneger mit Taillenschnürung, das afrikanische Gegenstück zum europäischen Korsett.

FIGURE 4.41. *Racial exotica. "Mekeo Negro with Drawstring Waist Belt, the African Counterpart of the European Corset,"* Das Leben des Menschen *(1927), 2:92. National Library of Medicine. Copyright Kosmos Verlag, Stuttgart.*

VISUAL SEXOLOGY

"The Influence of Lacing" performed a little erotic tease that was never declared in the text. But elsewhere Kahn directly addressed the topic of human sexuality. That too was not unusual; popular medical texts had for centuries discussed human sexuality and reproduction, and had provided advice on sexual and reproductive problems. Medical publishers and authors well knew that the treatment of such topics could drive sales. Sexology appealed to readers' urgent desire for reliable information about human sexuality—no easy thing to get. It also invariably

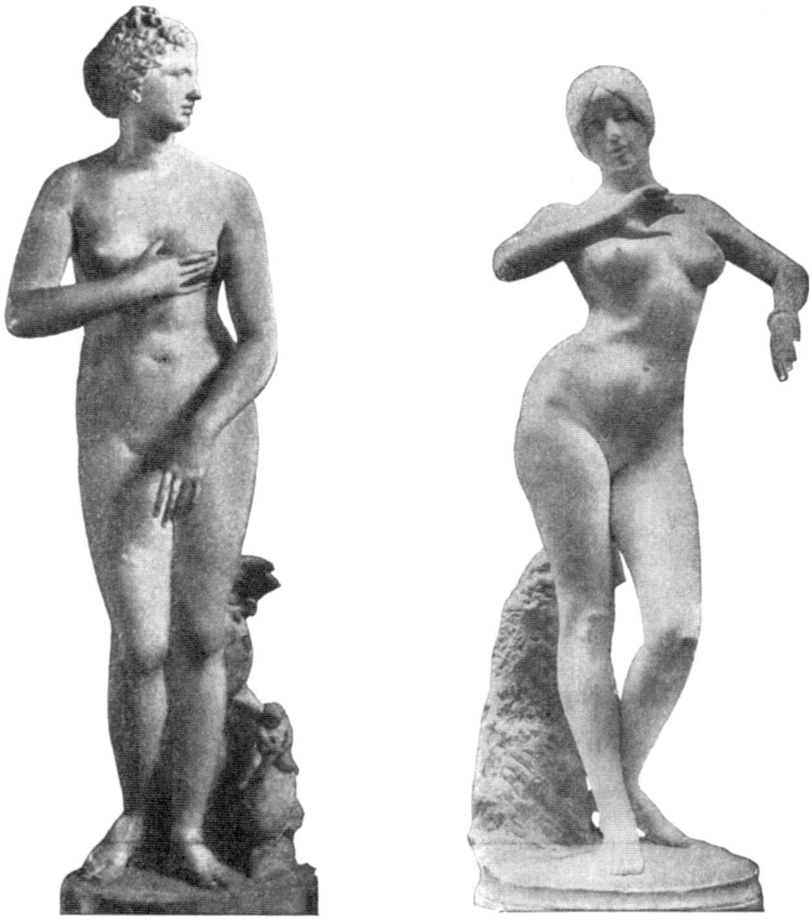

a b

FIGURE 4.42. *"The Influence of [Corset] Lacing on the Female Body. (From [Alexandre] Falguière)."* Das Leben des Menschen *(1927), 2:93. National Library of Medicine. (a) The Venus de Medici as a type of unspoiled female body. (b) The lifelike portrait statue of a dancer who was a world-famous beauty from the era when lacing was in fashion. Copyright Kosmos Verlag, Stuttgart.*

titillated readers, who may or may not have been comfortable seeking out overtly pornographic materials.

Kahn's medical specialty was gynecology and obstetrics, and while he never did any groundbreaking research, he could, with some justification, claim authority to speak on sexual topics. He included chapters on sexology in his human body books and contributed articles to newspapers and magazines. His book on the subject, *Unser Geschlechtsleben* (1937), was probably his best-selling work. Titled *Our Sex Life* in English (and translated into many other languages), it was respectfully reviewed and sold steadily until it was eclipsed by the works of Alfred Kinsey and other authors. Kahn also worked with sexologist Magnus Hirschfeld as a consultant on pharmaceutical promotional literature for drugs designed to help men who experienced difficulty with premature ejaculation or maintaining an erection.[62]

Human sexuality was controversial, a difficult topic. Authors had to walk a tightrope: they had to be tactful and inoffensive, but also clear and direct. Kahn's sexology focused on the usual topics: heredity, Mendelian genetics, embryology and pregnancy, sexual anatomy and

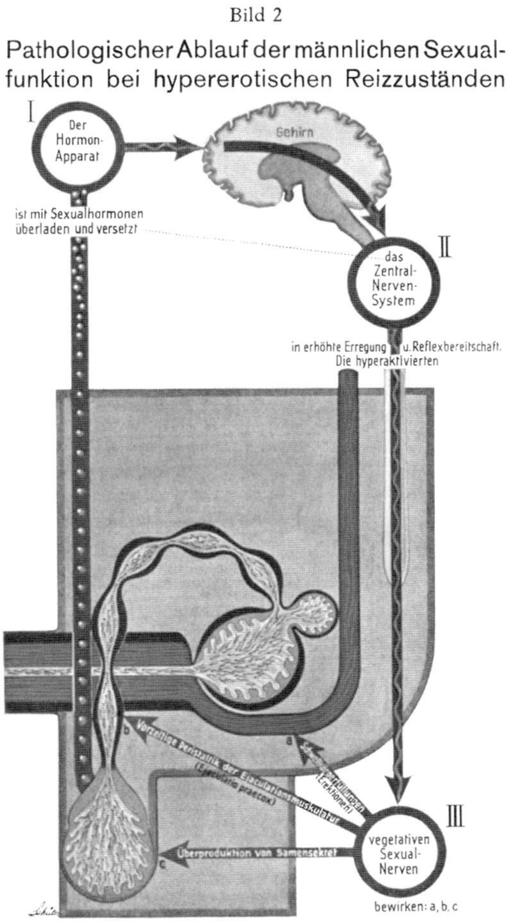

FIGURE 4.43. *"Therapy for Hypererotic Arousal Conditions." Artist: Fritz Schüler. In Bernhard Schapiro,* Präjaculin. Kombiniertes Epiphysen-Präparat gegen Reizzustände am Genitale und Hypererotismus *(Hamburg: Chemische Fabrik Promonta GmbH, 1932), pl. 3. Leo Baeck Institute, New York. In collaboration with sexologist Magnus Hirschfeld, Kahn worked on four-color illustrations for several pharmaceutical products, including this remedy for premature ejaculation.*

physiology, anomalies (cultural, physical, psychological), menstruation and fertility, contraception, virility, loss of desire, promiscuity, and venereal disease. As with his other work, a commitment to visual explanation was his calling card. His sexology featured close-ups, cutaways, body factories, modernist diagrams, and organicist bodyscapes.

Many of Kahn's sexological illustrations were of the same type, or even the same images, as those that appeared in the works of Hirschfeld and other contemporary sexologists. All of them featured typological, ethnographic, and pathological photographs and anatomical diagrams that could titillate as well as instruct. The topic of sex was itself in the register of modernity, and once again—whether he generated the visual material or took it from other sources—Kahn performed as a modernizer.

His most distinctive contributions to the field were his stylized illustrations of the erect penis. Within the universe of printed matter, and even within the domain of sexology, the male genitals in a state of erection was regarded as a difficult matter, what sociologist Erving Goffman called a "stigmatized," "spoiled" subject for representation. [63]

Kahn handled it by representing the phallus as both an anatomical illustration and as a machine-manufactured hydraulic apparatus—as something both medical and technological. This gave him something of a double license, and some protection against the accusation that his work was pornographic. That helped make *Unser Geschlechtsleben* and its many translations eligible for publication, even with the inclusion of the phallus, by respectable houses like Albert Müller and Alfred Knopf.

In "The Male Sex Apparatus," the erect penis looks like a U-boat or airplane fuselage sheathed in metal, powered by electrical cables (Plate 9). In the interior of the phallus, the structures, primed to shoot, look like the chambers of a revolver or a Gatling gun. Historian Cornelius Borck argues that Kahn's visual sexology had a masculine valence: "The female was represented," or not represented, "as that which resisted the very process of technological acculturation."[64] In Kahn's illustrations of female genitalia, the vagina and uterus are figured as passive, tropistic, almost aquatic creatures.

Beyond sexology, there was a strong masculinist drift to Kahn's image practice. Figures of men far outnumber figures of women. Kahn's most famous illustration, "Der Mensch als Industriepalast," conveniently leaves out the genitalia, but it still features a masculine profile, as do all of his other cutaway profile body factories. Yet there is also a queerness to Kahn's approach to the human body: its fragmentation, its technologism, and its ardent, excessive desire to pleasurably represent all of that. And no Kahn illustration was queerer than "The Male Sex Apparatus."

• • •

So that's the tour. More could be said about genres and tropes of conceptual illustration. Every paragraph of this chapter could be enriched by the addition of more illustrations from Kahn, his predecessors, and his contemporaries. But to include them would make this book even longer than the seven volumes of *Das Leben des Menschen* and *Der Mensch Gesund und Krank*.

5

KAHN'S TAKE AWAY

CONCEPTUAL SCIENTIFIC ILLUSTRATION'S ICONOPHILIC DIASPORA

OUTSIDE WEIMAR: *DER MENSCH*'S GLOBAL ITINERARY

We're tempted to treat Kahn as a made-in-Weimar story, or after 1940 as Weimar transplanted to America. But over the course of the twentieth century, his images and influence traveled far and wide. Kahnian pictures appeared all over the world, and Kahn's works were translated into French, Dutch, Spanish, Indonesian, Italian, Greek, Chinese, Hungarian, Danish, Farsi, Turkish, Czech, Swedish, Russian, and other languages.[1]

The most circulated image was "Der Mensch als Industriepalast." Early on, Kahn and his publishers recognized its commercial potential. Authorized versions appeared in French, Spanish, Portuguese, Romanian, Arabic, Chinese, Hebrew, and English.[2] It also appeared, in unauthorized versions, copied or redrawn with slight or significant modifications, in poster or other formats, or recast into something very different, with or without the modernist gloss that Kahn's artist Fritz Schüler supplied.[3]

Take, for example, a panel of a 1933 poster printed in Shanghai, as part of a series of health and hygiene posters (Plate 10).[4] It is not a direct translation of Kahn's 1926 poster but "Der Mensch als Industriepalast" reimagined by a Chinese artist. At the time, a small number of Chinese were traveling to Europe or America to study Western science, technology, and culture, and were bringing materials back to China. At the same time, Western printed material was entering into China through

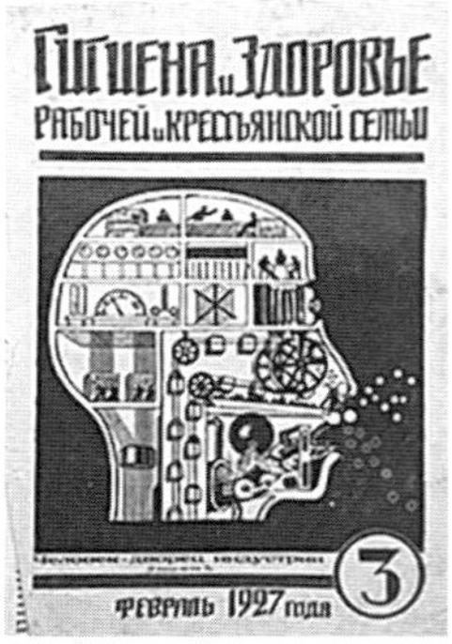

FIGURE 5.1. *An unauthorized Soviet "Der Mensch als Industriepalast": "Man [as] Palace of Industry,"* Gigiena i zdorov'e rabochei i krest'ianskoi sem'i *3 (February 1927): cover. Courtesy Nikolai Krementsov.*

the European-dominated "concessions." After the Nazis came to power in the 1930s, German refugees, many of them Jews, flooded into Shanghai's international settlement and other concessions. Did they bring "Der Mensch als Industriepalast" with them? Or did the activists of the Shanghai Physiology and Hygiene Association see the poster in an English-language version? (American missionary and secular health workers then were active in China, participating in projects to help modernize Chinese medicine and health practices, through direct provision of care, training programs and well-publicized exhibitions and poster campaigns.)[5]

Unlike "Der Mensch als Industriepalast," the Shanghai poster presents a double figure: an anatomical man stares at, perhaps studies, his industrial counterpart. It was made at a time when Chinese nationalists and socialists alike were adopting the rhetoric and methods of public health, personal hygiene, and scientific medicine. They believed that China, pressed by the Western powers and Japan, must modernize by turning to Western practices, ideas, and technologies. That meant the development of heavy industry, the adoption of Western forms of governance and education, and programs to implement Western nutritional science, hygienic practices, mass vaccination, and fitness regimes. To be a modern economic and military power, the Chinese people had to be made healthy and strong. And poster and pamphlet campaigns were seen as vital modern methods of educating the public.

In the body factory panel, the anatomical figure, a muscleman, exercises with a weight with his right hand as he demonstrates a powerful grip with his left. The parallel industrial figure has cranes in the place of arms, designed for a construction project: building the self, building the nation. One crane lifts an I-beam; the other is equipped with a steel claw that can grasp building materials. ("Der Mensch als Industriepalast," in contrast, has no arms; the figure only does the internal work of manufacturing himself.) "Der Mensch als Industriepalast" is revised, consciously or unconsciously, to make it distinctively Chinese. The color scheme is redone so that yellow and red dominate; the face of the anatomical figure is given Chinese features; and the respiratory system is equipped with a Chinese accordion bellows. The Shanghai panel is less complexly industrial, less synoptic, than "Der Mensch als Industriepalast." It only lightly suggests that the body is electrical and chemical in character (these themes are taken up more fully in other panels and posters in the series).

In Europe and America, early adaptations of "Der Mensch als Industriepalast" tended to follow Kahn more closely. A 1932 American school encyclopedia, *The Volume Library,* flips the profile and slightly modifies the body outline.[6] *The Miracle of Life,* a late-1930s British popular science volume, more deeply under the Kahnian influence, has several body factory figures, a "form suited to an essentially mechanical age."[7] Two of them are based on "Der Mensch als Industriepalast," following the concept closely, but reconceived. The profile is divided in

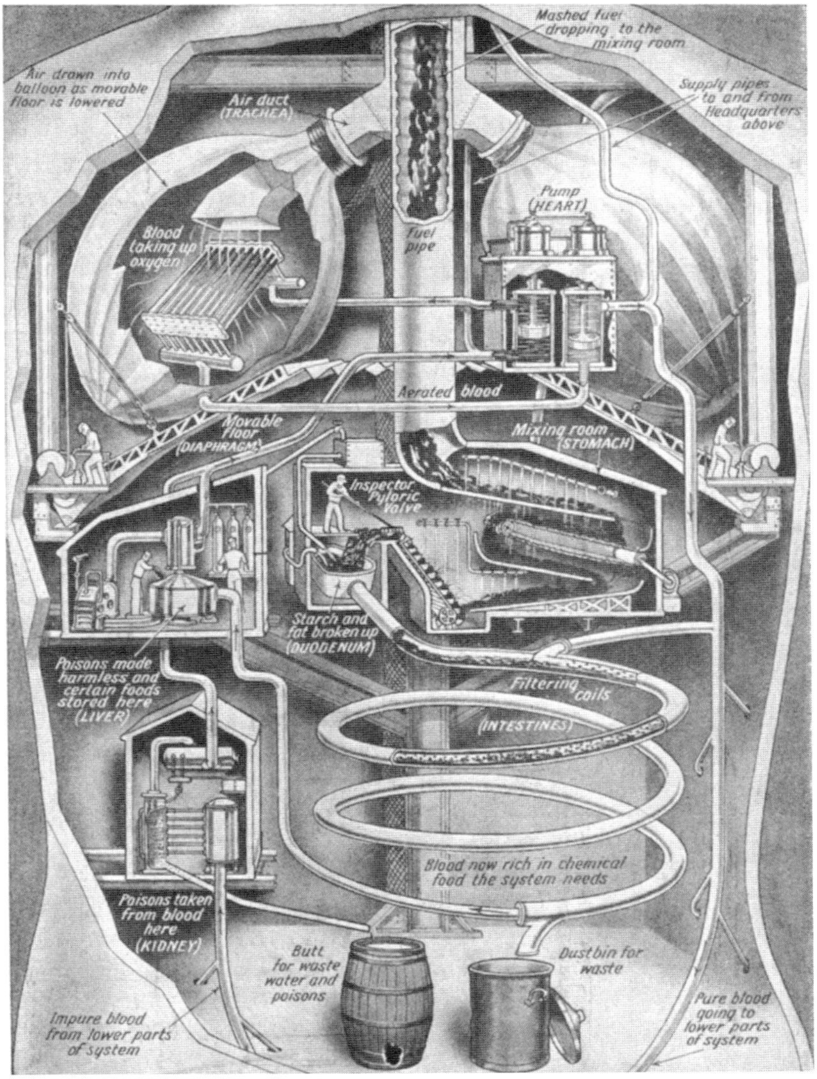

FIGURE 5.2. *"The Factory within the Human Body,"* The Miracle of Life *(London, ca. 1938), 341. Artist uncredited. National Library of Medicine. This British body factory looks like a cross between a futuristic city, pinball machine, and roller-coaster ride.*

two, one page for the head and the other for the torso. Both are given a gee-whiz treatment, made to look like a futuristic amusement park.

"Der Mensch als Industriepalast" borrowed many of its elements from dollhouse and mechanical and architectural display models. Scale models, with a side or top omitted to reveal the interior structure and functions, were often used in exhibitions. Sometimes these were working mechanical devices built to half or quarter scale, or scaled up, with openings (or no casing at all) to permit the viewing of operations. In the mid-1930s, the Buffalo Museum of Science's Hall of Man exhibited a display model version of "Der Mensch als Industriepalast" (without

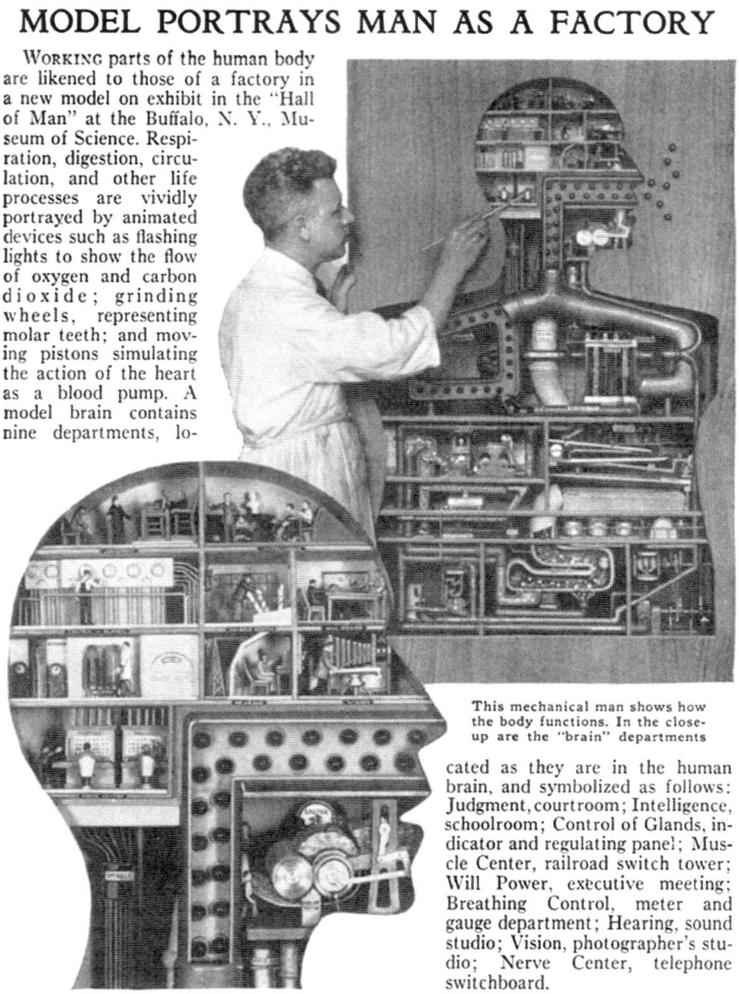

MODEL PORTRAYS MAN AS A FACTORY

WORKING parts of the human body are likened to those of a factory in a new model on exhibit in the "Hall of Man" at the Buffalo, N. Y., Museum of Science. Respiration, digestion, circulation, and other life processes are vividly portrayed by animated devices such as flashing lights to show the flow of oxygen and carbon dioxide; grinding wheels, representing molar teeth; and moving pistons simulating the action of the heart as a blood pump. A model brain contains nine departments, located as they are in the human brain, and symbolized as follows: Judgment, courtroom; Intelligence, schoolroom; Control of Glands, indicator and regulating panel; Muscle Center, railroad switch tower; Will Power, executive meeting; Breathing Control, meter and gauge department; Hearing, sound studio; Vision, photographer's studio; Nerve Center, telephone switchboard.

This mechanical man shows how the body functions. In the close-up are the "brain" departments

FIGURE 5.3. *A model based on "Der Mensch als Industriepalast," with moving parts and flashing lights, at the Hall of Man, Buffalo Museum of Science.* Popular Science Monthly *127 (December 1935): 30.*

credit to Kahn or Fritz Schüler). Like a pinball machine, it had "flashing lights to show the flow of oxygen and carbon dioxide; grinding wheels, representing molar teeth; and moving pistons simulating the action of the heart as a blood pump."[8] A few years later, curator Carlos E. Cummings commissioned a "twice life size" version for the Hall of Man exhibition at the 1939 New York World's Fair.[9]

THE CIRCULATION OF KAHNIAN IMAGES

SIX VERSIONS OF "THE TOTAL NUMBER OF BLOOD CELLS"

"Der Mensch als Industriepalast" was not the only Kahnian image to circulate. Kahn's publications, with their thousands of illustrations, served as sources for public health agencies and organizations, medical

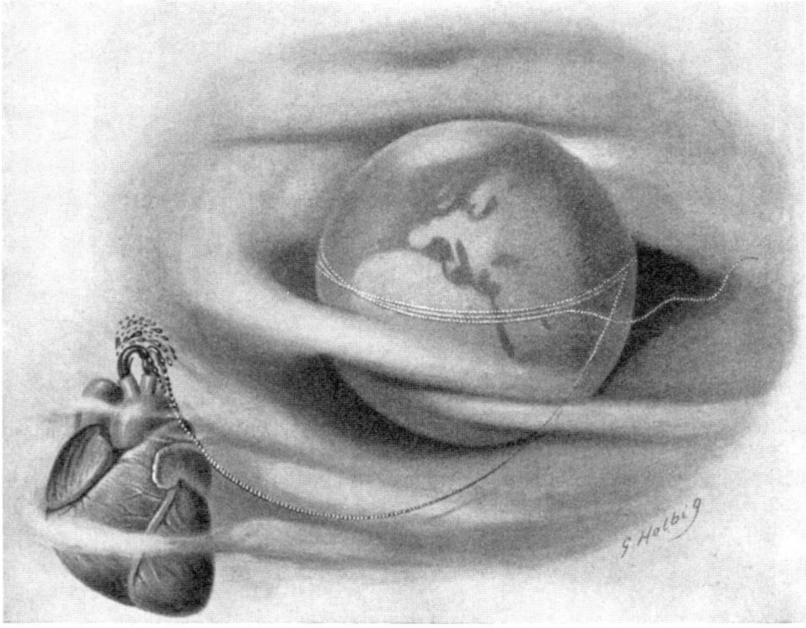

FIGURE 5.4. *The circulation of images 1: "The Total Number of Blood Cells,"* Das Leben des Menschen *(1924), 2:309. "The 22 trillion blood cells of a human being, placed end to end, would reach around the equator more than three times." Artist: Georg Helbig. National Library of Medicine.*

journalists, magazine illustrators, exhibition makers, poster artists, and cartoon animators. Closely or loosely derived from the Kahnian original, such images cropped up all over the place, in Germany and abroad. Copyright was impossible to enforce, and credit almost never given. Each step of the way, images were redrawn and revised for reasons political, practical, and contingent—or for no reason at all.

Consider, for example, "The Total Number of Blood Cells," an illustration that dramatized a numerical quantity through a global comparison: "The 22 trillion blood cells of a human being, placed end to end, would reach around the equator more than three times." It first appeared as a halftone pencil-and-charcoal drawing by Georg Helbig in 1924 in the second volume of *Das Leben des Menschen.*[10] A glistening heart, still beating as if recently extracted from a living body, spurts droplets of blood. Seen from an impossible extraterrestrial perspective, the disproportionately large organ floats in space, going around the planet like a satellite. The earth is tilted on its axis, enfolded in moody gray and black clouds. A string of cells, attached to the heart, loosely encircles the planet several times. It's an odd and not entirely successful image; it has a kind of eccentric sensuality that is entirely extraneous to its lesson.

The illustration was not a showpiece and never attracted any comment. Yet it traveled. In the year 1936, an uncredited version crops up in a Soviet health magazine. While closely following Kahn's image, the

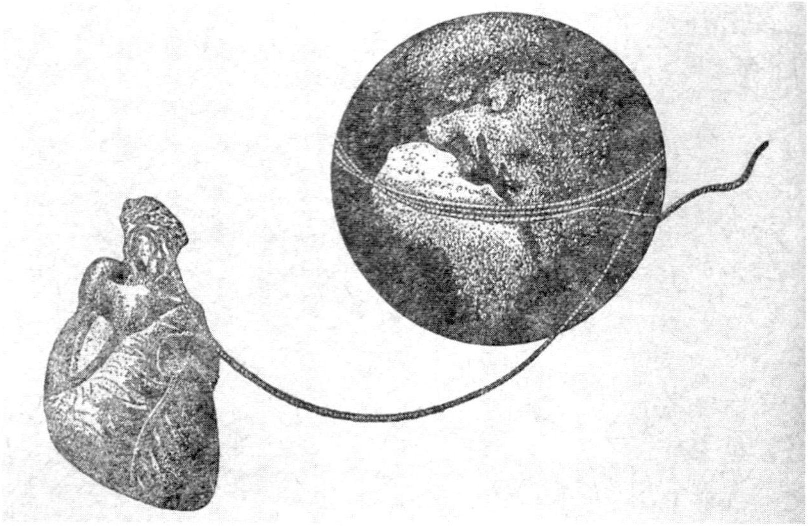

FIGURE 5.5. *The circulation of images 2:* Gigiena i Zdorov'e *7 (April 1936): 16. National Library of Medicine. This uncredited Soviet version of "The Total Number of Blood Cells" closely follows the original, but the stormy atmosphere and artist's signature have been removed, and droplets of blood no longer spurt from the heart.*

Soviet version deletes the moody atmosphere and spurting drops of blood. That may have been because the tone of the picture wasn't in keeping with the bright optimism of Soviet health propaganda. Or it may have been because, as was often the case with Soviet mass publications, limited resources meant that the ink, paper, and printing presses were of poor quality. Printed images were often made from a halftone of a photograph of an original. That especially degraded their appearance. The subtleties of Helbig's moody atmospherics would have been lost had they been retained.[11]

The same year, 1936, a variation on Kahn's image appears in a book version of a spectacular medicine and health exhibition, Das Wunder des Lebens, which had recently opened with great fanfare in Berlin. Reconceived by artist E. A. Mühler as a simple pen-and-ink drawing (and likely also used as a graphic on the exhibition floor), the picture is transformed, corrected. The idiosyncratic moodiness of Helbig's version gives way to a straightforward pragmatic rationalist rendering. The viewing angle is straightened. The spurting heart is deleted. The dark and textured clouds no longer trouble the viewer. All that remains is the planet earth, turned to include both western and eastern hemispheres, encircled by strings of straight-rule governed blood cell beads.[12]

In 1938 there comes yet another version, in a two-volume health encyclopedia designed to be a Nazi-approved successor to the banned *Das Leben des Menschen*. Taking off from the 1936 *Das Wunder des Leben* version, the image has its own peculiarities. The orbiting heart

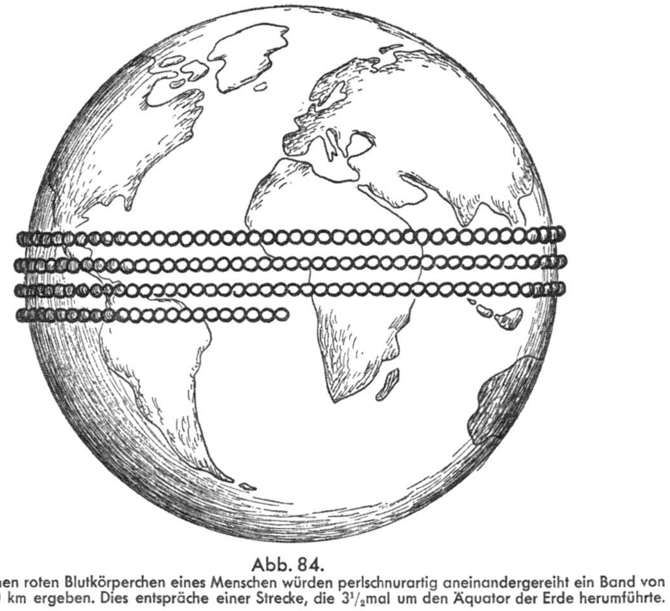

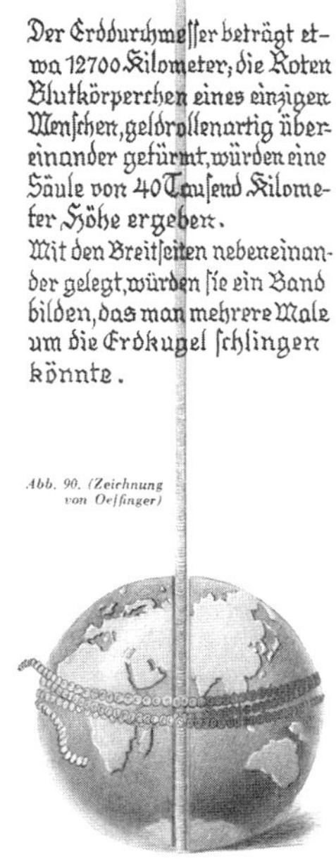

FIGURE 5.6. *The circulation of images 3: Bruno Gebhard, ed.,* Wunder des Lebens *(1936), 100 (exhibition catalog). Artist: E. A. Mühler. National Library of Medicine. The equatorial band and global numerical comparison remain (22 trillion changed to 25 trillion), but the image has been straightened, the heart and atmosphere deleted, and the globe turned to include the western hemisphere.*

FIGURE 5.7. *The circulation of images 4: The same visual comparison in a Nazi-era volume. The globe has been turned to show only the eastern hemisphere, and the modernist sans-serif caption of the Mühler version has been replaced by the Fraktur display type favored by the Nazis. Gerhard Venzmer et al.,* Hand- und Lehrbuch der Krankenpflege *(Stuttgart: Franckh'sche Verlagshandlung, 1938–39), 1:72. Artist: "Oeffinger." National Library of Medicine. Copyright Kosmos Verlag, Stuttgart.*

is again absent. The planet earth, now encircled in thick ropes of blood cells, is penetrated by a large measuring staff (which also numerically represents blood cells). A caption is now inserted inside the frame of the illustration in heavy, hand-drawn Fraktur lettering, instead of at the bottom or in the text, as previous versions had it. (Fraktur, a family of fonts derived from Gutenberg, was favored by the Nazis as a national German house style, in opposition to sans-serif and italic fonts.[13])

In exile from the Nazis, and perhaps responding to the pictorial revisions made in Franckh's new and now competing line of health books, Kahn worked with a new (and uncredited) artist to reconceive the picture for the 1939 *Der Mensch Gesund und Krank.* In the new version, the blood cells now spill from the back of a speeding train and are laid on railroad tracks that encircle the earth (instead of a string of blood cells spooled around the planet).[14] Kahn also added a lesson on the area of the surface of a blood cell, shown metaphorically as a plaza in front of a central train station. These revisions and additions made the image more dynamic—that onrushing train!—and no longer a mere variation on the original version. But it was now busy and hard to read.

In 1951, yet another version appears, as an animated segment in *The Wonder Engine of the Body,* an educational health film produced

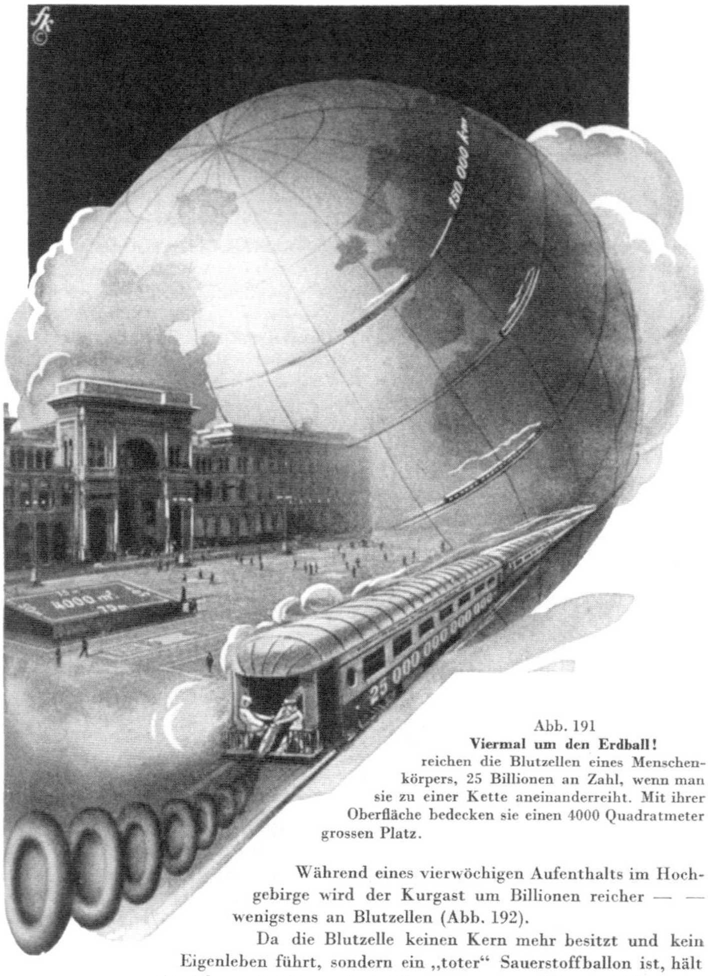

Abb. 191
Viermal um den Erdball!
reichen die Blutzellen eines Menschenkörpers, 25 Billionen an Zahl, wenn man sie zu einer Kette aneinanderreiht. Mit ihrer Oberfläche bedecken sie einen 4000 Quadratmeter grossen Platz.

Während eines vierwöchigen Aufenthalts im Hochgebirge wird der Kurgast um Billionen reicher — — wenigstens an Blutzellen (Abb. 192).

Da die Blutzelle keinen Kern mehr besitzt und kein Eigenleben führt, sondern ein „toter" Sauerstoffballon ist, hält sie sich nur 20 Tage. In 20 Tagen muss der Körper 25 Billionen Blutzellen neu ersetzen, d. h. in jeder Sekunde mehr als 10 Milliarden. Würden die Blutzellen das Knochenmark wie die Automobile die Fabrik am laufenden Band verlassen, eine Blutzelle dicht hinter der anderen, so müsste dieses Band mit einer Geschwindigkeit

223

FIGURE 5.8. *The circulation of images 5: Kahn's own revised version. "Four Times around the Earth!"* Der Mensch Gesund und Krank *(1939), 1:223. Artist uncredited. National Library of Medicine. The globe in this version is now turned to straddle the eastern and western hemispheres, showing the Atlantic Ocean and its coastlines.*

by Bray Studios of New York for the American Heart Association.[15] Based directly on Kahn's 1924 illustration, this too has key differences, apart from its conveyance into the medium of animated motion pictures. For American eyes, the globe is now turned to the western hemisphere. The heart is retained in the image, but as an iconic valentine, not the original blood-spurting anatomical organ. Both globe and heart are shiny, like the metallic finish of an American car.

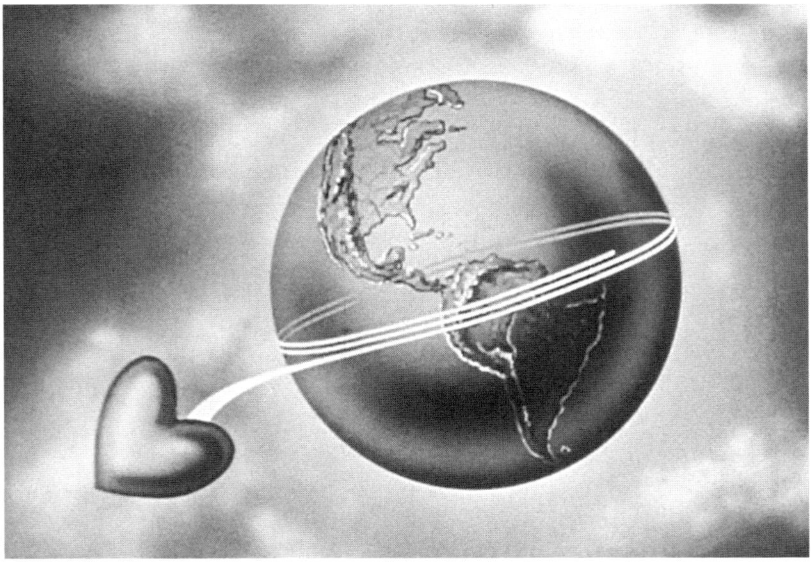

FIGURE 5.9. *The circulation of images 6: A frame from a postwar American animated cartoon version. Only the western hemisphere is shown; the heart is reinstated as a valentine instead of an anatomical organ.* Wonder Engine of the Body: The Human Heart *(American Heart Association and Bray Studios, 1951). National Library of Medicine.*

The atmosphere is sunny, suitable for the radiant American optimism of the postwar era, not at all like Helbig's gothic 1924 original.

"Der Mensch als Industriepalast" appeared in many versions: it delighted the eye, and its vision of man and machine was perfectly attuned to its moment. "The Total Number of Blood Cells" wasn't delightful or attuned to anything. Yet it too circulated. At first glance, that makes no sense. The scenario and analogies are contrived, the lesson not terribly important. Why, then, was it reused? It can only be because of the great enthusiasm for images as a modern tool of pedagogy (especially in science instruction) and the easy accessibility of Kahn's pictures. The adapters of "The Total Number of Blood Cells" didn't have a compelling need to present this particular lesson, but they did have a compelling need and desire to present pictorial science lessons that were in some way modern. Lacking the will or wherewithal to develop their own, Kahn's were available—off the shelf.

MEDICAL AND HEALTH EXHIBITIONS

Kahn was influential, but not the only show around. In the 1920s and 1930s, exhibitions and museums increasingly took on a prominent role as a modernizing venue for displays of the human body that made reference to, or enhanced, the experience of life in mass society. The Deutsches Hygiene-Museum (DHM), located in Dresden, was especially acclaimed as a model of modern medical and scientific education for the public. Its exhibits and traveling shows, as well as the circulation of objects manufactured in its workshops, stoked a global health exhibition movement. The 1935 Buffalo Museum of Science's Hall of Man, the 1938 New York Museum of Science and Industry, and the 1939 New York World's Fair's Hall of Man all fell deeply under its influence.[16] But the DHM was itself subject to influence. Kahn and the DHM were contemporaries with nearly identical agendas: to make life modern by bringing scientific medicine to the masses through prolific, substantive, and entertaining visuals. They both eclectically situated the body in modernity and modernity in the body, turned the body into a technological artifact, deployed or represented visual technologies of modernity (such as slide projections, and mechanical and electrical models), and, with no consistency, featured various versions of modernist aesthetics and architecture. Not surprisingly, then, there was overlap and traffic between them. Many of Kahn's illustrations and layouts looked like imaginative renderings of exhibition displays; some were based on actual displays at the DHM and GeSoLei.[17] Many of the museum displays looked like scaled-up book illustrations and page layouts; some were based on illustrations in Kahn. Das Wunder des Lebens, the DHM-produced Berlin exhibition of 1936, had a section titled "Das Leben des Menschen," and (as shown in Figure 5.6) some of its graphic displays were derived from illustrations that had appeared only a short time earlier in *Das Leben des Menschen.* (Kahn, by then persona non grata in Germany, received no credit or acknowledgment.[18])

The traffic between Kahn and the DHM was part of a larger traffic between printed illustration and exhibition display. Before the 1930 opening of the DHM, and even before the famous 1911 International Hygiene Exhibition that preceded and inspired it, magazines and newspapers were making illustrations that aspired to look like exhibitions on a page. Exhibitions returned the favor by trying to make their graphics look like print illustrations writ large.[19] That trend intensified in the interwar era with the rise of the international health exhibition movement. In 1934 the DHM came to America with an acclaimed but controversial traveling show, "Eugenics in the New Germany," which was set up in conjunction with the annual meeting of the American Public Health Association in Pasadena (then moved on to Oregon and then Buffalo).[20] More importantly, the cause of the health museum in North America was taken up by DHM curator Bruno Gebhard, who in 1937 fled Germany and emigrated to the United States. Gebhard went on to shape the 1938 exhibit at the New York Museum of Science and Industry at Rockefeller Center in Manhattan, the 1939 New York World's Fair Hall of Man, and later the Cleveland Health Museum. All of these were designed with a high modernist aesthetic, and all purchased exhibition materials in that style from the DHM's workshop (as did the Buffalo Museum of Science, which housed "Eugenics in the New Germany" until 1943!).

To make health museum displays with sans-serif type, evocative interactive display objects, negative space, and a playful, futuristic tone was at the time an intellectually and politically fashionable thing to do. Crowds poured in, and museum makers succumbed to a kind of exhibitionary delirium, fever dreams of Tivoli-style entertainments mixed with modernizing science instruction.[21] A *Popular Science Monthly* article described the displays at the New York Museum of Science and Industry as "the side show of a modern amusement park," a "scientific peep show" that "tells how our bodies work":

> By pressing buttons, turning cranks and pulling levers, visitors can test their strength, their lungs, and their voices and see for themselves how blood circulates, how their muscles work, and host of other interesting details. Controlled by a maze of motors, the exhibits provide a fascinating introduction to the mysteries of human anatomy and physiology.[22]

Many of the displays had no direct connection to health but focused on the human body. The environments and designs played with scale, volume, lighting, gradated texture, composition, decontextualization, and juxtaposition. Displays simultaneously referred to embodied life, the iconography and picture practice of popular science, and contemporary high graphic design. Visitors were presented with stylized giant ears, hands, blood cells, hair follicles, larger-than-life transparent or sliced anatomical figures, and so on. Some displays were interactive. These played on the contemporary fascination with meters, buttons, and dials that activated machines or mechanical displays that gave

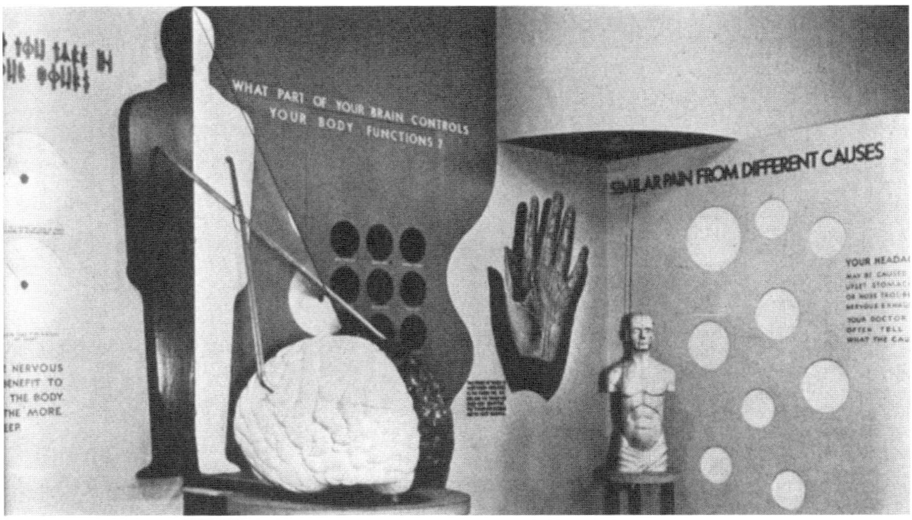

FIGURE 5.10. "*Where in the Brain the Most Important Functions Take Place,*" Man and His Health: A Guide to Medical and Public Health Exhibits at the New York World's Fair 1939 . . . *(New York: Exposition Publications, 1939), 25. National Library of Medicine.*

results. Museum visitors got to play the role of operant in the body factory, to mimetically become the homunculus.

Kahn's illustrations played games in representational space; the Hall of Man exhibitions gave viewers the chance to play games in physical space (or at least gestured toward that). The architectural-sculptural aspect of the exhibition hall was designed according to Bauhaus precepts of abstraction, negative space, clarity, and formalism. In the displays, visual metaphor and physiological dynamism were reduced to their most basic forms—heart as pump, nervous system as telephone, eye as camera, arm as mechanical device. The body was remade into a modernist graph, flowchart, schematic diagram, or composition.

In that way, the genre of public amusement merged with visionary modernism. The displays staged a surreal spectacle: humanity *designed,* like a mass-produced object, or a work of modern art, and housed in a designed "city of tomorrow." There was, then, no need to represent the

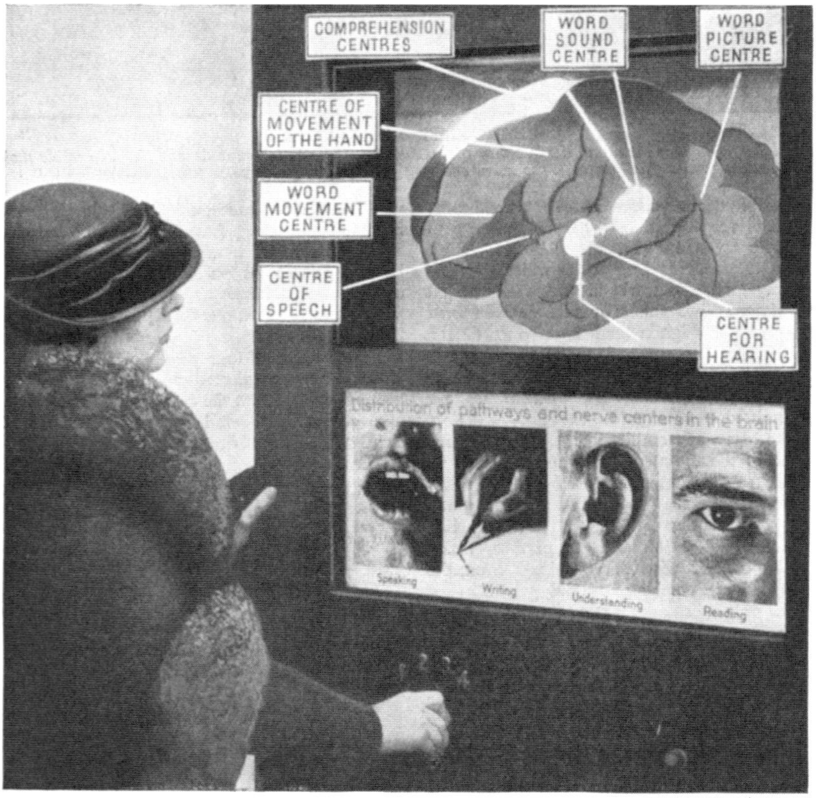

BUFFALO MUSEUM (MONDIALE)

PHYSIOLOGY MADE SIMPLE

A wonderful exhibit in an American museum. It indicates the distribution of pathways and nerve centres in the brain in relation to speaking, writing, understanding and reading.

FIGURE 5.11. *An interactive device in the Buffalo Museum of Science Hall of Man uses images derived from illustrated books to teach the structure and function of the brain. A picture of someone using the device is then returned to a book to serve as an illustration. "Physiology Made Simple," H. B. Wheeler, ed.* The Miracle of Life *(ca. 1938), 366.*

factory and the city in the displays, as Kahn and his artists did; the displays performed a purified version of factory and city. Made both bigger and smaller than embodied experience, the human body mimetically crossed the gap between life in mass society and the transcendent scientific human. It was a rosy vision of a future where the latest findings and technologies of the biomedical professions and the industrial age would abolish disease and deformity.

To get a feel of how that looked on the exhibition floor, consider "Mac the Mechanical Man," a three-meter-tall profile cutaway interactive exhibit, loosely based on "Der Mensch als Industriepalast." Mac was designed in high modernist style by Richard Huw, in consultation

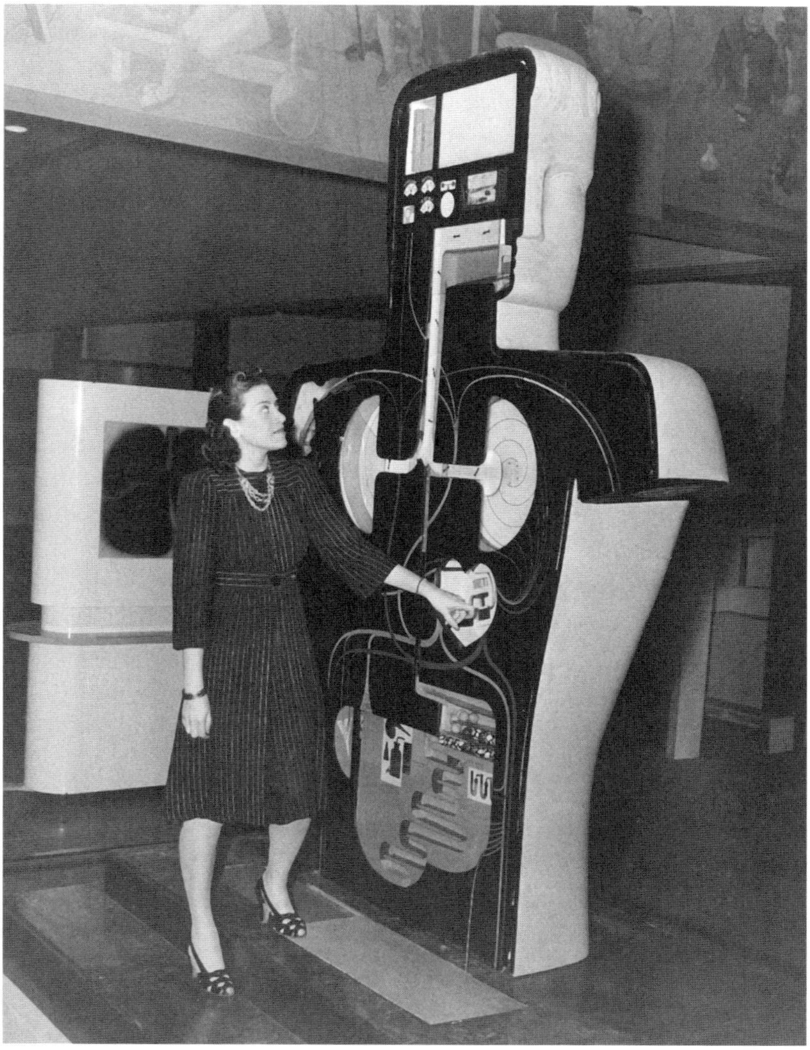

FIGURE 5.12. *"Mac the Mechanical Man," publicity photograph, Hall of Man, New York World's Fair, spring 1940. New York World's Fair 1939–50 records, Manuscripts and Archives Division, New York Public Library, Astor, Lenox, and Tilden foundations.*

with physiologist John Beattie of the Hunterian Museum in London, for the 1938 Empire Exhibition in Glasgow (where he was called "Godfrey"). In 1940, the exhibit traveled to New York and was shown to appreciative crowds at the World's Fair Hall of Man. After the close of the fair, Mac was sent back to Britain, where he continued to perform into the postwar era.[23]

NATIONAL SOCIALISM'S KAHN

The DHM was planned as a permanent successor to the famous 1911 International Hygiene Exhibition in Dresden. The outbreak of war in 1914, and then postwar political and economic difficulties, delayed its opening until 1930, when a second international exhibition was held. Like the earlier exhibition, the 1930 museum was a diversified effort that touched on many themes: the workings and biology of the human body, microbiology, the state of public health (represented in maps and statistics), eugenics, disease and disease control, cleanliness and preventative medicine, physical fitness, the progress of medical science, the health professions, and state bureaus of health.

The museum's natural political constituency was the liberal center and the left, in alliance with rationalist state-building technocracy (and the Bauhaus), and not National Socialism. The Nazis condemned technocratic neutrality, rationalist programs of demystification, Bolshevism, "Jewish" consumerism, "degenerate art," new architecture, free sexual expression, uncloseted homosexuality, and other signifiers of the modern. Even so, National Socialism had its own modernization program based on centralization of power, mass mobilization, efficiency, and militarism in the name of *Volk* and *Führer.* When the Nazis took control, the DHM, like other state institutions, succumbed. Modernity suddenly belonged to Hitler. While the museum retained its displays, it now answered to Nazi Party officials and was commanded to purge Jews and leftists from its ranks. The new regime wanted displays that emphasized racial health, eugenics, physical fitness, and the care of mothers and children in the service of the nation—National Socialist themes. The DHM complied, but in measured language that abstained from vulgar slurs.[24]

Such topics, of course, were not the sole property of National Socialism. Maternalism, physical fitness, and eugenics already had a place in the museum before the Nazi takeover—and also in the pages of *Das Leben des Menschen.* But the eugenics of Kahn and the DHM was not the eugenics of the Final Solution. Their eugenics was connected to a vision of social hygiene that included slum clearance, good nutrition, public education, preventive medicine, and so on. It was still eugenics—full of prejudice and condescension for the "feeble-minded" and racial others—but not a program of extermination. As the nature of the Nazi regime became evident and Kahn personally came under attack by polemicist Dietrich Eckart and other Nazis, that clarified things.[25] Kahn began to distance himself from eugenics and racial science. In early 1933, as the Nazis took power, Kahn and his family fled to Palestine, where he owned land. The mass book-burning events of May and June 1933 included works by Kahn.[26]

In this transitional period, the Nazis mainly treated illustrated popular medicine with contempt. This was partly because of its association with Enlightenment universalism (which was transnational and therefore un-German), partly because it overlapped with the field of sex education (which was condemned as morally corrupting), and partly because Jews and socialists were prominent in the field of popular medicine and were cited in popular medical texts as authorities on various topics.

However, by 1938, as the Nazi regime moved to consolidate its control, it embraced popular medicine and health, so long as it was purged of Jewish authors, refrained from sexology, and highlighted racial science, maternal and infant health, physical strength, and manly exercise. The emphasis changed, but many of the illustrations were not much different from what went before—indeed, many were exactly the same. The regime forced Franckh to "Aryanize," to promote non-Jewish authors, and to cease issuing publications by Jews. In a tense negotiation between Franckh and Kahn, Franckh took the German rights to Kahn's works and images, and retained possession of all of the original artwork.[27] Kahn had little bargaining power, but he was able to retain rights outside Germany. Gerhard Venzmer (1893–1986), a Nazi-approved author and editor from the *Kosmos* stable, was given permission to use Kahn's previously published images. In 1938, Venzmer, with two coauthors, put out the two-volume *Hand- und Lehrbuch der Krankenpflege* (Handbook and textbook of nursing), which included many illustrations from Kahn. Venzmer followed up in 1939 with a slender 180-page book (excerpted from the first volume of the *Hand- und Lehrbuch*), titled *Der Mensch und sein Leben* (Man and his life). *Der Mensch und sein Leben* loosely corresponded to the much longer *Das Leben des Menschen,* which had been taken out of print and circulation.[28] In much less detail, Venzmer covered many of the same topics that Kahn had covered, but he added material on the laws of heredity, racial science, and racial health. Roughly half of his images came from *Das Leben des Menschen* but were credited to "Bildarchiv Franckh" (Franckh image archive), and not "aus Kahn" (which had been Franckh's practice up to then).[29]

In the late 1930s, now living in Paris, Kahn had signed with Albert Müller Verlag, a Switzerland-based publisher, and put out *Unser Geschlechtsleben* (1937) and the two-volume *Der Mensch Gesund und Krank* (1939), both of which incorporated revised images and text from *Das Leben des Menschen,* although not any original camera-ready artwork. This was surely an urgent matter: Kahn and Venzmer were now in direct competition for German-speaking readers, using similar, and in some cases almost identical, images.[30] Banned in the domains of the

Third Reich, Kahn was at a great disadvantage. The only German-language markets open to him were Switzerland and Luxembourg, but Kahn assiduously began to secure translation deals for the Müller Verlag volumes, which soon appeared in Dutch, French, Danish, Swedish, and English.

With the Nazi conquest and occupation, Kahn was again forced to flee, this time to the United States. He arrived in New York in 1941, and once settled, he turned his attention to the English-speaking market via translations by physician-historian George Rosen. Kahn never became a household name in the United States, but did rate a seven-page spread in a 1943 issue of *Life* magazine and a human interest feature in the *New York Mirror*, a popular tabloid.[31] In both articles, Kahn denounced Venzmer as an opportunistic Nazi plagiarizer. That narrative, largely true, promoted Kahn as a heroic victim and advanced his goals, which were to gain the sympathy and confidence of the American public and resume his career in popular medicine.

The great irony is that Venzmer's appropriation of Kahn's illustrations pushed Kahn in turn to deny credit to his old artist-collaborators and also his new ones. As Kahn recycled material from *Das Leben des Menschen* to *Der Mensch Gesund und Krank* and its many translations, he erased his artists' signatures or covered them over with a trademark "©FK." The *Life* article, which focused on the "unique character" of Kahn's pictorial work, allowed that the illustrations were "actually rendered by various artists" but stressed that "they were roughed out by Dr. Kahn to the last detail"—something which artist Roman Rechn vehemently denied.[32] (See chapter 3.) None of the many illustrations in the *Life* magazine story bore an artist's signature (though many were

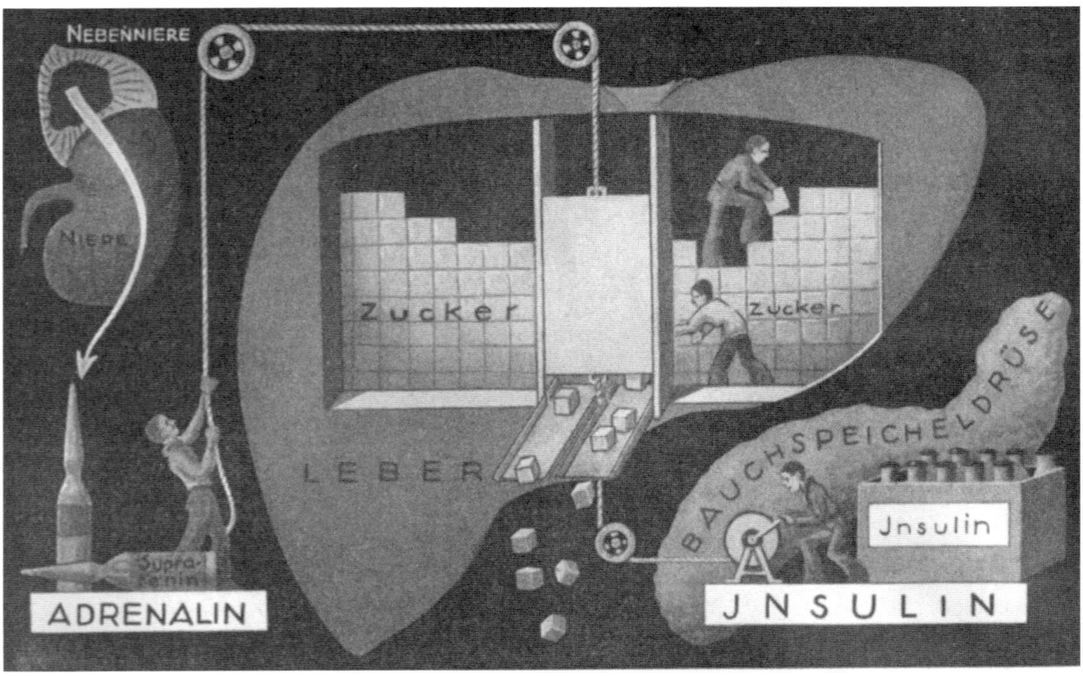

FIGURE 5.13. *"The Interaction of Adrenaline and Insulin in Glucose Metabolism," Gerhard Venzmer,* Der Mensch und sein Leben *(1939), 89. Artist uncredited. National Library of Medicine. While Venzmer took many illustrations from Kahn, he also featured original illustrations, like this one, in the style of Kahn. Copyright Kosmos Verlag, Stuttgart.*

adapted from *Das Leben des Menschen* illustrations signed by Arthur Schmitson). Kahn never publicly acknowledged by name or otherwise the crucial roles played by Rechn, Trester, Schmitson, and Fritz Schüler. (Their signatures all continued to appear in Venzmer's publications.) In exile, Kahn never again allowed an artist's signature to appear in any of his publications. Many readers undoubtedly assumed that Kahn was an artist, as do some readers even today.

CONSENSUAL MODERNITY

There was a larger paradox. While Kahn dismissed Venzmer as a plagiarist pure and simple, Venzmer's career suggests something more far-reaching: an emergent consensus. What Venzmer and other Nazi-approved authors stole from Kahn, they also affirmed; they advanced a common agenda.[33] The need to modernize self and society, the authority of science, the rampant pictorialism, and the pleasures of representing the human body as an artifact of modernity appealed to people and parties across the political spectrum.

Historian Jeffrey Herf, in *Reactionary Modernism,* his landmark monograph of 1984, argues that Nazism and Italian fascism, notwithstanding their critique of Jewish and Bolshevik modernity, had their own modernizing discourse and performance. In fact, nearly every sect and party in this period performed some version of the modern. The romance of science and technology, a delight in enchanted modernity, could be found left, right, and center. Both Hitler and Mussolini offered up modernizing visions of the state as a machine (organized on military principles), army (organized on mechanical principles), or body (organized on military-industrial principles), with the populace as troops or machine parts or cells, performing work guided by the supreme will of the leader (the builder, brain, spirit).[34] Body, state, and machine converged. Fascism—leadership by a charismatic dictator—was not just efficient, it was dynamic and heroic. Industrial technology hummed with primordial power, and Nazi rhetoric fetishized that power, as raw as they could make it. Economist Fritz Nonnenbruch, in the 1936 Nazi-published *Die dynamische Wirtschaft* (The dynamic economy), rhapsodized the experience of modernity as the apotheosis of romanticism:

> Technology is . . . a flaming illumination of reality. Flying in an airplane, driving in a car, the thunder of the elevated railway, the various landscapes of the battlefield, the glowing stream of flowing iron in the ghostly night filled with steel ovens—all of these things are incomparably more romantic than anything previous romantics could imagine.[35]

Like their enemies on the left, the Nazis, led by Joseph Goebbels, had a strong historicist sense of "the rhythm and hot impulses of our time." Goebbels proclaimed the new regime to be "the Reich of droning motors, monumental industrial creations": "The racing tempo of our century affects all areas of our life. . . . National Socialism never rejected . . . technology. Rather, one of its main tasks was to . . . fill it inwardly with soul, to discipline it and place it in the service of our people."[36]

Modernity, then, was never a stable or entirely coherent cultural category. All political actors and parties in this period were special cases. All were ambivalent and selective modernizers. The Nazis mobilized their core constituency on the basis of hatred of some aspects of modernity. But they did not, could not, refrain from embracing modernity in other ways. Their monopoly of state power meant that they also monopolized the role of modernizer of state and society and self, which became a leading theme in their propaganda and performance. (Eugenics was a perfect fit.) In contrast, the left began with a fervent identification with modernity (even while it criticized modernity as presently constituted). Both left and right sought to capture the state, centralize authority, build up bureaucracies, deploy new technologies, and mobilize society for action through modern measures and devices. Both left and right used the visual rhetoric of modernity to claim the role of a privileged modernizing agent; both sides claimed the emblems and methods of modernity. But both also articulated a kind of nostalgia for the world that had been lost. The left at times hearkened back to the idyllic condition of "primitive communism" before a voraciously modernizing capitalism wreaked havoc on workers and communities, and before labor was alienated. The right looked back to the primeval purity of the *Volk*, a time when masculine authority and female innocence was unchallenged and uncorrupted. Even so, to cure the evils of modernity, both sides advocated modern measures, and more modern measures still. Even avowed traditionalists felt obligated to perform traditionalism in modern ways.[37] That meant using modern media technologies, tactics, and themes—and visuality.

The embrace of signifiers was not limited to political parties. Visual methods and tropes of modernization circulated in the arts, sciences, economics, politics, administration, education, business, and everyday life. This was not because a historicist modern era had descended on the world but because there was a demand for materials that served as props for the performance of the modern and helped in the competition for the privileged role of modernizer. Modernizing visuality appealed to readers, educators, health advocates, publishers, authors, artists, political parties, and regimes across the spectrum, and around the globe.

Boosted by World War II, the modernizing imperative continued on into the postwar era. Competing modernization projects vied for control of the political apparatus and cultural terrain in Europe and North America, Asia, Africa, and Latin America. In such a milieu, Kahn's works, adapted to local conditions, had a modicum of success:

O Corpo Humano, a revised Brazilian edition of *Der Mensch Gesund und Krank,* went through six editions between 1943 and 1966. Kahn also exerted influence in Latin America and Spain indirectly through the health and sexological publications of the prolific "Dr. Adrián Vander," which circulated throughout the Spanish-speaking world from the 1920s to the 1970s. Vander modeled his pictorial approach closely on Kahn's. He used many of the visual tropes and genres developed by Kahn—body factories, the body in modernity, bodyscapes—accompanied by captions and inset into text, just as in Kahn. But Vander's illustrations, drawn and signed by the author himself, were his own idiosyncratic version of magazine and commercial illustration. They were entirely modern but innocent of any modernist aesthetic.[38] His work was as quirky as anything to be found in Kahn, but it couldn't match the rich variety, ingenuity, beauty—and modernism—that Kahn's artists contributed.

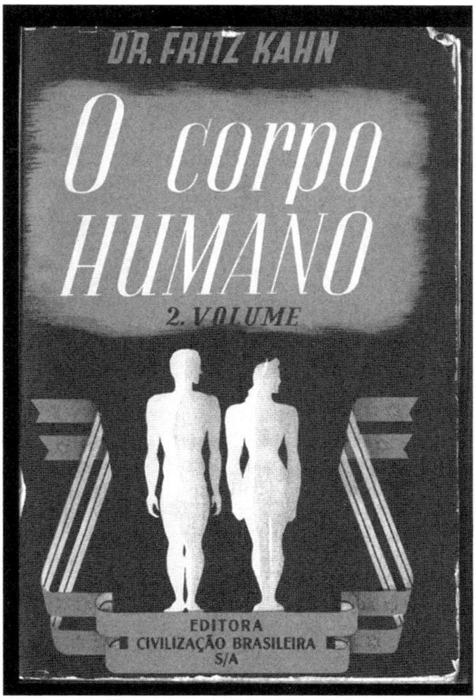

FIGURE 5.14. O Corpo Humano, *4th ed. (Rio de Janeiro: Editôra Civilização brasileira, 1960), vol. 1. National Library of Medicine. The fourth edition used a more modernist four-color cover based on "The Nerve Cell,"* Der Mensch Gesund und Krank *(1939), 2:134 (see Figure 4.38).*

CARTOONS, CHILDREN'S BOOKS, AND TRAINING MANUALS

Kahn's greatest influence was in the wartime and postwar United States. In the 1940s and 1950s, commercial artists and animators used the works of Kahn as something akin to a visual reference library. This section presents a small gallery of objects to show the diffusion, influence, and development of Kahnian tropes, genres, and methods. Not all of these productions arose from direct encounters with Kahn, although Kahn's images were then in wide circulation. But visual metaphor, dramatized statistics, modernity in the body—Kahn's trademark moves—were in the air and were enthusiastically taken up by American graphic designers, poster artists, cartoonists, illustrators, publishers, advertising agencies, and health organizations.

It was a time of unprecedented public confidence in the authority of science and professional experts—and belief in progress. There was a growing conviction that progress in science was paralleled by progress in art, in the form of new styles that expressed the spirit of the industrial age. Even more than the interwar period, it was an article of faith that innovative media technologies and tactics—potentially in alliance with some kind of democratic modernist aesthetics—could inform the public and shape opinion, perception, and mass behavior. The propaganda mobilizations of World War II and the ubiquitous print/radio advertising campaigns were irresistible demonstrations of the power of media. Governmental agencies, private nonprofit organizations, and for-profit companies all began to devote more resources to cultural productions that featured conceptual illustration. Kahn's pictures were there for the taking.

POST-KAHNIAN CARTOONS

During World War II, vast resources were allotted to the American military, which mounted an unprecedented effort to make medical and health films. Many of them were animated in whole or part, and many of them were made with top industry studios and talent. Animations were shown to millions of troops as well as to small groups of specialized personnel. The Navy, for example, commissioned the Walter Lantz Studio (the maker of Woody Woodpecker cartoons) to produce *Enemy Bacteria* (1945), a partly animated Technicolor training film for naval surgeons and operating room assistants. The aim was to dramatize the importance of methodical hand washing and sterilization of equipment before surgery. The film begins with live action: a surgeon, running late, skimps on the rigorous hand-washing protocol and gets careless during a surgical procedure. As a result, he infects the patient, a wounded sailor, with bacteria. Switching over to cartoon animation, the scene is the interior of the body, depicted as a cartoon world at war. Sadistic troll-like germ pranksters trek through a difficult tropical terrain of ponds, rivers, hills, and valleys. Along the way, they proliferate wildly, via cell division, and then mass in battle to overwhelm the body's defenses. They are beaten back, but at a terrible cost. Back to live action, the patient, lying in his hospital bed, is wracked with pain; a terrible fever rages. He survives but is crippled for life.

Though few people ever got to see *Enemy Bacteria*, Adolf Nichtenhauser, a public health doctor who spent his career promoting the use of film in medicine, singled it out as an "important

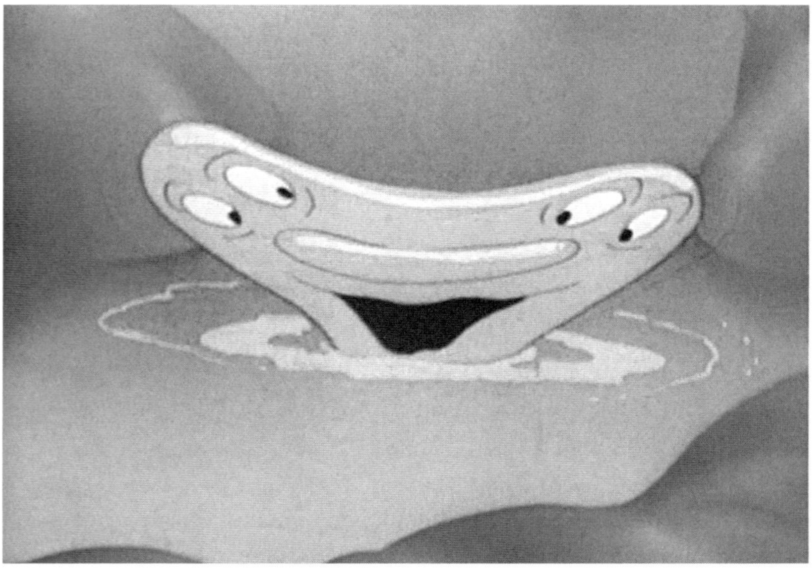

FIGURE 5.15. *A cartoon microbe divides in the bloodstream, adding to the bacterial force that is invading the body via the bloodstream.* Enemy Bacteria *(U.S. Navy; Technicolor; sound; Walter Lantz Studios, 1945). National Library of Medicine.*

signpost for educational films in general." He especially praised its animations, which, he observed, were "distinctly influenced by . . . the illustration technique developed by Dr. Fritz Kahn in his popular books."[39] Nichtenhauser's enthusiasm was misplaced—*Enemy Bacteria* doesn't delight; it has no brilliant visual devices or body factory or industrial metaphor. Its Technicolor body-scapes are beautiful to look at (though charmless compared to Schmitson's), but the production is marred by overblown narration, uninspired cartooning, and a woeful mismatch in tone between the grim live-action sequences (unrelenting, tedious) and cartoon sequences (cute, smart-aleck, slapstick). Shown only to small groups of military medics and then taken out of circulation, *Enemy Bacteria* was never a signpost of anything.

Much more influential was a round of wartime and postwar cartoons made by a group of brilliant young animators. Blacklisted in the aftermath of the Disney animators' strike of 1941, they banded together in small studios that survived on commissions from the military, advocacy groups, educational publishers, religious organizations, trade groups, and corporations. Their quirky medical-themed cartoons—*A Few Quick Facts about Fear* (Army–Navy Screen Magazine, 1945), *The Traitor Within* (American Cancer Society, 1946), *Swab Your Choppers* (U.S. Navy, 1948), *Rodney* (National Tuberculosis Association, 1950), *Multiple Screening* (Pennsylvania Department of Health, 1950), and *Man Alive* (American Cancer Society, 1952)—betray a group sensibility (Plates 11 and 12). Viewers were fed an absurd succession of visual metaphors, dramatized statistics, and fantastic voyages. Diagrams come to life. Flowcharts flow. Graphs graph themselves. Body factories hum, with busy production lines and conveyor belts rolling. Man-machine car engines fire (and backfire). Made in close collaboration with medical and health experts, the Kahnian influence was everywhere, but the tone and medium were entirely different: cheeky cartoon characters spoke snappy dialogue, suffered humiliations, took pratfalls, and played out absurdist gags and visual puns, with cacophonous sound effects and music.

That difference in tone was matched by a difference in style, a new aesthetic: frenetic oddball minimalism. The modernist cartoon dispensed with illusionistic perspective and detail. Instead, there were strong lines, bold shapes, dramatic composition, fast edits, speedy action, and disorienting motion.[40] By aesthetic preference and necessity—low budgets, fast deadlines, small studios with only a few animators—the cartoonists used less labor-intensive methods than the big studios customarily used. They made fewer drawings with fewer, but bolder, lines.

That modernist style came from the artists, but it also pleased the commissioning organizations and the people who staffed and funded films, who also felt compelled to perform the modern. Sponsors wanted to modernize themselves and their organizations and campaign materials and methods, just as much as they wanted to modernize the viewers who made up the intended audiences for their films. (Some of the wealthy philanthropic benefactors also collected modern art and sat on the boards of art museums.) Modernity was in the medium and style as well as the message. And so the new style was entirely in tune with the National Tuberculosis Association, American Cancer Society, and the American Dental Association, organizations that had resources to devote to films, but wanted to be economical while getting out simple but compelling messages plainly conveyed to the public and children. Be modern: get regular checkups, see your doctor if you suspect something is wrong, get an X-ray, brush your teeth, know the early warning signs, trust medical science, do regular exercise, celebrate scientific progress, donate money, and don't go to quacks.

That worked well with audiences. *Man Alive* (American Cancer Society–United Productions of America, 1952) was shown before the main feature in commercial movie houses and drive-ins, and in community centers and other venues. It received an Oscar nomination for best short film. *Hemo the Magnificent* (directed by Frank Capra; Bell Science Series, 1957), an hour-long episode on the physiology of the heart and circulatory system, and *Gateways to the Mind* (directed by Jack Warner; Bell Science Series, 1958), on the brain and nervous system, were initially broadcast on network television for a prime-time audience, then screened for decades after in classrooms throughout the country. Both were part of the acclaimed Bell Science Series, an initiative to bring well-produced, entertaining science education films, which mixed live action with lengthy cartoon sequences, to schoolchildren and the general public.[41]

Many of the animators—Zack Schwartz, Stephen Bosustow, David Hilberman, Lu Guarnier, William T. Hurtz, and John Hubley—were associated with United Productions of America, a small animation studio that initially lived off of commissions—some of its very best work—before achieving commercial and critical success with *Gerald McBoing-Boing* (1950) and the Mr. Magoo character.[42] Given the movement of artists through studios and projects, it seems as though everyone knew everyone else. Their medical and health films were particularly rich. A feeling of excitement runs through them. The cartoons didn't have to just suggest the dynamism of the body; they could perform it, in roller-coaster time and careening space. They could move in and out, stop and start, speed up and slow down, form and transform, stretch and squeeze, and play with shape and (if the budget permitted) color. Processes unfolded in snappy time and rhythm. In the space of a few minutes, they could never provide the same amount of technical detail that Kahn and his artists provided in text and illustration. So they kept it simple, making their points through radical condensation, ellipsis, allusion, substitution, metaphor, visual rhyming, and utter abstraction.

Kahn never theorized the modern or his image practice or the relation between the two, but under the influence of György Kepes's 1944 manifesto of modernist graphic design, *Language of Vision,* animators John Hubley and Zack Schwartz did. In a short 1946 article in the *Hollywood Quarterly,* they declared, "The medium of animation has become a new language" of visual forms and methods, using "dynamic symbols, images representing whole ideas":

> The artist and writer have at their command all the traditional means of graphic expression and the new means which grew out of moving symbols and sound. One of these is the concept of explanation through change from an object as it is to the thing it signifies. . . . Animation . . . is no longer the vaudeville world of pigs and bunnies. Nor is it the mechanical diagram, the photographed charts of the old "training film." It . . . encompasse[s] the whole field of visual images, including the photograph. . . . Line, shape, color, and symbols in movement can represent the essence of an idea, can express it humorously, with force, with clarity.[43]

As an example, Hubley and Schwartz pointed to *A Few Quick Facts about Fear* (1945), which they had recently made for the U.S. military. The assignment was to make an entertaining two-minute cartoon that reassured troops that fear was not a moral failing but instead an adaptive evolutionary psychophysiological response to danger, with well-defined and useful effects on the brain, heart, liver, glands, skin, and muscles.[44] But *A Few Quick Facts about Fear* did more than that. It fused the animator's traditional repertoire of gags, visual puns, wise-cracking vaudeville shtick, and impossible distortions and transformations with Kahn's repertoire of visual metaphors, tropes, and genres, in a kinetic style that drew haphazardly from Bauhaus, cubism, Matisse, Otto Neurath, Felix the Cat, and Merrie Melodies.

Citing Moholy-Nagy, Fernand Léger, and other high modernists, Kepes theorized that "the dynamic plastic nature" of modern "visual experience" should inspire artists and designers to develop "a dynamic iconography" that evokes the continuous motion and disorientation of "contemporary" life.[45] The language is familiar, not much different from the writings of Joseph Roth or Walter Gropius:

> Machines, motor cars, streetcars, elevated trains, aeroplanes, flickering light displays, shop windows. . . . Together with the new richness of light-effects from artificial light sources, the increased dimension of the landscape with the skyscrapers and their intricate spatial order above, and the subways underneath, [gives] an incomparably greater speed and density to the light stimulations reaching the eye than any previous visual

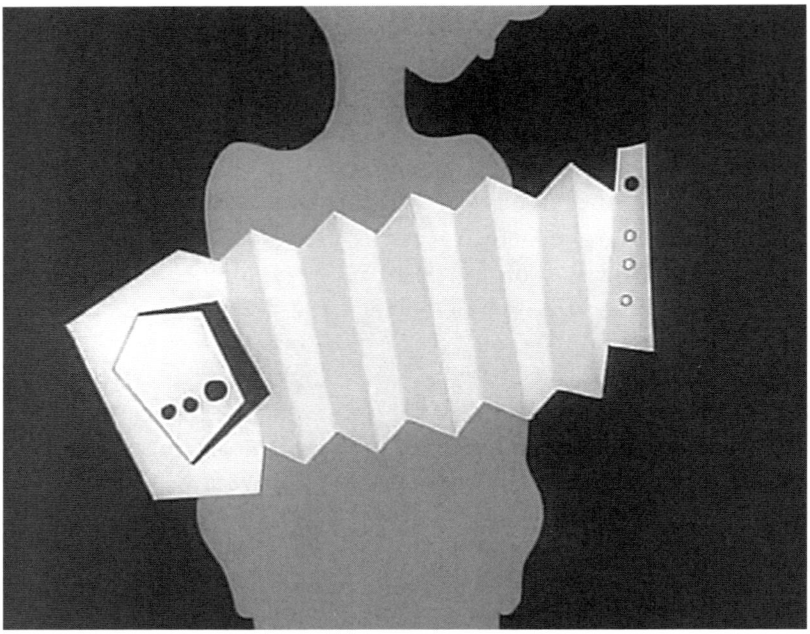

FIGURE 5.16. *Visual metaphor: when you feel fear, your lungs expand like an accordion.* A Few Quick Facts about Fear. *Animators: Zack Schwartz and John Hubley (U.S. Army–Navy Screen Magazine, 1945). National Library of Medicine.*

> environment. . . . There is no time now for the perception of too many details. The duration of the visual impacts is too short. To attract the eye and convey the full meaning in this visual turmoil . . . , the image must possess, like the traffic sign, simplicity of elements and lucid forcefulness.[46]

Kepes did not envision any special role for cartoon animation. But his manifesto stirred Hubley and Schwartz, who were already fooling around with everything that Kepes called for. Since the silent era, and on through the era of Betty Boop and Mickey Mouse, animators had been developing their own dynamic iconography, a vernacular aesthetic that could ignore gravity and physics, make bodies stretchable and partible, dissolve perspectival horizons into typographical symbols, and turn anything into something else and back again. With the paring away of decoration and the rejection of illusionistic perspective, there came a new aesthetic for the age of industrial technology and media, a synthesis of high and low: modern art meets the funnies.

TRAINING GUIDES AND SCHOOLBOOKS

Cartooning was, of course, not just limited to cartoon animation. The cartoon form seemed to speak directly to everyone, from the illiterate poor to the well-educated elite, and appeared all over the cultural landscape, in comic strips, advertisements, editorial pages, posters, and so on. The U.S. military and other producers of training guides and instructional books used it extensively. Take, for example, *Your Body in Flight* (U.S. Army Air Force, 1944), a training manual for airmen with cartoon illustrations by Eric Sloane. Before the war, Sloane specialized in drawings on aeronautical and meteorological topics.[47] *Your*

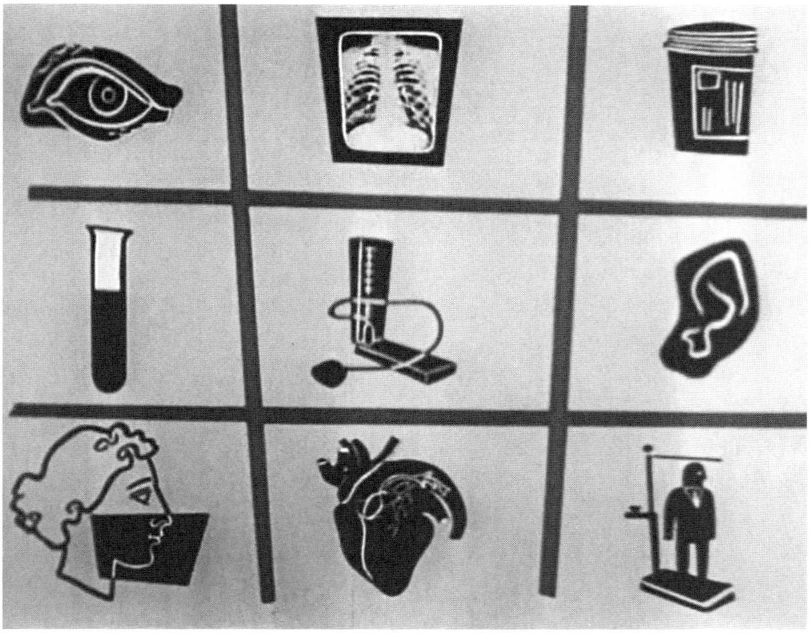

FIGURE 5.17. *A synoptic table in a modernist cartoon animation.* Multiple Screening *(dir. David Hilberman; Columbia University Press; Pennsylvania Department of Health; Tempo Productions, 1950). National Library of Medicine.*

Body in Flight combined those two fields with physiology. The subject was what happens to the human body, inside and out, under the stress of high-acceleration, high-altitude, unpressurized air travel. The assignment was to explain that visually. To do that, Sloane made crisply composed pen-and-ink cartoon visualizations that fooled around with the Kahnian tool kit: visual metaphor, sequential action, modernity in the body, physiological dynamism, and gracefully rendered flowcharts.

If soldiers were the principal target audience during World War II, over the long haul, children seemed to be the audience best suited for conceptual illustration. *Adventures in Science with Ruth and Jim* (1948), part of the Rainbow Series in Elementary Science, was, according to one reviewer, "one of America's most popular and widely used" textbooks for primary school readers because its illustrations had "great pupil appeal, challenging both his attention and interest."[48] Alongside halftone photographs, it featured playful cartoon elves conducting lessons on the workings of the human body, meteorology, earth science, botany, and other subjects.

For the most part, training manuals and schoolbooks only lightly took up modernist aesthetics. They figured modernity by using the cartoon idiom and by depicting industrial technologies, products, and scenes within the frame of the illustration. The informality and friendliness of the cartoon had a leveling democratic ethos, and readers were charmed into absorbing lessons on science, technology, work protocols, and so on.

Here too it bears emphasis: animated cartoons, training manuals, and children's books were collaborative projects. Animators and illustrators worked closely with educators, doctors, psychiatrists, public health experts, and advocacy organizations. Some of these collaborators were familiar with Kahn's illustrations, and perhaps even held them up to their artists as models of how to visually explain things. Others came to Kahnian ideas by osmosis. But in every case, the cartoonists brought their own idiosyncratic talents to bear.

THE LOOK OF MODERNITY: WILL BURTIN, LESTER BEALL, AND MODERNIST GRAPHICS

It didn't take long before the cartooning tradition was itself assimilated into modernist aesthetics, first in surrealism and then in the works of Roy Lichtenstein, Robert Rauschenberg, Andy Warhol, Joe Brainard, and other acclaimed artists. The modern was a capacious signifier, a moving target. Because it defined itself in opposition to the passé modernism of previous generations, there could never be any stability to

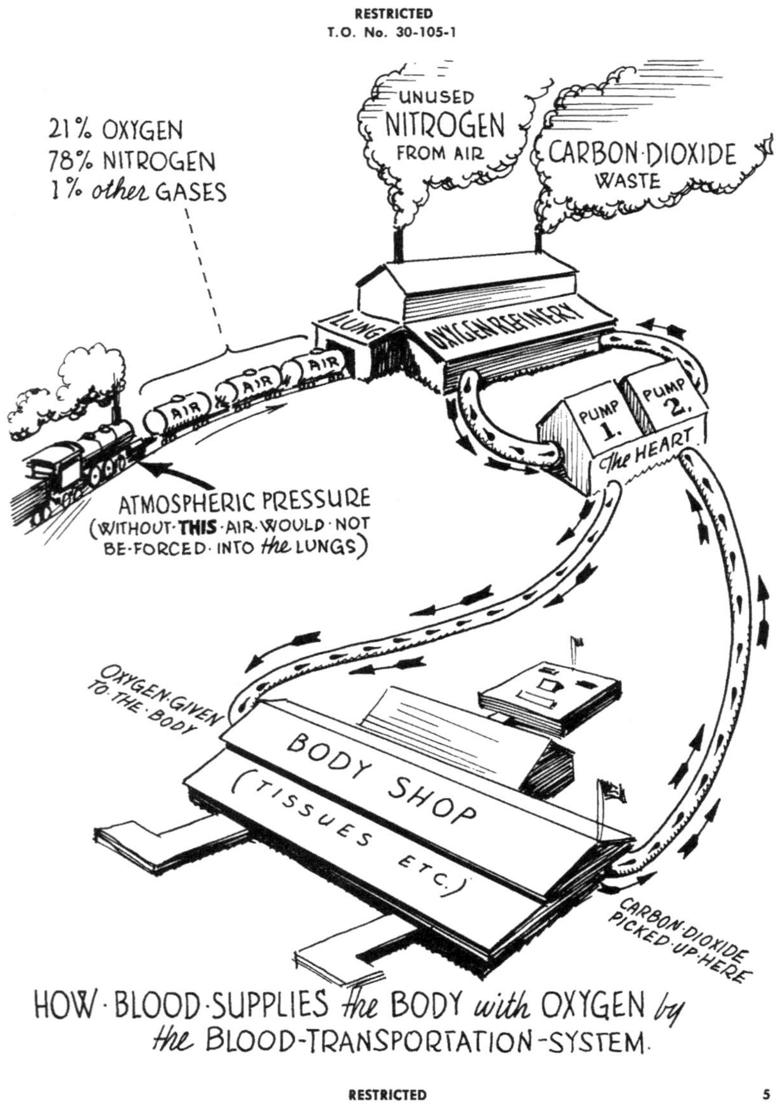

FIGURE 5.18 *Visual metaphor 1: The human circulatory system as rail transportation–refinery network.* Your Body in Flight *(TO-30-105-1; U.S. Army Air Force: Dayton, Ohio/ Washington, D.C., 1944), 3. Artist: Eric Sloane. National Library of Medicine.*

the concept. Eventually the modern aggregated so many different approaches and projects that the very idea of it began to seem incoherent, contradictory, and pervasive, incapable of sustaining the mood of rapture and futurist utopianism. That soon would lead to postmodernism, a critique that came from inside aesthetics, political theory, art history, and cultural criticism.

But before that, in the postwar/Cold War era, came a time of great confidence. In that moment, graphic design wielded an enormous and eclectic palette of moves, which were both functional—suited to almost

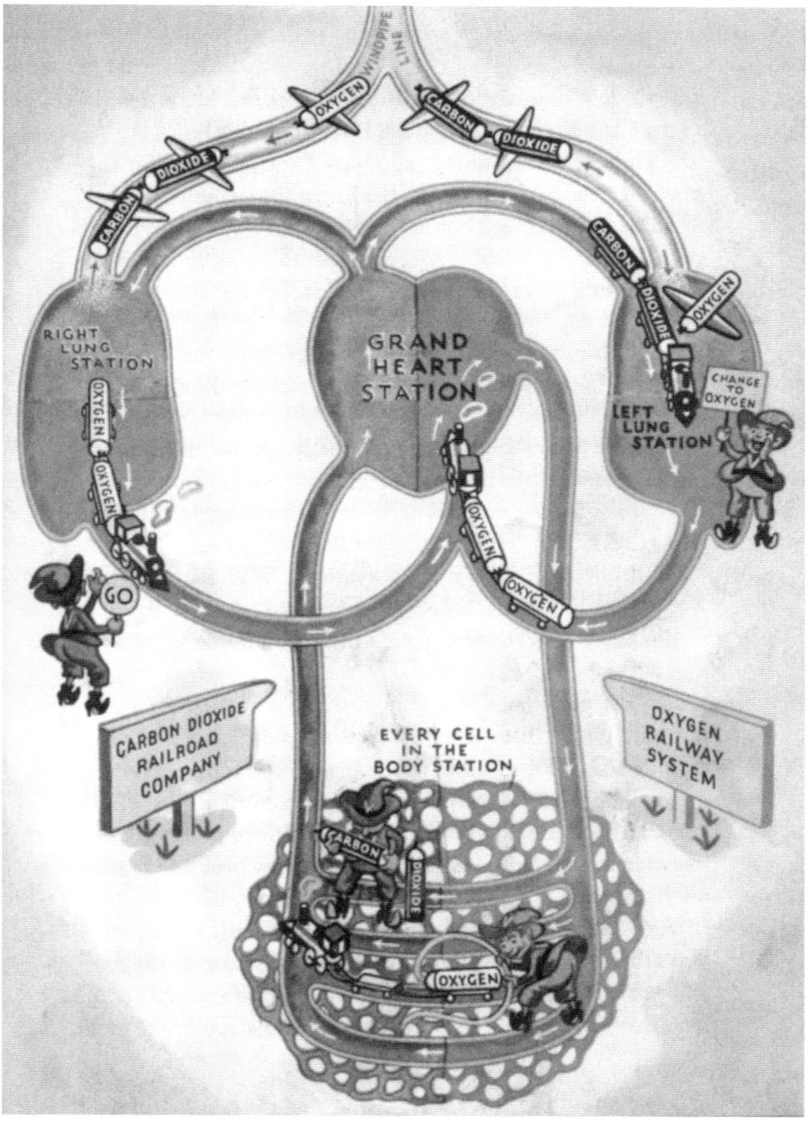

FIGURE 5.19. *Visual metaphor 2: The human circulatory system as an air–rail transportation system directed by elves. Four color. H. A. Carpenter, G. A. Bailey, and P. E. Smith,* Adventures in Science with Ruth and Jim *(Boston: Allyn and Bacon, 1948), 85. Artist uncredited. National Library of Medicine.*

any purpose—and transcended function. Science, modern art, and industrial and graphic design all occupied the same stratospheric register of cultural prestige and authority. Design was no longer yoked to the goals of advancing any particular narrative or lesson; rather, it belonged everywhere. Visual explanation was only one of many ways to get modern with scientific illustration.

And so this chapter ends by considering *Scope,* a promotional magazine of the 1940s and 1950s that was distributed to doctors. Sponsored by the Upjohn Company, a large pharmaceutical manufacturer, and under the direction of celebrated designers Will Burtin and Lester Beall (whose work was cited in Kepes's *Language of Vision*), *Scope* became a showcase for new medical products, sophisticated modernist graphics and the advances of postwar bioscience (Plates 13, 14, and 15).[49] More than that, Upjohn's sponsorship of the magazine branded the company itself as ultramodern, as much a vector of modernity as its innovative pharmaceutical products, biochemical research, and high-end graphics.

This was a giant step away from Kahn. For visual modernism to signify, it had to be new—and by the 1950s, as modernism radiated in every direction, Kahn's way was no longer that. The new thing in America, the new cultural capital of the modern world, was an abstractionizing, horizon-dissolving high modernism that seemed to be everywhere applicable. So long as text and image were surrounded by enough negative space to make a composition that was purified of the overdetermined chaos of everyday life, designers could use any image and any tactic to signify the modern. In a world crowded with images, negative space was a signifier of sumptuous luxury and a sign of cultural confidence. The image world annexed nothingness; it recruited the negative to serve as ballast or a scape or a frame. *Scope*'s page layouts were emancipated from the horizon and from illusionistic perspective. Images floated free, in zero gravity.

Kahn and his artists tried to give readers some feeling of the experience of life in modernity by presenting scenes of arrested motion within the picture. But their page layouts mostly featured rectangular illustrations planted on the page and image elements planted within the frame of the picture. Silhouetted images, even double-page spreads and splash pages, occasionally enlivened Kahn's books, but layouts were never the centerpiece of anything. *Scope*'s layouts were exactly that. Beall and Burtin made pages where image and text seemed to dance in space. If some pages didn't perform visual gymnastics, it was only because the designer was using a conventional page layout or a horizon as a counterpoint.

The design principle was dynamic compositional balance. The elements of a layout were not some fully conceived self-standing picture but rather a combination of photographic and drawn images, diagrams, lines, geometrical forms, organic forms derived from microscopy and other medical images, display type, and text. Collage assimilated drawing. The image was deployed as an allusive, citational, poetic, suggestive element. Visuality ate up pictoriality. That victory ratified everyday experience: immersive jumbled visuality was a remarkable fact of life in modernity that no single picture could contain.

Scope, like the kindred modernist medical exhibitions of the mid- and late 1930s, offered a purified version of that jumble; it brought order to chaos and chaos to order. Modernist graphic design was a sponge that could absorb any style. Even a seemingly intractable style like abstract expressionism could serve, as a chaotic element in a harmonious design. Electron microscopy, along with other modern imaging and staining techniques of organic structures, yielded dense patterns or patternless groupings that resembled the convoluted thickets of action painting. The action of action painting recorded the movement of the artist's hand wielding paint in a way that seemed to mirror the recordings of the movement of particles and waves by devices in the laboratory. By the same token, graphic design recorded the free hand of the graphic designer who took the liberty of tossing design elements onto a page, almost at random. Both action painting and graphic design mimetically suggested chemical and electrical actions occurring in and between tissues, cells, brains, and other organs. More than anything else, high modernism seemed to value the dynamic relationship between random chance and meaningful patterning.

If so, the self and its homuncular effects were dispersed. Identity in modernity, so vibrantly alive and at the center of Kahn's image practice, was no longer central or identifiable. The airy design of *Scope*—the principle of visuality—made a strong claim to be a rhetoric of modernity. But did so at a cost: it attenuated the mimetic power of figurative representation (the same loss of belief that fueled abstract expressionism and, later, conceptual art). By the mid-1950s, the tropes and genres that Kahn and his artists had developed, and the principles of visual explanation and allusion, had passed into general circulation. They could be found in magazine articles, advertisements for over-the-counter headache pills and indigestion tablets, schoolbooks, training manuals, funny cartoons, and health exhibits. But the more they circulated, the less salience they had as a visual rhetoric of modernity. Instead, they served merely as rhetoric. As images colonized every cultural nook and cranny, the significance of profuse illustration declined. Everyone was doing it. Pictorialism and the tactics of visual explanation no longer had any distinction.

6

"TO PICTURE THE BODY"

KAHN'S IMAGES IN THE POSTMODERN AFTERLIFE

> [*Der Mensch Gesund und Krank*] is a book dealing with *modern* man and written for *modern* people. It not only presents man in the light of *modern* science, but is also a *modern* book in text and illustration. It contains the basic facts of the *modern* science of man. . . . Anyone who owns the book . . . possesses a work which will teach him all that *modern* science has to tell contemporary man about himself. . . . It answers all the questions relating to health and a rational mode of life, with which *modern* man is confronted.
>
> —GEORGE ROSEN, PROPOSAL FOR AN ENGLISH TRANSLATION OF *DER MENSCH GESUND UND KRANK*, CA. 1941, GEORGE ROSEN PAPERS, YALE UNIVERSITY LIBRARY (EMPHASIS ADDED)

To take stock: in the early and middle decades of the twentieth century, there was a demand for cultural productions that performed the modern and offered answers to modernizing questions: how to get modern with machines, pedagogy, state administration, music, clothes, and so on—and how to get modern with printed pictures, charts, diagrams, and designs.

In some ways, this was just the continuation of a *longue durée*. From the sixteenth century (near the very creation of the idea of the modern) to the last decade of the twentieth century, the performance of modernity was especially identified with print culture and its proliferating pictures, which represented every facet of contemporary life. The printed image, its modalities, styles, captions, subjects, and technologies of reproduction, came to signify the modern, whether it depicted futuristic fantasies of life in space or nostalgic scenes of home and hearth. As everyday life was increasingly remade into a venue for immersive, saturating image/texts, there came new moves and movements to modernize image production with new styles and subjects, new forms and technologies. One of the great accomplishments of this hydra-headed multifaceted cultural project was the modernist conceptual illustration—a device that worked to modernize viewers with its visual explanations of the material and cultural world and the human body.

Fritz Kahn deployed and redeployed thousands of such illustrations. In collusion with artists, publishers, and readers, he developed the

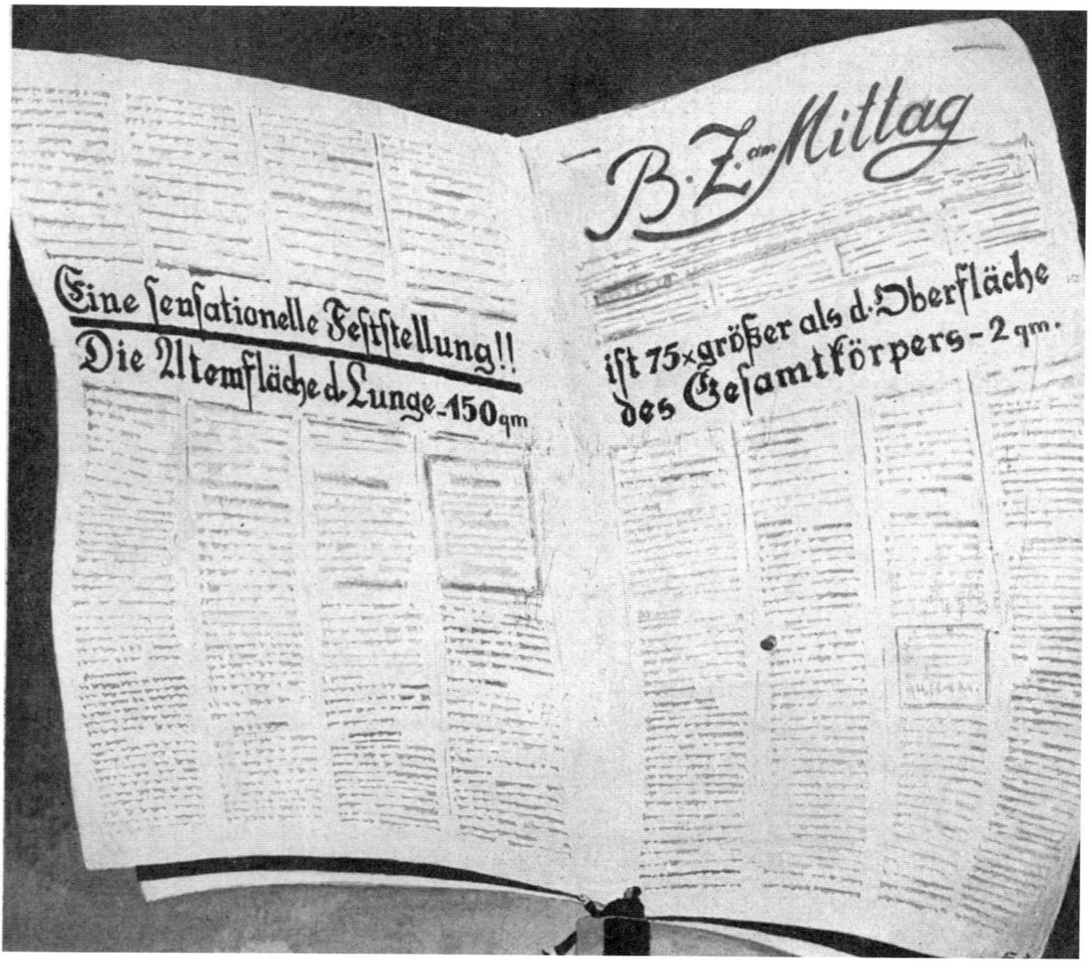

FIGURE 6.1. *Modern media is much bigger than its readers. "A sensational finding!! The respiratory surface of the lung, with its 350 million air sacs, is so large that, spread as a newspaper sheet in relation to a human, it reaches the huge proportion of this newspaper."* Das Leben des Menschen *(1926), 3:49. Artist: Fritz Schüler. National Library of Medicine. Copyright Kosmos Verlag, Stuttgart.*

image as a way of knowing and showing the human body—and a way of taking pleasure in the modern and the modern condition. The human body coursed with the machine rhythms of the combustion engine and turbine generator, glowed with the smoothness of polished metal, plastic, industrial glass, poured asphalt, and concrete, pulsed with light, electricity, and X-rays. Turned inward on the reader, the busy traffic of the industrial city and workplace and the modern home figured effigies of the self: homunculi that referred back to the familiar activities and conditions and technologies of everyday life. There was an unstructured, open-ended cascading loop of image correspondences.

MODERNITY'S DEPTHS AND SURFACES

The new visual vocabulary in turn revised presiding ideas of the modern. Like a Möbius strip, it brought the surfaces of modernity into the depths of the body, and the depths of the body out onto modernity's industrial, urban, and domestic surfaces. Interiority acquired its own industrial landscape and its own political economy.

This undermined the cultural valorization of depth in the evolutionary, historical, and psychological order of things. Nineteenth-century romanticism bequeathed to the twentieth century the belief that wisdom is attained by penetrating deeply through surfaces and layers that obscure the true inner workings of things.[1] The interior was mysterious, intractable, frightening, even menacing, but also beautiful and transcendent—sublime. But conceptual scientific illustration lit up the body, made its murky depths bright and legible. The inside story was almost too rational, too easy, too showy.

I have argued that Kahn's representation of agents within the body (the homunculus) modeled the reader's own complex relation to figurative representation. The philosopher of mind Colin McGinn rightly warns us that the homunculus is a delusion, a trap we continually fall into.[2] If so, it makes sense to wonder why that is. The partial answer offered here is that we should treat that delusion as a historical artifact, and think about what social and ideological work the delusion performs in its moment and setting. Our political, economic, and contingent circumstances (what Pierre Bourdieu terms a "habitus") continually generate homunculi in various genres and locations.[3] These figure the individual self as well as the collective work that constitutes the self. In this reading, "self" and "I" are homunculus words. The self is an amalgam of parts, glued together by training, mimesis, encouragement, discouragement, and routine, which converge to make a presentable, usable "I." Kahn represented the "I" as both factory and product of manufacture, the work of many hands and devices. In contrast, much of contemporary neuroscience treats it as sui generis, hardwired (albeit in a programmable, internally networked way). In neuroscientific accounts of subjectivity and agency, consciousness (shaped by and in evolutionary circumstances) is principally something

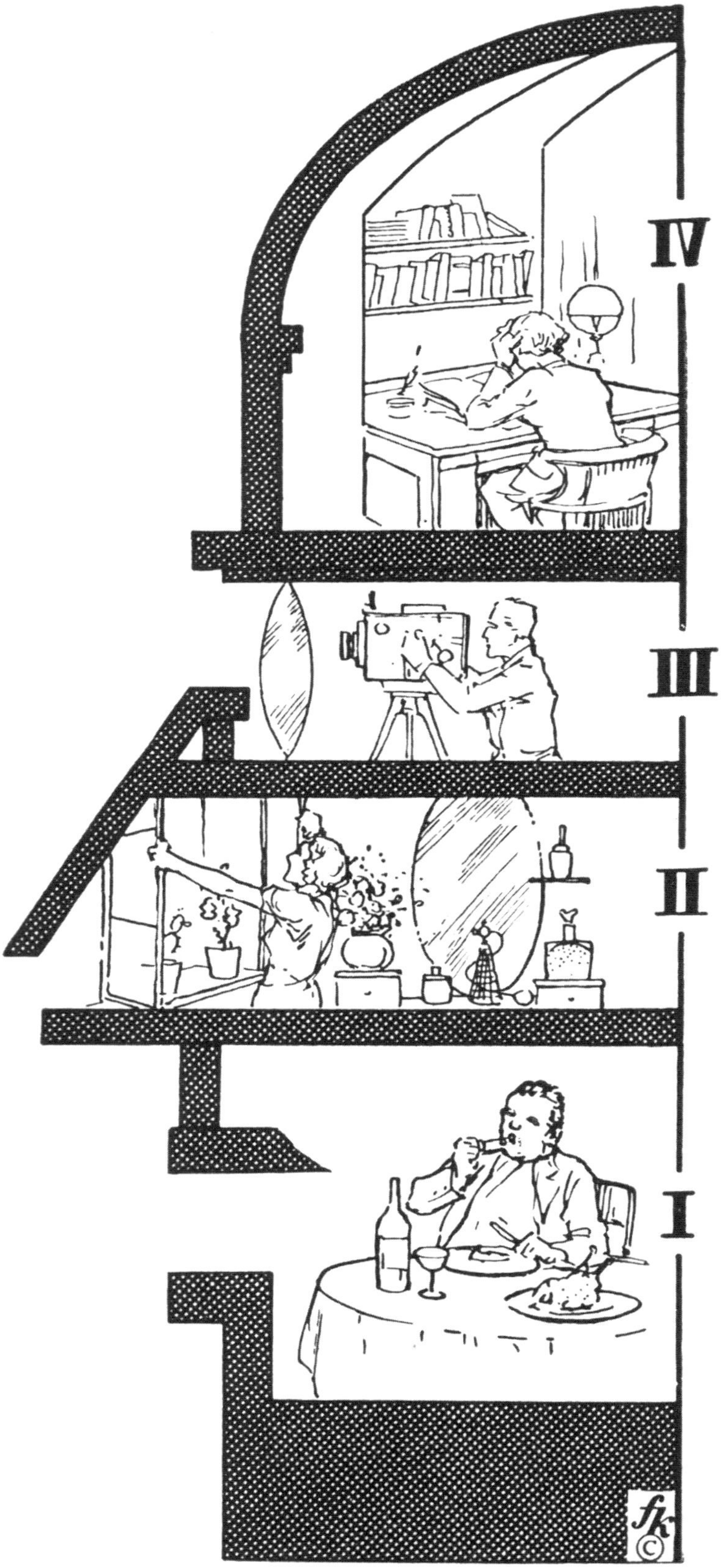

FIGURE 6.2. *Domestic modernity inside the head. "The Skull-House Has Four Floors,"* Der Mensch Gesund und Krank *(1939), 1:115. Artist uncredited. National Library of Medicine.*

that happens in individual brains. Subjectivity is a species-specific, genetically programmed biophysical device.[4] The dominant technologies of our times—the computer, the Web, the app—our metaphors of choice—inform neuroscientific understandings of consciousness, just as the factory, the automobile, the electrical circuit, and other technologies of Kahn's times informed his.

And just as productively. Or not. The recourse to metaphor is inevitable and necessary and problematic. According to psychologist-historian Douwe Draaisma, "Many metaphors, particularly in science, owe their existence precisely to the fact that they express what cannot be said literally—either not yet or in principle." In other words, there is no other way to speak about what happens in the body except by means of analogy. Metaphor is "not . . . decorative or superfluous, . . . there is simply no other choice." If "the domains which are brought into contact in a metaphor are rich in associations," Draaisma argues, then the metaphor can bring "a considerable heuristic yield," though nothing is guaranteed.

Given the centrality of metaphor to our story, we are obligated to consider the poetics of Kahn's image practice. Draaisma (following literary critic I. A. Richards and philosopher Max Black) explains that metaphors function by "substitution," "comparison," and, especially germane, here, "interaction": "In a metaphor the topic term and vehicle term are linked by a set of associations." These create "a new meaning which is given neither in the one nor in the other term separately."[5] The homunculus effect arises in a welter of metaphorical traffic; the topic term and its vehicle constantly reverse fields. Each refers back to and revises the other, and the oscillating pair together dynamically shape the individual recipient and the collective "us." Against our best efforts, no one in the chain of representation—not the publisher, author, artist, or reader—fully controls the meaning.

FIGURES OF THE MODERN

> What then is truth? A mobile army of metaphors, metonyms, and anthropomorphisms . . . which after long use seem firm, canonical, and obligatory. . . . Truths are illusions that we have forgotten the nature of, metaphors that are worn out and now impotent to stir the senses.
>
> —FRIEDRICH NIETZSCHE,
> "ON TRUTH AND LIES IN A NON-MORAL SENSE," 1873

In *The Order of Things,* Michel Foucault places a homunculus, or something very like it (a "doublet"), within the master narrative of modernity, as a defining practice of the modern: "The threshold of our modernity is situated not by the attempt to apply objective methods to the study of man, but rather by the constitution of an empirico-transcendental doublet which was called *man*."[6] Setting up his long account of the origin of "the gaze," Foucault tells us that eighteenth- and nineteenth-century anatomists and physiologists made that doublet by developing scientific practices that "operate within the space of the body" in order to study "perception, sensorial mechanisms, neuro-motor diagrams, and the articulation common to things." These practices worked "as a sort of transcendental aesthetic," unlike the tabular, flattening, analytical procedures of the "classical" period (that goes roughly from Descartes to the French Revolution). The resultant homunculus acquires "depth," multilayered volumetric qualities (or takes up residence within the layers). This applies both to the object under study and the investigator's discipline: "Knowledge has anatomo-physiological conditions," "is formed gradually within the structures of the body."[7]

Foucault's interest in the birth of the clinic and other medical practices arises out of his sense that, individually and collectively, each one of us lives these "anatomo-clinical" categories, institutions, and procedures (such as dissection, the gaze, the asylum, and the case). We are constituted by an ensemble of practices, of which doctors, scientists, jurists, teachers, and other authorities are the masters. Or: the ensemble lives us, structures and operates the self (there's that word again). In these and other practices, we develop and perform some (never entirely coherent or stable) version of personhood and agency. For Foucault, "modernity begins when the human being begins to exist within his organism, inside the shell of his head, inside the armature of his limbs, and in the whole structure of his physiology; when he begins to exist at the centre of a labour by whose principles he is governed and whose product eludes him."[8] This sense of psychic habitation inside an anatomical entity, governed by rules of physics, chemistry, microbiology, and a grab bag of hygienic "laws," is consolidated when "the anatomo-clinician's gaze has to map a volume[,] . . . deals with the complexity of spatial data which, for the first time in medicine, are three-dimensional."[9]

But there was another dimension: time. In the nineteenth century, with the rise of evolutionary discourse and the creation of biology as a disciplinary category, the doublet acquires a historical and developmental aspect that corresponds to spatial depth—"deep time."[10] The deeper one goes into the body and mind, and their sciences, the further one goes back into the historical and/or evolutionary past. (The maxim "ontogeny recapitulates phylogeny" collapses the two time streams.) In the twentieth century, this correspondence/collapse makes possible, and is in turn intensified by, ever more strenuous efforts to describe and mimetically represent the fine details of psychological experience (the unconscious or a stream of consciousness). This becomes the project of psychoanalysis, phenomenology, and all sorts of cultural productions (the novel, stage plays, symphonic music, visual art, and the movies).

There things go awry. The homunculus lives both deep within and at the surface. But even in the interior, it has a certain superficiality. Like a set of Russian nesting dolls, there's not much revelation to be had as you peel away the layers. The Kahnian image opens up a "gateway between interior and exterior" (Wolfram Schivelbusch's words), stages a platonic technological "unreality beyond . . . material existence" that is somehow very material.[11] The body factory is a border crossing between

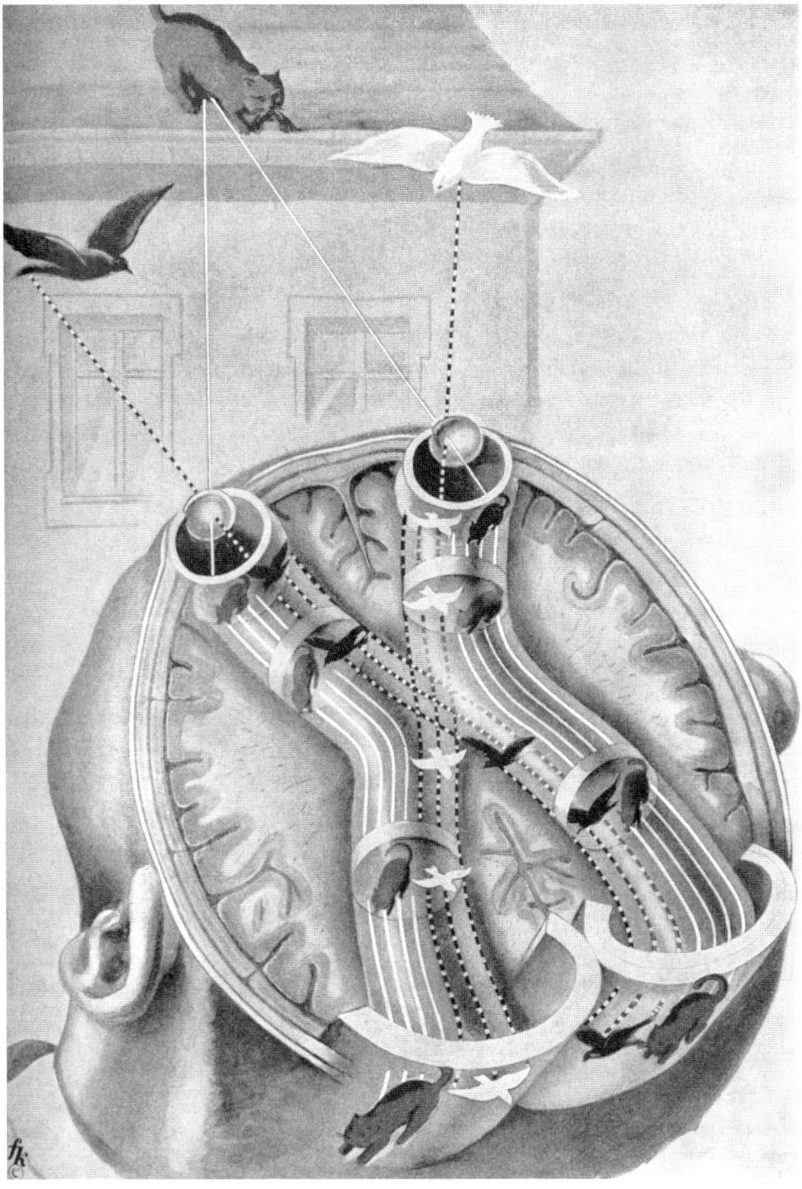

FIGURE 6.3. *Outside comes inside (and inside comes out). "How we see—the image-stream goes from the eyes to the visual cortex in the back brain like the movement of the image in a television set,"* Der Mensch Gesund und Krank *(1939), 2:195. Artist uncredited. National Library of Medicine.*

natural and social, real and imaginary. Even so, Kahn's pictures mainly figure that crossing as a day-to-day affair or, if they deal with growth and development, as something that lives in very shallow time. The idea of depth, which Foucault takes such care to explicate, requires a reserve of dark matter, an area of recalcitrant unknowability that challenges the artist or scientist to enter into unknown and obscure domains (which, if conquered, lead to still other domains of even greater obscurity). In contrast, there's no space for dark matter in Kahn's chatty little technical image/texts.

"THE TRICK OF METAPHORS"

Kahn's image work revised one other presiding idea of the modern: the attack on metaphor, allegory, and simile that was orchestrated in Francis Bacon's *Novum Organum* (1620) and Thomas Sprat's *History of the Royal Society* (1667), those foundational texts of the scientific revolution. Sprat, a key early member of the Royal Society, warned against "the trick of Metaphors." Samuel Parker, another founder, deemed metaphors to be "wanton & luxuriant fancies climbing up into the Bed of Reason," which "impregnate the mind with nothing but Ayerie and Subventaneous Phantasmes."[12] Almost two hundred years later, American medical professor John William Draper restaged that critique in his influential *History of the Conflict between Science and Religion* (1874), in which he acclaimed science as a defining "influence" on "modern civilization" and insisted that science requires empirical proof, not splendid arguments or showy spectacles. No metaphors are allowed: a "demonstration" cannot "be given through an illustration of something else."[13]

However, even as exponents of the new science denounced the macro- or microcosmic fantasies and correspondences of alchemy, astrology, and natural magic, they continued to use analogies. Metaphor was, despite everything, indispensable, an essential part of scientific model making and creativity.[14] Bacon himself analogized that "many actions" in the body—the digestion of food, "the heart and pulse beat," "the entrails"—are all "like so many workshops."[15] In medicine, too, it was often argued that reasoning from analogy led doctors and patients astray, but analogy played a vital role in formulating explanations and hypotheses. There was a kind of cognitive dissonance. Distinguished contributors to medical science used analogies—body as house, nerves as telegraph wires, muscles as pulleys—while at the same time denouncing the metaphorical reasoning of quacks, charlatans, and the common folk. Metaphor, it seemed, was only permissible when the scientific elite used it judiciously among themselves and when respectable popularizers and educators used it to address the uninformed public, colonial subjects, or schoolchildren.[16]

By the early twentieth century, some critics disavowed even these limited exceptions. Francis M. Walters, "professor of physiology at Missouri State Normal School" and a textbook author, complained:

> Too much latitude has been taken in the past in the use of comparisons and illustrations drawn from "everyday life." To teach that the body is a "house," "machine," or "city"; that the nerves carry "messages"; that the purpose of oxygen is to "burn up waste"; that breathing is to "purify the blood," etc., may give the pupil phrases which he can readily repeat, but . . . does not give him correct ideas of his body.[17]

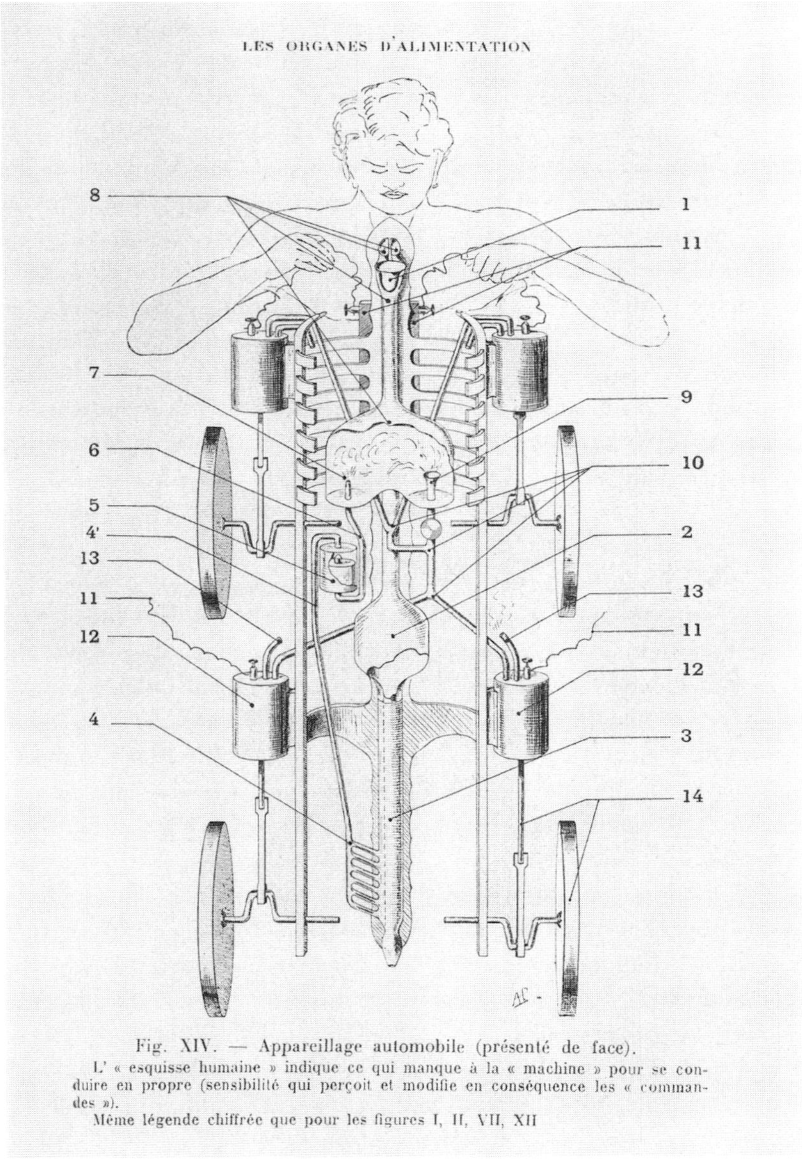

FIGURE 6.4. *Louis Chauvois (author/artist), "A Human Skeleton Is Like an Automobile Chassis,"* La machine humaine enseignée par la machine automobile *(The human machine, envisioned as an automotive machine) (1926), 35. National Library of Medicine.*

For Walters, distantly echoing Bacon, metaphor was unscientific and delusionary. The modern thing was to kill it off. Kahn, in contrast, promoted it to high visibility as a way to get modern—not to make a scientific proof but to explain science's proofs with methods suited to the modern age and masses. Once (seemingly) purged from the production of knowledge, metaphor could be safely recruited to play an indispensable role as a technology of explanation. Science, its epistemological authority secure, could embrace the use of metaphor to explain itself to the laity.

Kahn was not the only author to try his hand at illustrated metaphor. From the 1910s on, body metaphors that make reference to contemporary technologies, always a staple of popular medical textual rhetoric, began to crop up in a mixed bag of idiosyncratic illustrated medical publications for lay readers: James Cabell Minor's comical *The Plan o' the House o' Man, Sir! or, The Parts Water and Position Play in the Prevention and Treatment of Physical Disorders of the Body* (1912), Arthur Keith's dry-as-dust *Engines of the Human Body* (1919), Wilfred Grenfell's crudely charming *Yourself and Your Body* (1924), and Louis Chauvois's bizarre *La machine humaine enseignée par la machine automobile* (1926).[18] All of these works, written by doctors, used visual metaphor with some ingenuity to perform the modern and figure modern times. But they were largely unattended. Kahn's commitment to visual metaphor was far more extensive, and far more influential.

MODERNITY'S ICONOCLASTIC CONTROVERSY

In the first chapter of this book, I set aside "when does modernity begin?" as an impossible question. The modern appears as a privileged term and identity (and not just as a synonym for "contemporary") whenever people engage in a *querelle des anciens et des modernes.* Kahn's metaphor work joined together *querelles* in three fields: image making, pedagogy, and science. On behalf of his readers, he located that work in a grand narrative of progress, which was the summa of all *querelles.*

The twentieth-century development of conceptual scientific illustration and its metaphors was part of a larger cultural revolution that remade the relationship between text and image. A picture is worth a thousand words; so argued visual instructionists, filmmakers, advertising agencies, propaganda bureaus, and public health advocates. Pictures en masse could mobilize the masses. Shows of visual materials could seduce, instruct, motivate and entertain workers, children, students, consumers, soldiers—anyone, everyone.[19]

Enthusiasm for the image as a device of psychosocial engineering developed in tandem with emergent professions, disciplines, and institutions. As Germany underwent intensive urbanization and grew into a bureaucratic state, professionals increasingly claimed the intellectual authority to scientifically manage life and society. Hugo Münsterberg, a Harvard professor of psychology who was influential in both Germany and America, marked the moment with his 1916 proclamation, "We live in the day of the scientist and the expert."[20] In this milieu, there emerged a utopian politics of medicine and health fueled by new expectations about what science, the medical professions, and health bureaucracies could accomplish. The enactment of state and industrial health insurance legislation in late nineteenth-century imperial Germany, combined with the growth of great educational and research institutions, new bioscientific findings and technologies, and the proliferation of commercial advertisements for new remedies and therapies, all fueled public interest in medicine and the human body. Reformers, officials, authors, publishers, trade unionists, pharmaceutical manufacturers, and health enthusiasts worked assiduously to cultivate in the public a new kind of health consciousness.[21] These concerns were intensified by the calamitous experience of the Great War—and its dire consequences for the health of so many people, not least the maimed and wounded—and by the shocking collapse of the Wilhelmine regime, which opened the door to political unrest and violence as opposing groups vied to shape the new order.

That terrible loss presented opportunities for those who wanted to take giant steps into modernity. Alfred Grotjahn, a prominent health reformer, called on colleagues "to deepen, simplify and spread hygienic culture . . . , to take hygienic awareness into the dwellings of even the poorest workers." This work, he predicted, would become "the ruling star of the twentieth century."[22] Credentialed by science, the voice of the expert would become generalized, universalized, recited, and heeded by both the individual and the masses. And to be heard, expertise had to be seen via the image.

By long familiarity, Kahn's readers were primed to see themselves and their world in images. Pictures were ubiquitous. There were illustrated newspapers and magazines, posters, museums, movies, exhibitions, framed pictures hanging in houses, buildings, and workplaces, all around. The image marked, archived, and authenticated experience. The twentieth century, Otto Neurath famously announced, was "the century of the eye."[23] For Martin Heidegger, modernity and the conversion of experience into image were mutually defining givens: "The fact that the world becomes picture *[Bild]* . . . is what distinguishes the essence of the modern age."[24] Ludwig Wittgenstein, whose oddly stripped-down and numbered philosophical prose was almost the textual equivalent of Bauhaus functionalism, used "picture" as a foundational term: "A picture held us captive. And we could not get outside it, for it lay in our language and language seemed to repeat it to us inexorably."[25] For Wittgenstein, picturing, like language, was a universal practice of human subjectivity; language and picturing stand in some ambivalently reciprocal but hard-to-characterize relation to each other. These pronouncements were responses to the experience of life in a built environment of images. Mass media technologies and tactics had uniquely historical cognitive effects. Heidegger's world picture *(Weltbild)* was conceived—could only have been conceived—in a world of pictures, a *Bildwelt.* We look at the world and see a picture because we look at pictures and see the world. This is so even if world and picture exceed each other and never neatly match up (despite our efforts to make them match).[26]

For even the most profound philosophers, the image was irreducible, a given, and therefore undertheorized. For modernizers like Kahn, Neurath, and Waldemar Kaempffert, the instrumental uses and pleasures of the image were cause for celebration. Kaempffert, upon assuming the editorship of *Popular Science Monthly* in 1916, remade the magazine almost overnight into a mass-circulation publishing success by cramming "as many illustrations as can be crowded" into its pages, photographs, and drawings, theatrically framed or silhouetted to simultaneously entertain and instruct readers. Introducing the change, he celebrated magazine pictoriality as "mechanical vaudeville": "The picture," he exulted, "is the quickest, surest way of communicating ideas."[27]

Cultural critics were less enthusiastic. They noticed the proliferation of images (and media technologies), and deplored them. Gustave Le Bon, in his widely read *The Crowd: A Study of the Popular Mind* (1895), argued that "a crowd . . . thinks in images. . . . It accepts as real the images evoked in its mind though they most often have only a very distant relation with the observed fact."[28] French critic Louis Haugmard, writing in 1913, sarcastically predicted the future of the century: "The bewitched masses will learn to not think anymore, to resist all desire to reason . . . ; they will know only how to open their large and empty eyes, just to look, look, look."[29] Under the spell of printed pictures, movies, and other visual displays, critics argued, the masses were incapable of understanding a scientific worldview. Science's sober epistemological procedures would fail, Le Bon worried, when put up against the crowd's "collective hallucinations."[30]

If so, those who would appeal to the masses with images were pandering. Martin Jay has shown that twentieth-century critics of mass culture maintained a well-studied hostility to the image.[31] The relation between image and text was thus "not merely a technical question," W. J. T. Mitchell argues, "but a site of conflict, a nexus where political, institutional and social antagonisms":

> Even something as mundane and familiar as the relative proportion of image and text on the front page of the daily newspaper is a direct indicator of the social class of its readership. The real question when confronted with these kinds of image–text relations is not "what is the difference (or similarity) between the words and the images?" but "what difference do the differences (and similarities) make?" That is, why it does matter how words and images are juxtaposed, blended or separated.[32]

REVIEWING KAHN

For Kahn, the point was to juxtapose and blend pictures and visual tropes that appealed to, incited, and satisfied the psychological needs, desires, and capabilities of the collective and individual viewer. How well did he do? Pretty well, according to reviewers. In his heyday, the critical response in both Europe and America was mainly favorable and focused on his images. Reviewing *Man in Structure and Function* (the 1943 American revision/translation of *Der Mensch Gesund und Krank*), Logan Clendening, a Kansas City physiologist, syndicated health columnist, and author of books and articles on medicine for the public, recalled how "years ago I saw on the shelves of a Munich bookstore a set of volumes called 'Das Leben des Menschen' . . . and was fascinated by the illustrations":

> They were scientifically accurate and yet infused with a power of imagination . . . to discern what parts and activities of the body the non-technical layman would be curious about, and . . . to infuse into the picture a point of view which lifted it to the level of magic—to catch the body . . . at the very split second when it was most interesting. . . . It is as if [Gustav] Doré and Salvadore Dali and the editors of Fortune had combined their powers to illuminate this text.[33]

Other reviewers emphasized the power and modernity of Kahn's pictures. Mabel S. Ulrich, a doctor and contributor to *Scribner's* and the *Saturday Review of Literature,* connected Kahn's "beautiful, graphic designs" to the most advanced imaging technology of the time: "the new electronic microscope." She also linked Kahn's images to modern art: some pictures were "influenced by a Dali-esque surrealism," while others resembled "the rhythmic abstractions of a Bracque."[34] It was as if the haphazard formations and deformations of nerves, veins, tissues, proteins, cells and crystals were naturally occurring works of art (an idea that painter Pavel Tchelitchew played on in his anatomical interior landscapes of the 1940s).[35] Our bodies are both scientifically and aesthetically modern.

An article in the tabloid *New York Mirror* marked the difference between Kahn's illustrations and the illustrations that typically appeared in public school textbooks:

> A striking feature of Kahn's lifework is its accidental kinship with some of the latter-day trends of the non-scientific graphic arts. Anyone will be struck by the resemblance of some of these sketches to the so-called abstract or sur-realist schools of "City of the Future" art. Certainly they will have the general effect of making the observer feel much more uplifted about his insides than has been his miserable lot since the awful day when his

> eye first fell upon the meat-market art of the plates in the "Phizzology Book" back in the Seventh Grade.[36]

We should not take the comment too seriously: the *Mirror* reporter, like Kahn, had to entertain readers. Kahn's illustrations provided the writer with an opportunity to indulge in a little bit of comedy. Even so, the *Mirror* piece gets to the essence of Kahn. His illustrations were a radical departure from the anatomical tradition (and its rhetorical connection to the dead body); were oddly entertaining in some kind of futuristic way; and in some ways resembled modern art.

But Kahn also staked his authority on a close identification with current science. Some medical reviewers were uneasy with that. Wendell J. S. Krieg, a professor at Northwestern University School of Medicine and the author of a standard textbook on neuroanatomy, credited Kahn for originality, accessibility, and impact: "The accounts are not academic descriptions, but visions. Dr. Kahn pulls no punches. One amazing fact succeeds another, and vivid analogies strike blow upon blow." But Krieg also faulted Kahn for visual excess:

> Many of his drawings are so far from the merely representational as to be actually surrealistic. . . . Kahn oversteps the possibilities of a picture to present a concept.

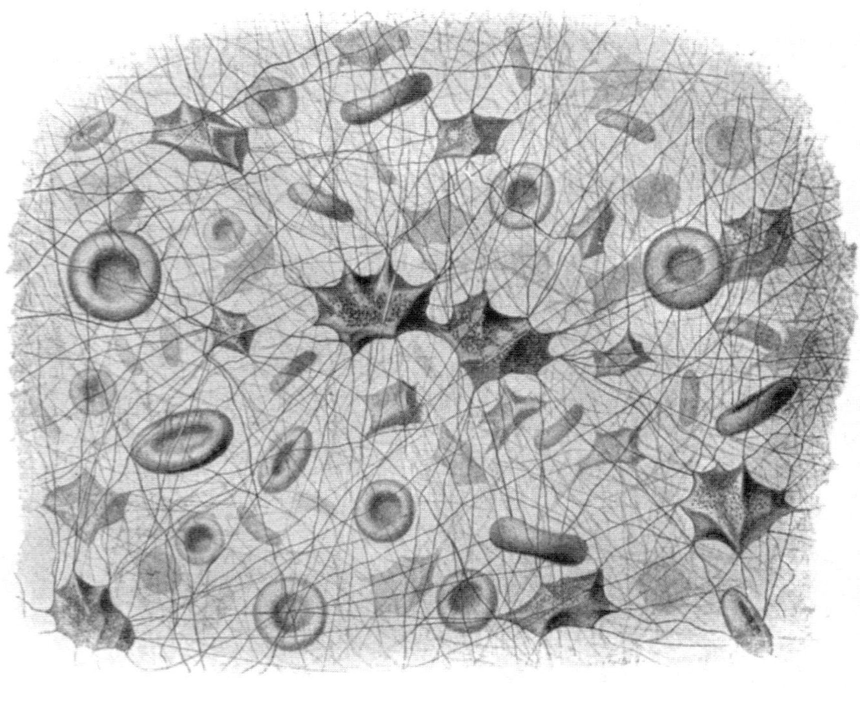

FIGURE 6.5. *"Blood Crystals,"* Das Leben des Menschen *(1924), 2:320. Artist: Georg Helbig. Rhythmic abstractions in the human body. National Library of Medicine. Copyright Kosmos Verlag, Stuttgart.*

> It is fatuous to illustrate the effects of a cigar by depicting a brace of fresh corpses. . . . Elaborately wrought reconstructions . . . illustrate the structure of organs, frequently to good avail, but the method is foredoomed to failure when detail is attempted in a large block of tissue which has a fine grained structure. A drawing of a microscopic section would have been better.

Krieg also faulted Kahn for errors of fact and bias:

> The human embryology is very out of date, the heart does not have the skeleton represented . . . , the adrenal cortex does not secrete adrenalin, the brain does not float in blood, and the account of the effective stimuli of the semicircular canals is quite wrong. Unwarranted

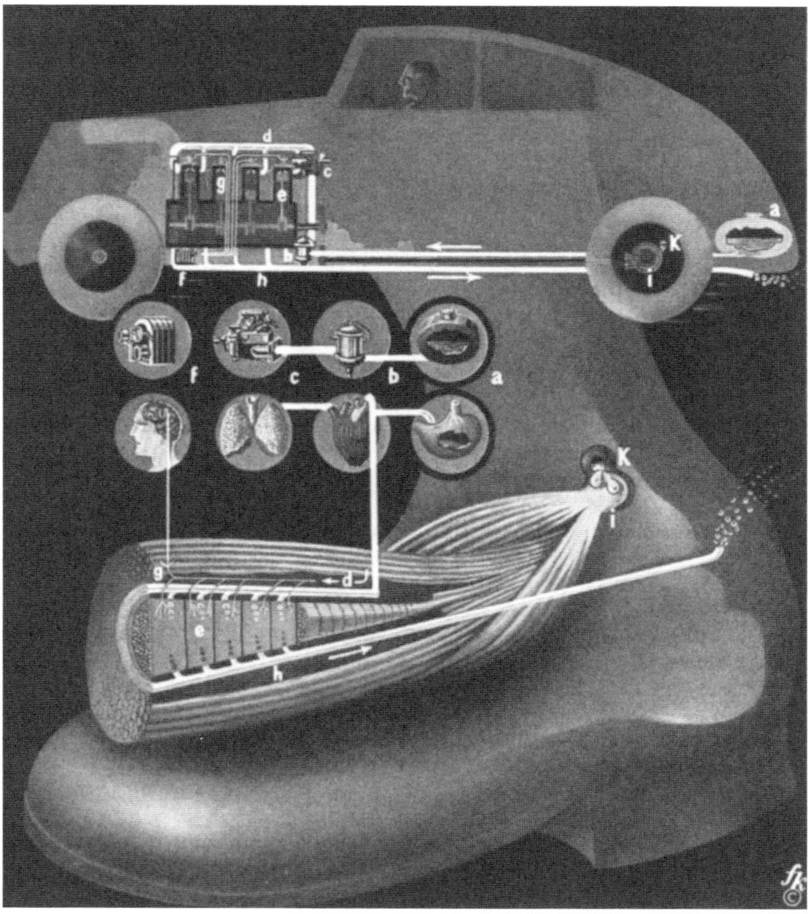

FIGURE 6.6. *"Muscle Fibers and Automobile Engine Show Substantial Similarity in Structure,"* Der Mensch Gesund und Krank *(1939), 2:141. Artist uncredited. National Library of Medicine. An anonymous medical critic complained, "The desire for originality has gone to almost absurd extremes."*

> generalizations are occasionally made, as that "Woman's chief biologic function is the bearing and raising of children while man is destined by nature for intellectual activities," and that "only animals with a cerebrum can be trained."[37]

Dr. Ralph W. Gerard, a University of Chicago physiologist, also had qualms. *Man in Structure and Function* "will probably be a success—lovely pictures, positive answers to the sort of questions laymen are wont to ask, an aggressive promotion . . . almost insure it. But the reader will be entertained more than instructed."[38] What made the pictures accessible to the general reader made it unsuitable to the scientist, who by training and inclination saw himself as a higher authority, the ultimate arbiter of the content of public scientific discourse. In the same vein, an anonymous reviewer in the *Quarterly Review of Biology* acknowledged that Kahn's illustrations were "novel," "stimulating," "highly imaginative," and "instructive to a rare degree," but also complained that in places "the desire for originality has gone to almost absurd extremes and some illustrations interpret the facts in a distorted fashion."[39]

THE ICONOPHILIC SELF

The distortions were not entirely random, mistaken, or accidental. Kahn tried to be accurate and up-to-date, and he certainly wanted the endorsement of scientists and physicians (which he usually got), but first of all, he needed to persuade and entertain readers. And to do that, he had to be ingenious with his images and analogies.

As we have seen, Kahn's illustrations narrated and interpreted the "life of man," "man's structure and function," and "man in health and sickness" by means of "metaphors to live by."[40] In aggregate, the images provided source material for a universalist cultural and political identity (or anti-identity). They affirmed the reader's filiation with modernity—located the reader's body in the order of science and reason. Politics received no representation, and for the most part neither did history. The "tense" of Kahnian illustration was the present tumbling over into the future, in some cases, just the future.

Kahn's pictures naturalized industrial technology and its products, naturalized capital's continual need for novelty in marketing, and naturalized the web of sociopolitical powers that deployed industrial technology. Inside the body, modernity was simply the order of things. There was no partisan strife, economic competition, or brute force. Modernity could be discovered within; it didn't have to contend against political forces and obdurate matter. Kahn's images (along with the words of his text) instated biomedicine as a privileged, disinterested authority over the body and everyday life—and they instated Kahn as a representative of that authority, a go-between. Kahn's image work thus contributed to the formation of a self-surveying panoptic gaze, on and within the subject. Identification with the images prompted readers to identify themselves with the technoscientific order that governed and operated the body.

That sounds manipulative, even coercive. Writing in 1925, Fritz Giese noted that for workers, "*Psychotechnik* has become . . . a mechanization of spirit, an oppression of will," but *Psychotechnik*'s very success as a rhetoric of modernity took it out of the hands of manufacturers, marketers, and efficiency experts, moved it beyond the instrumental uses that made it so disagreeable.[41] Unlike the psychotechnicians, Kahn needed his subjects' consent to proceed. He had to give readers something they wanted. Kahn's *Psychotechnik,* then, was not a top-down attempt at domination. It responded to the desires of his public: the "salaried masses," industrial workers, teachers, students, "new women," and so on—people who wanted to make themselves into moderns, to live a new kind of life, to be a new type of human. His images offered to assist in that transformation.

More than that, the sheer pleasure of Kahn's images, their wayward indexical and metaphorical references, exceeded any rhetorical

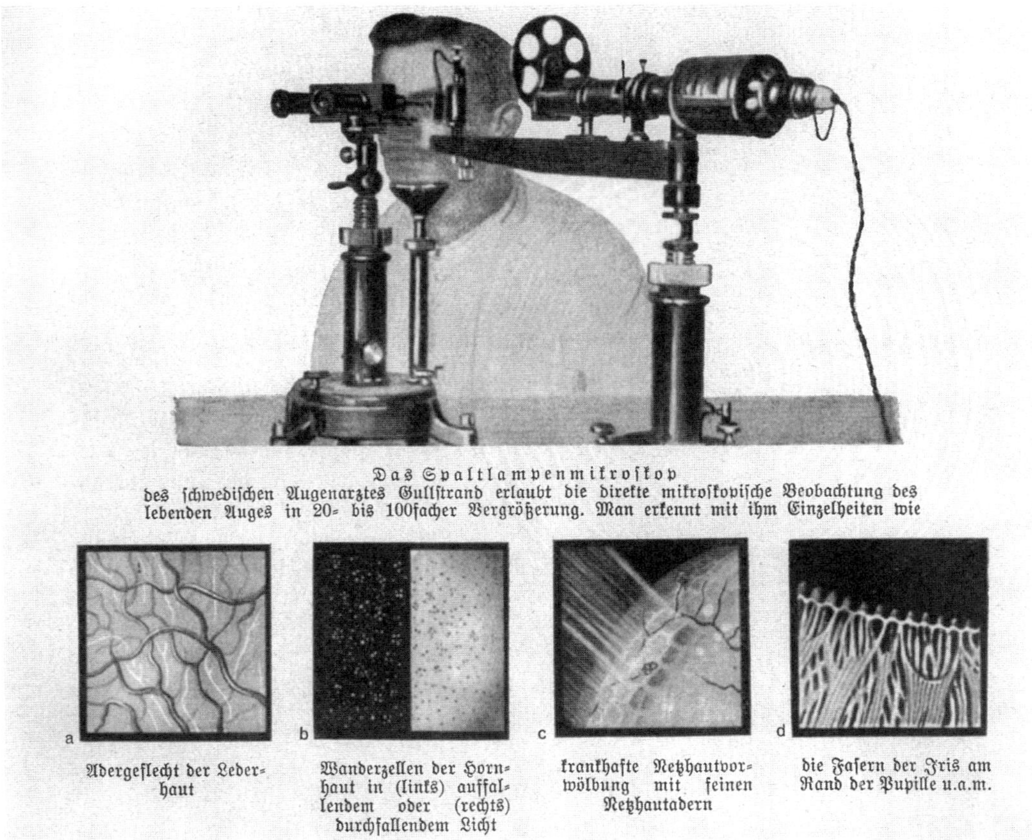

FIGURE 6.7. *"The Slit-Lamp Microscope,"* Das Leben des Menschen *(1931), 5:23. Artist uncredited (for subillustrations a through d). National Library of Medicine. Copyright Kosmos Verlag, Stuttgart.*

design and subverted any project of self surveillance and rationalization. Embodied life is dirty and disorderly; it produces urine, feces, blood, sweat, mucous, semen, pus; it generates passions, emotions, and sensations. Bodies are recalcitrant, impossible to control. In discourse and iconography, the body and its interior inevitably come to stand for the mess of embodied experience and its opposite. In contrast, on the printed page, bodies are cleaned up, put under the governance of some real or imagined medical, legal, moral, or technological order.

But, just as embodied life always exceeds regulation, so too does representation. Catherine Gallagher and Stephen Greenblatt argue that "the body functions as a kind of 'spoiler,' always baffling or exceeding the ways in which it is represented." If so, representation can also function as a spoiler, baffling or exceeding the thing it purports to represent. This is perhaps what most troubled Kahn's critics: the tension between his avowed purpose—to use images to control his readers, and teach self-control through an induced identification with the scientific order of things—and his own evident loss of control over his images (and text), his showboating iconophilia. What Roland Barthes wrote about the plates of Diderot's *Encyclopédie* is even more true of Kahn's illustrations: every figure "vibrate[s] well beyond its demonstrative intent."[42] And that, of course, could be a very good thing.

LAST WORD: HOW TO GET MODERNITY, VISUALITY, SUBJECTIVITY

At stake in all this are the mutually affecting relations that link representation to experience and subjectivity. Experience stocks representation with tropes, guidelines, and problems—and vice versa. The how of it depends on circumstance. But merely to assert the reciprocal relation, though inarguably true, tacks around the difficult (impossible?) issue of agency: whose experience? Who or what is the recipient of experience? Where is experience made or stored? How is it mobilized? Continually groping toward some originary agent—the self—we end up reinstating as a given the very object in question. We are tempted to quarrel with selfhood, to debunk it as a nonentity, an artifact of philosophizing, a delusory existential convention. Or we defer to neuroscience and its tendency to treat the self as a neural homunculus, a natural operating system distributed in networked brain clusters within a singular mind, which dynamically assembles itself in evolutionary time, memory time, and the phenomenological moment. Or we can treat it as a historical/cultural property, a contingent, emergent identity effect: the Western self, non-Western selves, non-Western nonselves, and smaller, less hemispheric units. Or we can be pragmatic: the self is a necessity of embodied life.

We've circled around again to the question: What is the ontological status of the self? Does it have a secure ontological status as one of those conditions that defines us as human? Are people in every culture assigned a name and identity that becomes destiny? What kinds of destiny precede naming? Do nonhuman animals also have an inborn notion of "me/not-me," a sense of selfhood based on self-recognition of characteristics that differ from other creatures, other members of their species?

If so, maybe the thing (or one of the things) that makes us human is the fractured, partible, and only partly sutured-together self. If self is made, taught, enforced, assigned, purveyed, assaulted, defended, performed publicly, kept private, inspected, or revised, then it takes a lot of work, both on the part of individuals and collectives, to keep it together. It also takes a lot of work to keep it stable, to fix it as human or modern or German or masculine or some other identifying label. We live under the eyes and at the hands of workmates and classmates, parents, authorities, police, teachers, friends, strangers, and so on. Those assessors confer subtle and sometimes not so subtle rewards; they mete out punishments that work to enforce self-coherence or undo it. The more complex and contradictory a society is, the more fraught its identity formations.

And so never-articulated tensions afflict Kahn's images: they are jammed up next to each other, or separated by many pages. The self is an internal combustion engine or a factory or an elevator building or a car or a diagram or an anatomical illustration or an interior landscape or something else. The style is Bauhaus or surrealism or Art Deco or commercial realism or . . . Outside Kahn's pages, his images bump up against the images of advertising, government propaganda, and the cinema. Even more incoherence.

"What sort of man comes out of modern technics?" critic Lewis Mumford troubled to ask in his 1934 jeremiad *Technics and Civilization.*[43] Kahn's images reply: every human, and every human part. The modernizing conceptual illustration effects a kind of smoothing and polishing of overlaid inner surfaces, a kind of flattening and fragmentation through iteration and reiteration. So agency is real, but depends on a mirage: you read pictures, pictures read you. "You" are an image bank. Otto Neurath, in his 1926 manifesto *Pictorial Representation of Social Facts,* almost breathlessly described the new, immersive, built visual environment: "Today, frequent changes of visual environment belong to the characteristics of modern urban life which is also penetrating into rural areas. Wall posters call out to us from the streets and hallways; exhibitions are inviting us; millions of people are watching the motion picture screens every evening; a growing number of magazines and booklets are bringing new pictures."[44]

In Neurath's account, the visual calls out, and invites. Displays and pictures penetrate, colonize, mobilize, transform, constitute; they have an aggregate effect. Cultural theorist W. J. T. Mitchell, in a characteristically provocative gesture, poses the question, "What do pictures want?"[45] The answer, of course, is nothing and everything.

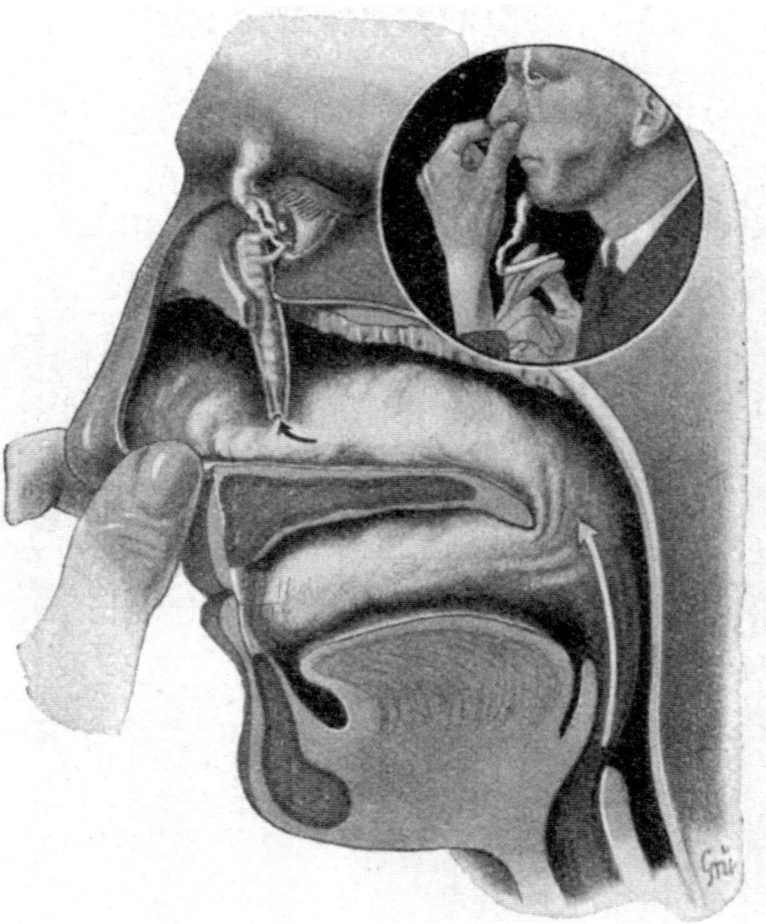

FIGURE 6.8. *A trivial scene of modern life, a trick scientifically explained. "Smoking through the Eyes,"* Das Leben des Menschen *(1931), 5:142. Artist: Grü (Helmut Grandner). National Library of Medicine. A cross section with an interior inset representation of the outer scene that the cross section explains. Copyright Kosmos Verlag, Stuttgart.*

We are operators who are shaped and worked on; configurations of things form us and operate us. Agency, like subjectivity, is distributed and exchanged and made; it is the result of multiple acts of collusion and resistance. It emerges in the puzzle pieces of words and images that we (collectively and in one bag of skin) try to fit together to make a unified self. That's not easy because images, accompanied by a surrounding immersive cloud of metadata (image tropes, genres, styles, media), refer back to, swarm around, and bounce off each other. There is an infinite regression of "echo objects" (to use Barbara Maria Stafford's phrase) and "mirror stages" (Jacques Lacan).[46] This mirroring and echoing turns out to be perhaps the most evocative, uncanny theme of Kahn's image work.

As we have seen, the conceptual scientific illustrations of Kahn and his studio are a lost body of modernizing cultural productions, a small subset of a very large and still poorly attended body of vernacular image work. The images swarmed together in cheap publications, and (to us) remote and isolated places, but were not devoid of contact with the higher modernism of Gropius, Moholy-Nagy, Le Corbusier, Bayer, and their colleagues.

Because of our unfamiliarity with Kahn, his images can serve as echo objects for us. This book-length meditation on the history of Kahn's image practice is itself implicated in those practices. In narrating the origin story of conceptual scientific illustration, I have relied on the principles set forth in *Pictured Knowledge* and contemporary advertising manuals, and practiced in the works of Fritz Kahn. I have made an illustrated argument that tries to pleasure the eye with visual poetry.

In a 1929 radio address, looking back on the long history of image in relation to text, Walter Benjamin remarked that "the most obvious thing to do was to emancipate visualization as much as possible from the word."[47] That is now our standard operating procedure: the proliferation and showing of images and a corollary demotion of text has become our order of things. (The practice of showing is now so pervasive that it only serves as a marker of ordinary social and intellectual competence.) That order is naturalized in these pages, as it is in our apps, websites, video games, operating systems, television shows, direct mail, textbooks, and software interfaces—and also naturalized in our moral, psychological, behavioral, and cognitive principles and methods.[48]

So this book tells what I have learned of, and from, Fritz Kahn. Immersive visuality—call it hegemony or consensus or an epistemic regime or a scopic regime—is real and tangible, and in the current moment, inescapable. It is an ongoing story, to which Kahn, the makers of early twentieth-century picture encyclopedias, and post–World War II animators and graphic designers all made telling contributions. In developing so many different modernizing visual tropes and strategies, Kahn and his artists ended up deconstructing the very thing they aimed to construct: an atomic individual for industrial modernity. Their thousands of pictures contributed to the blizzard of representations that now make selfhood nearly impossible. How fitting, then, that Kahn's trademark image—"Der Mensch als Industriepalast"—ends up gracing the cover of the latest English translation of Deleuze and Guattari's deranging "schizoanalysis" of the universal human subject in late capitalism: *Anti-Oedipus*.[49]

We live in a technoeconomy that every day disarticulates us by breaking down our behavior into clicks, patterns, and finely delineated market segments, then spews out a continual stream of figures and products for us to identify with, scripts and props for our performance of self. That is our way of being. But this is not to say that our homuncular delusions—"I," "self," "the body"—are only conjured from the conspiracies of corporate and political interests that continuously rearrange our pieces (the digital sequel to early twentieth-century *Psychotechnik*). That would be too easy, and in any case, the

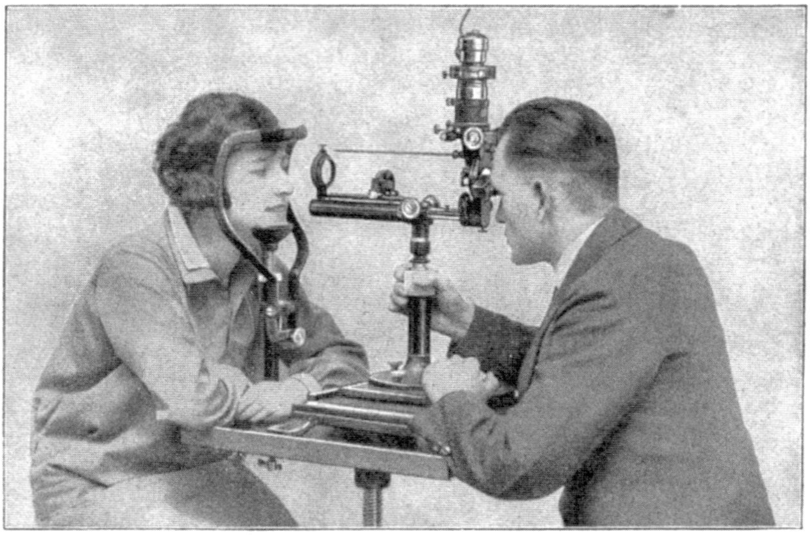

FIGURE 6.9. *"Examination of the Back of the Eye II: The 20th-Century Method,"* Das Leben des Menschen *(1931), 5:65. National Library of Medicine. Copyright Kosmos Verlag, Stuttgart.*

same reductive suggestion could easily be applied to a structuralist/poststructuralist critique of the unitary self. Global capital is equally content with heteronormativity or queer normativity, critical theory or free market ideology; it happily exploits every need and preference (and flows around every obstacle placed in its path). Does the incoherent power of it all make the unitary "I" a useless singularity? Or a necessary fiction? "You" and "I" have to hang on something.

Fritz Kahn's lost images were once offered as a visual lexicon of a new utopian biopolitics. In showing what we imagined we were, they now suggest something else. We are not (never were) atomic individuals. The personal, insofar as it has existed, has always been impersonal: political, economic, theological, medical, technological. That impersonality now intensifies. Increasingly, we are quantized selves—subatomic image/text packets, waves of meaning. Biotech, finance capital, big pharma, medical-industrial complexes, media mobilization machines, Silicon Valleys—they make us and play us.[50] Of course we also play back, with consequences that no one fully intends or anticipates. Some remainder of us always falls outside the totalizing penumbra (even if that doesn't amount to emancipation). Simultaneously associating and dissociating, Roland Barthes says "to write the body" (but we will say "to picture" it): "Neither the skin, nor the muscles, nor the bones, nor the nerves, but the rest: an awkward, fibrous, shaggy, tangled thing, a clown's coat."[51]

EPILOGUE

TOWARD A THEORY OF THE HOMUNCULUS

ARCHIVING THE SELF

Why was archiving the self as figures on paper so valuable to Kahn and his readers? In a seminal article, Bruno Latour discusses the "advantages of the 'paperwork.'"[1] Images on paper, he argues, are "mobile," can be easily transferred from medium to medium, place to place. At the same time, they are "immutable"; they retain their content and are easily retrieved and read. Pictures allow you to capture and fix—make flat—and then circulate some selected aspects of three-dimensional life. The advantage, Latour asserts, is that "there is nothing you can *dominate* as easily as a flat surface; there is nothing hidden or convoluted, no shadows, no double entendre."[2] Domination, yes, but not always easy, and not always total in the way that Latour makes out. Think of what an artist may hide in the shadows of a picture, or pictorial double meanings that may slyly mock a patron, or how a commissioned artwork may provide the artist the chance to play some private game. Even the most determined illustrators cannot intend, or know, all of the meanings generated by their own pictures. If the picture is the result of collaboration—as are all of Kahn's pictures (and as are all pictures in a sense)—then inevitably there will be some power struggle within the frame of the picture. No single agent entirely controls the production. The reception of a work is even more difficult. A picture may be read in ways that its producers would not assent to, cannot agree on, cannot control. Yet by virtue of its fixity, as well as the larger set of conventions and ideological assumptions that govern its production and use, Latour has a point: an image is inscribed with some particular vision of how the world works. Some images, and some genres, do a better job of that than others. Kahn and his artists did a pretty good job of dominating their pages with pictures. Who or what dominated within the borders of any particular Kahnian picture is more difficult to ascertain.

Of course, domination goes beyond the picture plane and extends out into the material world, where it is political and personal. Latour tells us that the making of pictures is in part an "agonistic" struggle for power between rivals over interpretation, patronage, use, and profit.[3] (Think of Kahn's critics and rivals, the other authors trying to win over readers to their articles and books on medicine and science.) The picture is a bid for some kind of governance over, and exploitation of, readers. If readers want to see and enjoy the picture, then they have to accept some terms and conditions of use. While this may seem coercive, there is also mutual benefit and collusion. Readers ask to be dominated. The act of reading is a kind of willing submission to the book, article, or illustration. The author and illustrators know they have to please the reader and satisfy reader demand, creating a feedback loop. In that way, readers call at least some of the shots.

Within the picture plane, all sorts of powers are at work. Using the examples of mapping and micrography, Latour emphasizes the advantages of "immutability," of keeping things as they are, then limiting the mutation to one register or scale, which "may be modified at will, without any change" in the "internal proportions" of pictures.[4] But changes in internal proportions may also be toyed with for effect. Kahn's image practice depended largely on continually playing with the scale of bodies and body parts. (Alfred Döblin praised Kahn's "unusually vivid magnifications.")[5] Microscopic parts are enlarged as photographic transcriptions or worked up as imaginative presentations; macroscopic parts, whole bodies, are reduced to the size of an illustration on a page. Reading the images with some degree of imaginative identification, readers scale themselves up or down, in part or in whole.

Mobility is the other principal virtue that Latour assigns to picture making. From the fifteenth century on, print technology made it possible for images to be "reproduced and spread at little cost, so that all the instants of time and all the places in space can be gathered in another time and place." Because images "are mobile, flat, reproducible, still, and of varying scales, they can be reshuffled and *recombined*," "images of totally different origins and scales can be superimposed"—both within the frame of a single picture and through overlays—and also sequenced and resequenced.[6] Latour is principally concerned with knowledge production and circulation among producers, but Kahn's profile cutaways, synoptic pictures, and photomontages recombine images with the aim of explaining science and capturing lay viewers as a form of pleasurable play for both image maker and viewer, as a contribution to the viewer's identity, sense of self, and personhood.[7]

Latour brings all of these ideas together in the concept of immutable mobiles. He insists, "Innovations in graphism are crucial . . . only insofar as they allow new two-way relations to be established with objects (from nature or from fiction) and only insofar as they allow inscriptions either to become more mobile or to stay immutable through all their displacements."[8] However, Kahn's illustrations retain "optical consistency" and immutability for only part of their itinerary. They are transferred mostly intact from finished artwork to the printed page, with some loss and some revision in successive editions and titles. But their ultimate destination is to go from the printed page to the eye, and then the complexly mediated thought processes of the viewer, where they must inevitably decay, ending up as a murky contribution to a murky gestalt of self and collective visualization, rehearsal, and explanation. That murk cumulatively amounts to something: the thousands of printed image/texts in books on the shelf are a retrievable, shareable archive of self-words and self-images that displace each other and dissolve into each other, nucleate, and grow around each other. Subjectivity, then, does not only take place in the mind of the individual; it is not only confined to the self. Instead, it is generated in distributed networks of interactions within selves and among selves, and in objects and collections of objects, including printed and illustrated texts.[9] "What goes on in you" only makes sense if we include the onlooking audience for whom the illustration is intended, and the instructional diagram itself. "You" is in both the reader and the image that is being read.

SELF AS HOMUNCULUS

Latour is concerned largely with the translation of images from place to place and medium to medium, scaled up and down, combined and recombined, formed and transformed. All of those things happen in the making and circulation of Kahn's illustrations (and everyone else's). But what gets black-boxed in Latour is a consideration of the terminus. "Is the body a medium?" asks historian Philipp Sarasin. "We usually think of media as technical devices that 'mediate' far-away things by bringing them closer, by which we mean 'close to our bodies.' So the body might thus seem to be the exact opposite of a medium."[10] Sarasin shows, however, that nineteenth-century popular anatomy and hygiene discourse treated the body precisely as a medium, as a communications system with built-in devices that transmit messages from outside to part to mind and back again—messages that can only truly be decoded in collaboration with medical authority and scientific knowledge. That only some of those messages are consciously apprehended breaks down the idea of a unified self. If the human body and mind are really just circuits wired together, who or what is the ultimate receiver?

Kahn—so anxious to lead readers to the contemplation of their own inner workings—so anxious to modernize the illustration and reader—takes it several steps further. Mediation happens inside the body just the way it happens outside, on its way to the body. The actor-network reaches in and extends out, breaches the surfaces and sense organs of the body, and travels through nerves and glands and lungs and intestines and everything else. The body is colonized by external powers or, constructed from within, reaches out into the media environment. (Connectivity is not much of an issue in Kahn: textual and pictorial discourse connects everything.)

Following Latour's lead, the next step is to expand the notion of homunculus, to scale it up. What appears to be an agent, a person with an identity who acts in the world, is really a complex node, part of a network of homuncular relays. If so, "homunculus" can serve as a key term in a descriptive account of identity and, more broadly, subject formation. The self is a hybrid of performed visual and narrative elements, practiced over and over, but never fully consolidated. Kahn and his readers lived in a media-saturated ocean of cultural productions that teemed with homuncular figures that could be retrieved and worked up into self-models (if not consumed raw). And so do we. We live in a built environment of homunculi. We are a built environment of homunculi. We can be retrieved to serve as a homunculus for others. Homuncularity is viral, distributed.

The homunculus thus has an itinerary, or many itineraries; it behaves like something that sociologists Susan Leigh Star and James Griesemer call a "boundary object." Such objects are

> both plastic enough to adapt to local needs and constraints of the several parties employing them, yet robust enough to maintain a common identity across sites. They are weakly structured in common use, and become strongly structured in individual-site use. They may be abstract or concrete. They have different meanings in different social worlds but their structure is common enough to more than one world to make them recognizable, a means of translation. The creation and management of boundary objects is key in developing and maintaining coherence across intersecting social worlds.[11]

The homunculus is exactly that: an identity figure that crosses back and forth between people and groups—crossings that are continuously replayed and revised in the mind of a recipient (one of the "several parties") for future presentation to others as self. In the webs of negotiated, regulated, and contested social relations, self acquires generic meaning, and selves acquire particular idiosyncratic meanings and identities. What makes this even more complicated is that the identity and self-conceptions of "the several parties" are themselves only partly stable boundary objects.

So we have an "I" that moves from a social network to model and provisionally unify internal subject homunculus fragments within a single person. These are collated and condensed, via interaction with the social world, and made into a semistable composite that is then returned back to the social network as a homunculus—a meta-homunculus. Collectively and individually, this "I" is an ongoing work in progress, an artifact of multiple mirroring and echoing and actions and experiences that provoke mirroring and echoing. It is always idiosyncratic and unique, but it is also a generic figure of a person and a kind of a person. The "creation and management of boundary objects," Star and Griesemer argue, is "a key process in developing and maintaining coherence."[12] The "I" is always governed by social conventions and formal regulations, which are more or less effectively tutored and policed in family life and at work, school, and elsewhere. This process is carried on in what Michel Foucault called "technologies of the self": the classroom, the interrogation, the confessional.[13] But the category of "technologies of the self" is much larger than that. Our immersion in the multivalent images of Kahn teaches us that every structured human activity and tool of human devising—the alphabet, the family dinner, farming, the printing press, the steam pump, the electromagnetic induction coil, the assembly line—is a technology of the self or figures a technology of the self (and the figuration is itself a technological device).

Because an "I" is also an agent that acts on the world—something that takes shape in the process of formation through the actions it performs and the reactions it elicits—it is also a *boundary subject.* All of those slippery homuncular actors that both act and exist, that actively shape the world and are shaped by it—"my body," "my mind"—are boundary subjects. With all of that, the traditional injunction of the popular medical text, "know thyself," becomes complicated. We are busy and disunified, like Kahn's homuncular figures and pictures, a pileup of commodity exchanges, friendly greetings, manufactories, traffic intersections, anatomical illustrations, primordial jungles, and everything else.

NEUROSCIENCE

How, then, does this account of Kahn's image work and the homunculus square with current neuroscientific ways of doing business? The protagonist of our contemporary neuroscience is the sui generis mind, which cobbles together the disparate elements reflected in the array of mirror neurons to make an emergent consciousness that is assigned the role of operant according to an as-yet-unspecified program. The telos is still the atomic individual, the unified self—but that now becomes a complex system, constituted by involuntary neural processing mechanisms and procedures. Under the influence of computer technology and the world of hardware, software, programs, devices and networks, neuroscience mostly engages in a kind of conceptual reverse engineering. Neural software and hardware constitute the individual mostly by bracketing off ideational content from neural mechanics—and bracketing off what happens inside our skulls from what happens outside our skulls, the traffic between brains and everything else in our environment.[14]

My discussion here argues something different: subjectivity is intersubjectivity. Human beings are social creatures and have a collective (and hence political and historical) existence. For us, the material world is artificial. We enclose even the mutest and most desolate environments on earth, the most natural regions of nature, in linguistic categories, and boundaries and regulations and disputes. Nothing is innocent of language or politics or history, broadly conceived, even if nonhuman materiality haunts us, exceeds us, is much more than a residue—a kind of cognitive, experiential dark matter (out of which Burke and the romantics conjured the sublime, Freud the uncanny, and Kristeva the abject).

We are material that changes over time. We come in units of one bag of flesh (except in the case of conjoined twins). Statute and common law—and human rights ethics—require that one unit of subjectivity and one corresponding unit of civil rights be allotted to each bag of flesh, a one-to-one correspondence. That seems morally right, a precious historical moral accomplishment, but each of us is a self among selves. Our mirror and nonmirror neurons receive stimuli from the outside world of oceans and forests and deserts, as well as billions of humans, cities, highways, assembly lines (still!), fiber optics, flat screens, cyberfinance, agribusiness, genomics, and container ships—the whole postmodern industrial/environmental palace. We only know ourselves—come to be ourselves—through avatar selves and avatar others. All categories are ineluctably contaminated—structured and restructured—by human social, political, cultural, technical, and environmental interactions. These take place over historical experiential time and are formed, reformed, sustained, neglected, or destroyed by negotiations, exchanges, contests, and, always, accidents—and all of it with plenty of unintended consequences.

None of this invalidates neuroscience. That ever more influential field is now opening up the black box of cognitive formations and subjective experiences and identities that we designate as the mind by posing empirically testable questions about the workings of human and nonhuman minds. One of its most interesting topics it studies is how representations are constituted, and how internal representations (or the illusion thereof) relate to external views. But to make good on its promise, neuroscience will have to traverse the distance from the interior of the brain to the outer world and back again, what John Sutton and other theorists of cognition term an ecological approach. This would entail some kind of thoughtful dialogue with philosophy of mind, history, and political economy, based on the assumption that neural/cognitive mechanisms are connected to long chains of mediations that traverse the boundaries between body and environment and that connect mutually affecting, provisionally constituted agents, both human and nonhuman. John Sutton and Nicholas Keene, following Andy Clark, ask us to put "brain, body and world together" in "distributed cognitive ecologies."[15] If so, neuroscience must at regular intervals reverse the field: the focus on the neurologics (or naturalization) of subject formation must be matched by the sociopolitical logic (or socialization) of neural object formation. Social networks shape neural networks—and the very questions that neuroscientists pose.

I would start with the proposition that human consciousness is socially distributed, constituted, and performed, and not just an emergent effect of a biologically programmed individual psychoneurological growth cycle. A second premise: human consciousness and identity are fundamentally divided. We come to believe that we have unified selves through a dialogical interactive process.[16] The unified self—the "I" effect—is produced by naming, registration, marking practices, laws, parenting actions, governmental regulations, and image production, all working to shape familial, legal, and peer expectations that the individual must be the bearer of a stable, bounded identity. These expectations, suggested and reinforced and inflected by mimetic/experiential models, are developed and enforced by rewards and punishments. They become cognitively hardwired in formative stages (or successive stages) as subjectivity. Identity, originally plastic and subject to arrangement and rearrangement, becomes fixed (to a greater or lesser degree). This hardened wiring makes the structured self in some way durable and persistent. It moves us to recur to thematics, obsessions, procedures, vocabularies, grammars, moods, and desires that were established at the time of fixity. In the above account, identity formation moves from the social to the neuroscientific (plasticity and fixity)—the opposite trajectory of most current neuroscientific research.

Even after fixity, there are historical/experiential times and places when crises of identity—power matched against power—result in rewirings, self-fracturing revisions, countervalent expectations, and so on. Induced by war, elections, legislation, religious fervor, hypnotic trances, drugs, literary fashion, sexual obsession, disaster, consumer electronics, the World Wide Web, economic depression, and other contingencies, contests occur in different social structures and genres, such as spirit possession, life-cycle transformation, role reversal, the invention of alter egos, transgender.

All of this turns the question of identity upside down. Instead of wondering why some people have multiple identities and split personalities, we can wonder why most do not—or why we think that most people are singularities. Neuroscientist Michael Gazzaniga, who works on the "sense of unity" problem, posits that human beings have a left-brain narrating system, which he dubs "the interpreter." Gazzaniga theorizes that the interpreter creates the illusion of a meaningful script as well as a unified coherent self.[17] But this leads to conceptual difficulties: meaningful to whom? Following what narrative structure? To whom (or what) does the interpreter speak? The right brain? Other sectors of the left brain? Other parts of the body? Other left brains? If there is a (proto?) homuncular interpreter in the brain, then is there also a (proto?) homuncular brain audience that receives the interpretation or gets to see/hear the script performed? Or is it (externally) other people who hear the interpretation in language or implicitly in social performances (such as dress, gesture, or food preference)? Or some internalized restaging of other people? All of the above? And beyond all that, how does the interpreter itself stabilize and become consolidated as a singular entity? Could someone have several competing or disagreeing or uncoordinated interpreters?[18]

Going further: where does Gazzaniga's "interpreter" originate? What is its source material? In what dialects, registers, voice, syntactic modes, or genres does the interpreter narrate and interpret? Is the development of the interpreter related to language acquisition and other mirror-stage effects? And how is it related to other deep cognitive acquisitions, such as personality, sexual preference, or erotic attachments? Do dogs, cats, monkeys, dolphins, and so on, also have a mirror stage and a coherent self? Rats? Octopuses? Snails? Jellyfish? How low can you go? What roles do various sensory experiences play? (Which returns us to the arguments of early twentieth-century academic psychologists about the evolutionary priority of the visual and other senses over language.) Autointerpretation, or self-narration, must have a cognitive logic, a grammar, vocabulary, scenography, a habitus that, like language, is continually practiced and repracticed, rehearsed and revised and corrected, conditioned by historical configurations, social position, cultural practices, and political contests, and always an unaccountable drift.[19] In naturalizing the interpreter, we are always subject to slippage. Our internal narrators turn out to resemble the omniscient narrator of the nineteenth-century novel or the late twentieth-century film director's edited pastiche of dialogue, camera movement, acted characters, evocative settings, and voice-overs. There is a shifting montage of voices and figures: mother, father, brother, sister, teacher, best friend, television star, and, always, something else.[20] These productions play back to us individual and collective human experience; they help us to understand our bodies, our selves. But they also deceive, because as we go about our everyday lives, we want to assume, indeed almost have to assume, that these contrivances, genres, and technologies are our natural (neural) order of things—which is also how the images and works of Fritz Kahn helped their readers to get modern.

ACKNOWLEDGMENTS

This book was written in stops and starts over the past decade, disrupted by work commitments, dire illness, and crazy love. I often doubted that it would ever see the light of day. Now daylight has come and I'm moved to give thanks. Like every work of scholarship, *Body Modern* is the work of many hands—a collaborative effort. I want to thank everyone who has contributed; I have tried to name you here and in the notes. If I neglected to list you, please forgive my sin of omission.

I want especially to offer thanks to my wife, Eva Åhrén, who gave me love and intellectual companionship and consoled me in dark hours, and to my dear daughters, Eve and Nell, who somehow grew to adulthood despite my Fritz Kahn obsessions. I also especially want to thank Jeff Reznick, chief of the History of Medicine Division (HMD), National Library of Medicine, who warmly supported this book and my work as division scholar. My colleagues at the National Library of Medicine have helped me in uncountable ways.

A number of scholars heard presentations or read parts of the project in various stages. I am entirely indebted to them. Miriam Eilers, a young Kahn scholar, was especially generous in sharing with me some of her research findings and ideas. I feel honored to have her as a trusted colleague, and I urge readers of this book to consult her smart scholarship on Kahn. Mary Cappello, Claudia Stein, and Carsten Timmermann all read portions of the book. Their insightful comments greatly improved the text, as did University of Minnesota Press peer reviewer Matthew Biro and a reviewer who chooses to remain anonymous. Claudia also helped me with translations, brilliantly teasing out nuances that I had entirely missed. I also owe thanks to Sita Reddy and Mark Dery for their friendship, brilliance, and willingness to indulge in zonked-out intellect play. Mark encouraged me to submit *Body Modern* to the University of Minnesota Press, for which I am ever grateful. I thank my exemplary editor, Pieter Martin, as well as everyone else at the University of Minnesota Press, for putting my baby to bed.

I want to give a shout-out to the grandchildren of Fritz Kahn—Daniel Turban, Joseph Turban, David Turban, and especially Dror Kahn, who invited me into his home and broke bread with me. They generously gave me their permission and blessing to use Fritz Kahn's images and words. I hope they are pleased with the results, and that this book helps them and their families to get better acquainted with Kahn, a complex and brilliant figure with whom they share family histories and secrets.

There are still more people and institutions to thank: Sam Alberti, Emewedish Bekele, Laurie Block, Cornelius Borck, Robert Brain, Julie K. Brown, Liping Bu, Karen Bucky, David Cantor, Anne Carter, Sterling and Francine Clark Art Institute, Roger Cooter, Rachel Core, Milton Corn, Duyan Dang, Kristi Sweet Davenport, Thilo von Debschitz, Aruna D'Souza, Thiago de Sousa Brandão, Erica di Benedetto, Dittrick Medical History Center (Cleveland), Joanna Ebenstein, Jim Edmonson, Sarah Eilers, Elizabeth Fee, Deborah Fehr, T. Kenny Fountain, Sander Gilman, Liliya Gusakova, Hagströmerbiblioteket (Stockholm), Richard C. Hall, Michael Hau, Bill Helfand, Holly Herro, Walter Hickel, Michael Ann Holly, Nick Hopwood, Institute for the History of Science and Technology (Deutsches Museum, Munich), Miriam Intrator, Mark Jarzombek, Margaret Kaiser, Jonathan Katz, Ken Koyle, Leo Baeck Institute for the Study of German Jewry (New York), Laura Lindgren, Tom McDonough, Philippa Mole, Morbid Anatomy Museum, Library, and Observatory, Elizabeth Mullen, Erika Naginski, National Institutes of Health Library, National Library of Sweden, Alex Nemerov, New York Academy of Medicine, Caitlin Newman, Jennifer Nieves, Michael North, Lisa O'Sullivan, Jennifer Burek Pierce, Karen Pitts, Lawrence Principe, Susan Roeper, Corey Ross, Ginny Roth, Lutz Sauerteig, Thomas Schlich, Arlene Shaner, Crystal Smith, Thomas Steller, Mike Stoffel, Kathy Stuart, Akihito Suzuki, Swedish Collegium for Advanced Study, Ken Swezey, Paul Theerman, Helmuth Trischler, Kristian Tvedten, Michaela Ullmann, Thilo von Debschitz, Uta von Debschitz, Jessica Wang, Bernadette Wegenstein, and Claude Zachary.

Over the years, I have presented aspects of the work of this book to the National Library of Medicine, the New York Academy of Medicine, the School of Visual Arts (New York), the Corporealities and Medicine and the Media workshops, University of Warwick (U.K.), the Smithsonian Institution, McGill University, the University of Uppsala, Johns Hopkins University, the University of Tokyo, the University of British Columbia, the Morbid Anatomy Museum and the Observatory (Brooklyn), and at the annual meetings of the American Association for the History of Medicine, the American Society for Bioethics and the Humanities, the College Art Association, and the Society for the Social History of Medicine. Small portions of *Body Modern* have appeared, in modified form, in the National Library of Medicine's *Circulating Now* blog.

This book exists in this form only because the National Library of Medicine employed me in the unique position of staff scholar in residence and gave me the intellectual freedom to work on projects of my own choosing, in my own way. I feel lucky to have had that job. Most of the materials and images presented here can be found in the library's amazingly rich collections. Research for this book was supported in part by the Intramural Research Program of the National Library of Medicine, the National Institutes of Health (NIH), and by a grant from the Sterling and Francine Clark Art Institute. Opinions expressed in this book are my own and do not reflect the view of the NIH, the Department of Health and Human Services, or the United States government.

FRITZ KAHN

A CHRONOLOGY

1888 Born 29 September, Halle an der Saale, Germany

1889 Father, Arthur Kahn, a physician, emigrates to Hoboken, New Jersey, leaving family behind

1893 Arthur Kahn brings wife, Hedwig, and three children to United States; family moves to New York City

1895 Hedwig Kahn and children move back to Germany (Hamburg, Halle, Bonn)

1900 Arthur Kahn returns to Germany, moves family to Berlin

1912 Fritz Kahn graduates medical school, University of Berlin. Dissertation: *The Care of Pregnant Women in Popular Belief and Literature*

1914–20 Serves in World War I German army medical corps. Begins contributing to *Kosmos*, *Berliner Illustrierte Zeitung*, and other periodicals. Pamphlets include *The Milky Way* (1914) and *The Cell* (1919)

1918–20 Recuperates from war in Italy, Algeria, London

1920 Returns to Berlin, remarries, resumes medical practice. Writes *Der Juden als Rasse und Kulturvolk* (The Jews as a race and culture).

1921	Travels to Palestine, buys property. Gynecological/ obstetrical practice in Berlin.
1922–31	*Das Leben des Menschen,* 5 vols., 1,200 pictures; sells about 70,000 sets
1926	Consultant to GeSoLei exhibition (Dusseldorf) and exhibition on nutrition (Berlin)
1930	Goes on geological expeditions to Palestine and the Arctic
1932	Travels to Sahara
1933	Nazis come to power, burn Kahn's books, force him to close his medical practice. Kahn emigrates with family to Palestine (Haifa, then Jerusalem)
1937	Moves to Paris. *Unser Geschlechtsleben* (Our sex life), which is translated into French, English, Dutch, other languages
1938	Kristallnacht. Kahn's books placed on list of "harmful and undesirable works"; *Unser Geschlechtsleben* banned
1939	*Der Mensch Gesund und Krank*
1940	Moves to Bordeaux. Interned by the French government as enemy alien but gains release
1940–41	France falls. Aided by Varian Fry and Albert Einstein, escapes to United States via Spain and Portugal, settles in New York
1942	*First Aid: Popular*
1943	*Man in Structure and Function*
1950s	*The Book of Nature*; *Design of the Universe.* Returns to Europe (Switzerland, Denmark)
1968	Dies in Lugano, Switzerland

NOTES

INTRODUCTION

1. In current parlance, "infographics" and "data visualization" mostly refer to images generated with computer software to create new kinds of diagrams that may reveal patterns not previously evident, with the suggestion that this is leading to a new epistemic regime based on pattern recognition rather than linear reading or narrative. See Lev Manovich, "What Is Visualization?," *Poetess Archive Journal* 2 (2010), http://manovich.net/index.php/projects/what-is-visualization; Edward Tufte, *Visual Explanations: Images and Quantities, Evidence and Narrative* (Cheshire, Conn.: Graphics Press, 1997); and other works by Tufte and Manovich. See also John Grady, "Edward Tufte and the Promise of a Visual Social Science," in *Visual Cultures of Science: Rethinking Representational Practices in Knowledge Building and Science Communication,* ed. Luc Pauwels (Lebanon, N.H.: Dartmouth College Press, 2006), 222–64.
2. The abundant scholarship on scientific illustration mainly focuses on the role of images and diagrams in the production, rather than circulation, of knowledge and largely neglects twentieth-century conceptual illustration. See, e.g., Brian Baigre, ed., *Picturing Knowledge: Historical and Philosophical Problems Concerning the Use of Art in Science* (Toronto: University of Toronto Press, 1996); John Bender and Michael Marrinan, *The Culture of Diagram* (Stanford, Calif.: Stanford University Press, 2010); Robert Brain, "Representation on the Line: The Graphic Method and the Instruments of Scientific Modernism," in *From Energy to Information: Representation in Science, Art, and Literature,* ed. Bruce Clark and Linda D. Henderson (Stanford, Calif.: Stanford University Press, 2002), 155–78; Klaus Hentschel, *Visual Cultures in Science and Technology* (Oxford: Oxford University Press, 2014); Lorraine Daston and Peter Galison, *Objectivity* (New York: Zone Books, 2007); Martin Kemp, *Visualizations* (Oxford: Oxford University Press, 2000); Caroline A. Jones and Peter Galison, eds., *Picturing Science Producing Art* (New York: Routledge, 1998);

S. H. Poggenpohl and D. Winkler, "The Frame of Reference: Diagrams as Tools for Worldmaking," *Visible Language* 26 (1992): 252–71; Barbara Maria Stafford, *Body Criticism: Imaging the Unseen in Enlightenment Art and Medicine* (Cambridge, Mass.: MIT Press, 1991); and Pauwels, *Visual Cultures.*

3. *Das Leben des Menschen; eine volkstümliche Anatomie, Biologie, Physiologie, und Entwicklungsgeschichte des Menschen* [The life of man: A popular anatomy, biology, physiology, and evolutionary history of man] (Stuttgart: Franckh'sche Verlag, 1922–31).
4. Cornelius Borck, "Communicating the Modern Body: Fritz Kahn's Popular Images of Human Physiology as an Industrialized World," *Canadian Journal of Communication* 32 (2007): 495–520. Kahn's publisher used the "over 1,200 illustrations" claim in its promotional campaign for *Das Leben des Menschen.*
5. Kahn's artists were active collaborators, not just instruments. Uta von Debschitz (personal correspondence, 2 March 2012), following Cornelius Borck, characterizes the relationship as a "blurred . . . anonymous collective authorship." Kahn's artists and the nature of their collaboration with Kahn, as well as the tropes and genres they invented, are more fully discussed in chapters 3 and 4.
6. Apart from *Das Leben des Menschen,* Kahn's most notable books are *Unser Geschlechtsleben* [Our sex life] (Zurich: Albert Müller Verlag, 1937); *Der Mensch Gesund und Krank* [Man in health and sickness], 2 vols. (Zurich: Albert Müller Verlag, 1939); and *Das Buch der Natur* [The book of nature] (Zurich: Albert Müller Verlag, 1952).
7. Borck, "Communicating the Modern Body," and Miriam Eilers, "Urbane Kultur und Natur: Das Berlin der 1920er Jahre im populärmedizinischen Werk Fritz Kahns," in *Visiten: Berliner Impulse zur Entwicklung der modernen Medizin,* ed. J. Bleker, M. Hulverscheidt, and P. Lennig (Berlin: Kulturverlag Kadmos, 2011), 197–212, are thus far the best articles on Kahn. See also Borck, "Electricity as the Medium of Psychic Life: Psychotechnics, the Radio and the Encephalogram in Weimar Germany" (Berlin: Max-Planck-Institute for the History of Science, 2000); "Living Ambiguity: Speculative Bodies in Weimar Culture," in *Weimar Culture and Quantum Mechanics,* ed. C. Carson, A. Kojevnikov, and H. Trischler (London: Imperial College Press, 2011), 453–74; Michael Hau, *The Cult of Health and Beauty in Germany: A Social History, 1890–1930* (Chicago: University of Chicago Press, 2003), 139–44; Douwe Draaisma, *Metaphors of Memory: A History of Ideas about the Mind* (Cambridge: Cambridge University Press, 2001), 214–15; Uta von Debschitz and Thilo von Debschitz, eds., *Fritz Kahn: Man Machine Maschine Mensch,* with C. Borck and M. Eilers (Berlin: Springer Verlag, 2009); Uta von Debschitz and Thilo von Debschitz, *Fritz Kahn* (Los Angeles: Taschen, 2013); Miriam Eilers, "Lebensbild(er) von Fritz Kahn" (Ph.D. diss., Ruhr-Universität Bochum, Institute for Medical Ethics and History of Medicine, 2014).

8. Fritz Kahn, "Der Mensch als Industriepalast," *Das Leben des Menschen* 3 (Stuttgart: Franckh'sche Verlagshandlung, 1926); insert poster in back inside cover pocket, offset chromolithograph, 48 × 94 cm; also sold separately in several editions. In this chapter, "1st ed." denotes the poster from the initial press run, which bears the credit "Beilage zu [supplement to] Kahn"; "2nd ed." denotes the poster with the credit, "Aus [from] Kahn" with some revisions. After that were other iterations, some of which also bear the credit "Aus Kahn." After its initial printing, the poster was relocated to the back pocket of volume 4 or 5.
9. Gilles Deleuze and Félix Guattari, *Anti-Oedipus* (New York: Penguin Classics, 2009). The animation is Henning M. Lederer's *Der Mensch als Industriepalast* (2010), http://www.industriepalast.com.
10. *Modernism: Designing a New World: 1914–1939*, Victoria and Albert Museum, 6 April to 23 July 2006; also at MARTa Herford Museum, Cologne, 16 September 2006 to 7 January 2007; and Corcoran Gallery of Art, Washington, D.C., 17 March to 29 July 2007. The poster can also be seen in the exhibition catalog: Christopher Wilk, ed., *Modernism, 1914–1939: Designing a New World* (London: V and A Publications, 2006).
11. Poul Henningsen, "Tradition and Modernism" (1927), in *Nordic Architects Write,* ed. Michael Asgaard Andersen (New York: Routledge, 2008), 19: "Modern artists in all countries struggle to ensure that the objects we surround ourselves with, as well as the buildings we live in, should look completely different than before the revolution brought on by social problems and new technology. The modern attitude seems to be that most new problems quite naturally demand new forms. . . . These new products will nurture a new aesthetic culture."
12. My account of the self aligns with Daniel Dennett, "The Self as a Center of Narrative Gravity," in *Self and Consciousness: Multiple Perspectives,* ed. F. Kessel, P. Cole, and D. Johnson (Hillsdale, N.J.: Erlbaum, 1992). Dennett argues that the self is fictive, but a very real, socially made fiction with profound effects and consequences. For a nuanced discussion, see Annemarie Mol, *The Body Multiple: Ontology in Medical Practice* (Durham, N.C.: Duke University Press, 2002), 33–50 (bottom), 129–49 (bottom).
13. Calvin N. Kendall and Eleanor Stackhouse Atkinson, eds., *Pictured Knowledge* (Chicago: Compton-Johnson, 1916, and many editions thereafter).
14. See W. J. T. Mitchell, *Iconology: Image, Text, Ideology* (Chicago: University of Chicago Press, 1986), and *Picture Theory: Essays on Verbal and Visual Representation* (Chicago: University of Chicago Press, 1994). More precisely, Kahn's publications situate images of bodies within and around text, and textual material within and around images of bodies. For a smart take on recent scholarship on image/text relations, see Jan Baetens, "Image and Visual Culture after the Pictorial Turn: An Outsider's Note," *Visual Studies* 28 (2013): 180–85.
15. Fritz Kahn, *Man in Structure and Function* (New York: Knopf, 1943), trans. of *Der Mensch Gesund und Krank* (Zurich: Albert Müller, 1939)

by George Rosen, 1:5. Rosen's translation/adaptation changes the year from 1940 to 1943 (the date of American publication). In this quotation and below, I have changed it back to 1940.

16. *Man in Structure and Function* (1943), 1:vi, from *Der Mensch Gesund und Krank* (1939), 1:5.

17. Fritz Kahn, "Das Versehen der Schwangeren in Volksglaube und Dichtung" [Mistaken ideas about pregnancy in popular belief and literature] (medical diss., University of Berlin, 1912).

18. Horace B. Samuel, *Modernities* (New York: E. P. Dutton, 1914), vii. Samuel translated Nietzsche, Schnitzler, and Strindberg into English.

19. Bruno Latour, *We Have Never Been Modern,* trans. Catherine Porter (Cambridge, Mass.: Harvard University Press, 1993), 10, 72; see also Richard Bauman and Charles L. Briggs, *Voices of Modernity: Language Ideologies and the Politics of Inequality* (Cambridge: Cambridge University Press, 2003). Latour's overcomplicated argument is usefully synopsized and reinterpreted in Jane Bennett, *The Enchantment of Modern Life: Attachments, Crossings, and Ethics* (Princeton, N.J.: Princeton University Press, 2001), 95–98.

20. Lisa Gitelman, *Paper Knowledge: Toward a Media History of Documents* (Durham, N.C.: Duke University Press, 2014), 9.

21. I am bracketing here all of the usual questions about when modernity begins, what its defining principles are, its practices and technologies, and so on, and focusing solely on the modern as an identity. If so, then we can begin to locate the origins of the identity (but not modernity as a historicist period) in the sixteenth and seventeenth centuries, when the *querelle* of moderns and ancients became the generic subject of debate in print and in salons. For a crisp introduction to the history of the *querelle,* see Peter N. Miller, "A Passionate Master," review of Marc Fumaroli, *La République de lettres,* in *New York Review of Books,* 9 July 2015, 61–63.

22. The modern generates its own antimodernism, and perversely, antimodernists contribute to modernization and perform some version of the modern (which they can never acknowledge).

23. See Christina Cogdell, *Eugenic Design: Streamlining in America in the 1930s* (Philadelphia: University of Pennsylvania Press, 2010); and Oliver Botar and Isabel Wünsche, eds., *Biocentrism and Modernism* (Surrey: Ashgate, 2011).

24. Samuel, *Modernities* (1914).

25. Charles Knowlton used the term in *Elements of Modern Materialism* (1829) to mark the difference between his radical materialism (informed by contemporary anatomy and physiology) and the materialism of La Mettrie and other eighteenth-century philosophers. The use of *moderne* to refer to the condition of contemporary urban industrial life is especially associated with the circle of Charles Baudelaire. Baudelaire is said to have coined *modernité* in the 1860s, but Roberto Galasso, *La Folie Baudelaire* (New York: Farrar Straus and Giroux, 2012), trans. Alastair McEwen, shows that Théophile Gautier used it

in correspondence in the 1850s. Kahn's modernity was not the same as the modernity of Baudelaire and his circle—no willful ambivalence, no defiant decadence, no defiant challenge to classicism—no defiance!—but the two modernities have some overlap. See also Luc Sante, "In Baudelaire's Dream Brothel," review of Galasso, *La Folie Baudelaire,* in *New York Review of Books,* 21 March 2013, 21–23.

26. The phrase comes from Walter Gropius, "Idee und Aufbau des Staatlichen Bauhauses Weimar," in *Bauhaus, 1919–1928,* ed. Herbert Bayer, Walter Gropius, and Ilse Gropius (New York: Museum of Modern Art, 1938), 27. My starting point for this discussion is Marshall Berman's classic *All That Is Solid Melts into Air: The Experience of Modernity* (New York: Penguin, 1988).

27. Uno Åhrén, "Turning Points," review of Exposition Arts Décoratifs, Paris, 1925, in Andersen, *Nordic Architects Write,* 313.

28. For early twentieth-century European visions and experiences of modernity, see Stephen Kern, *The Culture of Time and Space, 1880–1918* (Cambridge, Mass.: Harvard University Press, 1983); Hans Ulrich Gumbrecht, *In 1926: Living on the Edge of Time* (Cambridge, Mass.: Harvard University Press, 1998); Janet Ward, *Weimar Surfaces: Urban Visual Culture in 1920s Germany* (Berkeley: University of California Press, 2001). Marinetti is quoted in Gumbrecht, *In 1926,* 359.

29. Joseph Roth, "Affirmation of the Triangular Railway Junction," *Frankfurter Zeitung,* 16 July 1924, in *What I Saw,* trans. Michael Hofmann (1996; repr., New York: Norton, 2003), 105.

30. See Robert Michael Brain, "The Pulse of Modernism: Experimental Physiology and Aesthetic Avant-Gardes Circa 1900," *Studies in the History and Philosophy of Science* 39 (2008): 393–417. Brain disputes "one of modernism's founding myths": the assumption that "modernism is either a deterministic effect of modernity or a reaction to or negation of it" (394). He argues that "the relation of modernism to modernity, or aesthetics to technology," is not external nor an adjunct to or a comment on modernity and technology, but instead "immanent," part of a decades-long "combined effort" of artists, philosophers, physiologists, and other scientists, "quite deliberate, to remake the human body, especially the human sensorium" (394).

31. Siegfried Kracauer, *The Salaried Masses: Duty and Distraction in Weimar Germany* (1929), trans. Quintin Hoare (London: Verso, 1998), 92. For theories of "mass psychology" and the "mass/individual" distinction, as well as their deployment in Weimar marketing discourse and practice, see Holm Friebe, "Branding Germany: Hans Domizlaff's *Markentechnik* and Its Ideological Influence," in *Selling Modernity: Advertising in 20th-Century Germany,* ed. Pamela Swett, Jonathan R. Zatlin, and S. Jonathan Wiesen (Durham, N.C.: Duke University Press, 2007), 78–101.

32. Kracauer, *Salaried Masses,* 94–95.

33. Joseph Roth, *Neue Berliner Zeitung* ("12-Uhr-Blatt"), 24 February 1923, in *What I Saw,* 88.

34. Detlev J. K. Peukert, *The Weimar Republic: The Crisis of Classical Modernity,* trans. Richard Deveson (New York: Hill and Wang, 1992), argues that the triumph of modernizing agendas in Weimar Germany led to a "crisis of classical modernity." Recent scholarship has complicated that judgment by taking apart the concepts of both crisis and modernity; see J. B. Shank, "Crisis: A Useful Category of Post–Social Scientific Historical Analysis?," *American Historical Review* 113 (2008): 1090–99; Peter Fritzsche, "Landscape of Danger, Landscape of Design: Crisis and Modernism in Weimar Germany," in *Dancing on the Volcano: Essays on the Culture of the Weimar Republic,* ed. T. W. Kniesche and S. Brockmann (Columbia: University of South Carolina Press, 1994), 29–46; and K. Canning, K. Barndt, and K. McGuire, eds., *Weimar Publics/Weimar Subjects: Rethinking the Political Culture of Germany in the 1920s* (New York: Berghahn, 2010). My interpretation is mainly influenced by Ward's *Weimar Surfaces.* According to Victoria de Grazia, "The Arts of Purchase: How American Publicity Subverted the European Poster, 1920–1940," in *Remaking History,* ed. Barbara Kruger and Phil Mariani (Seattle: Dia Foundation, Bay Press, 1989), "there was a kind of unanimity in favor of economic modernism" (225), but disagreement about whether Europeans should adopt the American road to modernity. Carsten Timmermann, "Constitutional Medicine, Neoromanticism and the Politics of Antimechanism in Interwar Germany," *Bulletin of the History of Medicine* 75 (2001): 717–39, and "Rationalizing 'Folk Medicine' in Interwar Germany," *Social History of Medicine* 14 (2001): 459–82, argues that Germans were torn between revulsion for what science and technology had wrought in the Great War and attraction for the emergent mass culture of the interwar period. Nick Hopwood, "Producing a Socialist Popular Science in the Weimar Republic," *History Workshop Journal* 41 (1996): 117–20, argues that the industrial killing, maiming, and destruction of the war engendered widespread opposition to science and technology. In his account, the Marxist left was the principal dissenter from antiscientism and an advocate for utopian scientism. See also Alfred Kelly, *The Popularization of Darwinism in Germany, 1860–1914* (Raleigh: University of North Carolina Press, 1981); Matthew Biro, *The Dada Cyborg: Visions of the New Human in Weimar Berlin* (Minneapolis: University of Minnesota Press, 2009), 84; and Anne Harrington, *Reenchanted Science: Holism in German Culture from Wilhelm II to Hitler* (Princeton, N.J.: Princeton University Press, 1996).

35. For the politics of "masses" and "mass society" as descriptive terms, sociological categories, and identity markers, see Stefan Jonsson, "Neither Masses nor Individuals: Representations of the Collective in Interwar German Culture," in Canning, Barndt, and McGuire, *Weimar Publics/Weimar Subjects,* 279–301.

36. For modernism as a programmatic attempt to reject historical influences and national differences in favor of a new universalist culture, see Peter Galison, "Aufbau/Bauhaus: Logical Positivism and Architectural Modernism," *Critical Inquiry* 16 (1990): 709–52.

1. READING KAHN AND THE HOMUNCULUS

1. Walter Gropius, "Aims of the Program of the Staatliche Bauhaus in Weimar" (1919): "The Bauhaus strives to bring together all creative effort into one whole, to reunify all the disciplines of practical art. . . . The ultimate . . . aim of the Bauhaus is the unified work of art—the great structure—in which there is no distinction between monumental and decorative art." Such goals fit in perfectly with the historicist assumptions of modernizers like Finnish architect Eliel Saarinen, who proclaimed, "There is something fundamental in . . . the power of a cultural epoch, which directs the whole life." Address to the annual meeting of the American Institute of Architects (1931), in Andersen, *Nordic Architects Write,* 117.
2. Kahn, *Man in Structure and Function,* 1:vi.
3. See, e.g., Fritz Kahn, *Die Juden als Rasse und Kulturvolk* [The Jews as race and culture] (Berlin: Welt-Verlag, 1920). Kahn also wrote (and roughly illustrated) an unpublished natural history of Palestine; Arthur and Fritz Kahn Collection, Leo Baeck Institute, New York.
4. Fritz Kahn Scrapbook, "Wissenschaftliche Studien—Gemeinschaft Mittweida—Fritz Kahn Abend," 50, Arthur and Fritz Kahn Collection, Leo Baeck Institute, New York. Thanks to Miriam Eilers, who came across this document in the course of her research and generously shared it with me.
5. A full list of authors of epigrams: Friedrich Rückert, Goethe, Emerson, G. C. Vogt, Darwin, Schopenhauer, Angelus Silesius, the Bible (Genesis 2), C. L. Schleich, Pascal, Pierre Bayle, C. M. Hufeland, Buddha, Leonardo, Plato, Nietzsche. (Some of these make more than one appearance.) In *Der Mensch Gesund und Krank* (1939), the quotations and cultural references are stripped away and the ratio of text to image is greatly reduced. This is evidence perhaps that Kahn (and his editors) were modernizing the modernizer, working to intensify his rhetorical commitment to science and technology to make his books more appealing to general readers who might be put off by ostentatious displays of *mitteleuropäische Kultur.*
6. Nick Hopwood, "Producing a Socialist Popular Science," 122–23. The Gesellschaft der Naturfreunde was one of many book clubs in the period between 1890 and 1930 that aimed to appeal to the self-improving aspirations of some segments of the German working and middle classes; Michael Naumann, "Germany, by the Book," *Nation,* 18 June 2012, 23–24.
7. Quoted in Hofmann, *Die Lektüre der Frau* (Leipzig: Quelle and Meyer, 1931), 29–30, in Kerstin Barndt, "Mothers, Citizens, and Consumers: Female Readers in Weimar Germany," in Canning, Barndt, and McGuire, *Weimar Publics/Weimar Subjects,* 102.
8. Edgar Eichbaum, *Born Jewish: The Force of Destiny* (e-book; Xlibris, 2008), 39–40; phone conversation with Dr. Eichbaum, 10 October 2013. Eichbaum lived in a youth village for refugee children.
9. Anna Fischer-Dückelmann (1856–1917), *Die Frau als Hausärztin: Ein ärztliches Nachschlagebuch der Gesundsheitspflege . . . in der Familie*

mit besonderer Berücksichtigung der Frauen- und Kinderkrankheiten (Stuttgart: Süddeutsches Verlags Institut, 1901, and subsequent editions; continued after 1929 in new editions revised by E. A. Müller).

10. Fritz Kahn, *Die Zelle,* illus. Georg Helbig (Stuttgart: Kosmos, Gesellschaft der Naturfreunde, Franckh'sche Verlagshandlung, 1919). Kahn's first publication was *Die Milchstrasse* [The Milky Way], illus. Rudolf Oeffinger (Stuttgart: Kosmos, Gesellschaft der Naturfreunde, Franckh'sche Verlagshandlung, 1914). Kahn was in the German Medical Corps during the war. In the 1910s, his topics were astronomy, physics, natural history, and medicine. All of these publications featured original illustrations as well as images taken from other sources. The cover photograph, originally made in 1897, was taken from Martin Heidenhahn, *Plasma und Zelle* (Jena: G. Fischer, 1907–11), 1:103. Some of the material in *Die Zelle* was later reused in *Das Leben des Menschen* and later works.

11. In the late 1920s and early 1930s, *Berliner Illustrierte Zeitung* was distributed nationally and had a weekly circulation of one to two million copies—a big number in a country with only sixty-five million people. (The publication also had readers in other German-speaking areas of Europe.) *Uhu* had a monthly circulation of more than 100,000, and sometimes more than 200,000. See Julia Silvia Feldhaus, "Between Commodification and Emancipation: Image formation of the New Woman through the Illustrated Magazine of the Weimar Republic" (Ph.D. diss., Rutgers University, 2010), 10n20; Maud Lavin, *Cut with the Kitchen Knife: The Weimar Photomontages of Hannah Höch* (New Haven, Conn.: Yale University Press, 1993), 56; W. J. Freyburg, and Eva Noack-Mosse, *Uhu*, in *Hundert Jahre Ullstein, 1877–1977,* ed. W. J. Freyburg and Hans Wallenberg (Berlin: Ullstein, 1977), 256 and 186. According to Lavin, a significant portion of *Berliner Illustrierte Zeitung*'s income came from advertisements aimed at women (Feldhaus, "Between Commodification and Emancipation," 10n20).

12. Thomas Nipperdey, *Deutsche Geschichte, 1866–1918* (Munich: C. H. Beck, 1995), 1:797, trans. and quoted in Arnulf Kutsch and Andy Wagner, "Expenditures for Media Consumption in Germany: An Explorative Study for the Years between 1870 and 1930 Testing the Principle of Relative Constancy," *Media History* 20 (2014): 182.

13. GeSoLei is an acronym for *Gesundheitspflege* (hygiene), *Soziale Fursorge* (social welfare), and *Leibesubungen* (physical fitness). See Egon E. Albrecht, *Das Deutsche Hygiene-Museum und sein internationaler Gesundheitsdienst* (Dresden: Deutsches Hygiene-Museum, 1931); Klaus Vogel, "The Transparent Man: Comments on the History of a Symbol," in *Manifesting Medicine,* ed. Robert Bud (London: Taylor and Francis, 1999), 31–61, especially 36–47. Kahn served in a minor role as a consultant for GeSoLei exhibits in Dusseldorf and Berlin.

14. Walter Benjamin, "Children's Literature," broadcast on Southwest German Radio, Frankfurt, 15 August 1929, in Lecia Rosenthal, ed., *Radio Benjamin,* trans. Lisa Haries Schumann, (London: Verso Books, 2014), 253.

15. Otto Neurath, "Statistische Hieroglyphen," *Österreichische Gemeinde-Zeitung* 3 (1926): 328.

16. Loose translation by Claudia Stein and Michael Sappol; Alfred Döblin, "Metapsychologie und Biologie," review of *Das Leben des Menschen,* vol. 1, *Neue Rundeschau* 12 (1923): 1231: "Das Buch muss aus allen vorliegenden ähnlichen volkstümlichen Werken hervorgehoben werden, wegen der Leichtigkeit und Originalität des Vortrags, besonders wegen der Originalität und Schlagkraft seiner bildlichen Darstellung. Man kann nicht gut fasslicher und eindringlicher darstellen, aus welchen Teilen ein Mensch besteht, als auf dem Bilde: 'Es wird ein Mensch gemacht,' geschieht: zwei Hände werfen Substanzen auf einen Tiegel und giessen Flüssigkeiten hinzu; man sieht Kohle in Form von unten in der Retorte bildet sich der Homunculus. Der Wassergehalt der menschlichen Körpers wird durch Schraffierung am stehenden Körper gezeigt; das Wasser reicht vom Kopf bis fast an die Knie. Der Autor gibt ungewöhnlich plastische Vergrösserungen und systematische bildliche Darstellungen. Die Bilder sind zum Teil abenteürlich; so der Kampf im Zellenstaat zwischen eindringlichen Bakterien und den Wanderzellen; es sieht aus wie eine urweltliche Schlacht. Kahns Buch will ich sehr empfehlen. Es ist keine Populariesung biologischen Wissens, sondern eine ganz originelle Durchfühlung und Darstellung dieses Wissens. Die Belebung dieses Wissens, das Heranbringen des Menschen an die Natur, die Fortnahme der Natur den Wissenchaftlern, scheint mir zu den wünschenswertesten Dingen von heute zu gehören." Claudia Stein (e-mail to author, 10 January 2012) comments, "*Durchfühlung* is a difficult word to translate. It means: obtaining something through 'feeling' (but not the way advertisement works). The term, which goes back to German idealism, was an important concept. It belongs to the epistemological toolkit of the *Geisteswissenschaften* of the time (which aimed to find *Verstehen* rather than *Erklären* in the *Naturwissenschaften*). So the last sentence is key, because Döblin argues that Kahn is not merely giving us biological knowledge but bridging the gap (perceived as widening at the time) between two models of understanding human nature, the natural [physical] sciences way and the human [social] sciences way. Kahn's visual representations, the author suggests, bridge this gap. They allow us to 'feel' ourselves into the wider understanding (*Verstehen*) of what it means to be human (the domain of the *Geisteswissenschaften*)."

17. Fritz Kahn, "Man as Industrial Palace: Supplement," *Das Leben des Menschen* (1926), vol. 3. In later editions, the supplement, included in a pocket on the inside back cover, was moved to volume 4 (1929) or 5 (1931).

18. The term "homunculus" continues to be used in debates over the nature of human consciousness. Theories that assign agency to functional structures or rules within the brain are said to commit the homuncular fallacy. To further complicate matters, in cognitive psychology and neuroscience, the cortical homunculus refers to a bizarrely disproportionate figure that serves as a map of sensory experience.

19. "Der Kreislauf des Sauerstoffs: Was geht in dir vor, wahrend du dieses Bild beschaust?" (The oxygen cycle), *Das Leben des Menschen* (Stuttgart, 1926), vol. 3, pl. 11. For a dialogical interpretation of the image, see Michael Sappol and Eva Åhrén, "The Strange Spaces of the Body: Two Dialogues," in *Strange Spaces: Explorations in Mediated Obscurity,* ed. Amanda Lagerkvist and André Jansson (Aldershot, U.K.: Ashgate, 2009), 73–100.
20. Roth, "Affirmation of the Triangular Railway Junction," 106–7.
21. For a discussion of the homunculus and how it is always figured metaphorically with reference to a leading technology, see Draaisma, *Metaphors of Memory,* chapter 9. For a succinct explanation of the homuncular fallacy, see Colin McGinn, "Storm over the Brain," *New York Review of Books*, April 24, 2014, 62.
22. For a brief but useful discussion of the concept of identification in a Weimar context, see Biro, *Dada Cyborg,* 13–14.
23. Heinrich Eduard Jacob, *Blut und Zelluloid* (Blood and Celluloid) (1930; repr., Bad Homburg: Oberon Verlag, 1986), 23–24, trans. and quoted in Ward, *Weimar Surfaces,* 172. The phrase is uttered by the character Benno, a Berlin film producer.
24. Quoted in Debschitz and Debschitz, *Fritz Kahn,* 78.
25. See "Hans Surén der bahnbrechende Führer in Sport- und Gymnastik-Erziehung schreibt über Kahns 'Das Leben des Menschen'" (Hans Surén, the pioneering leader in sports and gymnastics education, writes about Kahn's *Das Leben des Menschen*), promotional pamphlet, ca. 1930, in Series II/1/C/Scrapbook I.22, Arthur and Fritz Kahn Collection, Leo Baeck Institute, New York; Döblin, "Metapsychologie und Biologie."
26. Ward, *Weimar Surfaces*; Monika Wagner, "Berlin Urban Spaces as Social Surfaces: Machine Aesthetics and Social Texture," *Representations* 102 (2008): 53–75. My argument here is greatly indebted to Ward.
27. At the same time, German nationalists selectively identified with the modernity of German military organization, tactics, and technology. For many Germans across the political spectrum, the modernity of industrial science and technology, even if it had universalistic aspirations and implications, was in some measure a German accomplishment as well as a resource for the construction of German identity. Modernity could be seen as the fruition of the *longue durée* of advances from antiquity to the present and future world civilization, with special attention paid to the contributions of German *Kultur, Wissenschaft,* and *Technik*—Gutenberg, Bach, Kant, Humboldt, Goethe, Beethoven, Koch, Roentgen, Diesel, etc. (German Jews could construct a parallel Jewish identity in modernity through the *longue durée* of Jewish *Kultur, Wissenschaft,* and *Technik*: the Bible, Spinoza, Heine, Mendelsohn, Freud, Einstein, etc., which converged with the German national narrative of world civilization.)
28. See Ward, *Weimar Surfaces*; Wagner, "Berlin Urban Spaces," especially 55–62.

29. The penis and rectum are outside the frame. For the sake of modesty, these parts were often suppressed in images designed for public viewing, but as convention often dictated, *Der Mensch*—Man—stands for universal humanity, not just the male body (obviously a sexist, parochial symbol of the universal human). In such situations, it was customary to omit genitals.
30. Fritz Kahn, *Der Mensch als Industriepalast: Versuch einer technischen Darstellung der wichtigsten Lebensprozesse* [A technical description of the main processes of life] (supplement to poster) (Stuttgart: Franckh'sche Verlagshandlung, 1926). Diagram matches the first edition poster ("Beilage zu Kahn").
31. Kracauer, *Salaried Masses*, 40–41.
32. Schlesinger headed the Institute for Industrial Psychotechnics at the Technical University, Berlin-Charlottenburg (the same quarter of Berlin where Fritz Kahn lived). Influenced by Hugo Münsterberg, F. W. Taylor, and Henry Ford, he described and analyzed industrial workers as "working machines"; quoted in Andreas Killen, "Weimar Psychotechnics," *Osiris* 22 (2007): 57.
33. Alfred Döblin, "Der Geist des naturalistischen Zeitalters," *Die neue Rundschau* 35, no. 2 (1924), repr. in Döblin, *Schriften zu Ästhetik, Poetik und Literatur*, ed. Erich Kleinschmidt (Olten: Walter-Verlag, 1989), 189, trans. and quoted in Ward, *Weimar Surfaces*, 228.
34. For Weber and disenchantment, see Harrington, *Reenchanted Science*, especially the introduction and chapter 1.
35. Roth, "Skyscrapers," *Berliner Börsen-Courier*, 12 March 1922, in *What I Saw*, 112–13. Roth finishes the piece by connecting that technological accomplishment of the age to jazz with this happy thought: "Oh—and already you hear that the first skyscraper in Berlin is to contain a great entertainment palace, with cinemas, dance hall, bar, Negro bands, vaudeville, jazz" (113).
36. According to Killen, "Weimar Psychotechnics," 57, some German psychotechnicians made a point of distinguishing German psychotechnics from American Taylorism and Fordism.
37. See Corey Ross, "Visions of Prosperity: The Americanization of Advertising in Interwar Germany," in Swett, Zatlin, and Wiesen, *Selling Modernity*, 52–77; Mary Nolan, *Visions of Modernity: American Business and the Modernization of Germany* (New York: Oxford University Press, 1994), especially 3–8.

2. "MUCH BETTER THAN WORDS"

1. *Wunder in Uns: Ein Buch vom menschlichen Körper für Jedermann* [The wonder in us: A book on the human body for everyone] (Zurich: Rascher and Cie, 1921). Quotation is from Hans Hauri, "Mensch als Maschine," *Wunder in Uns*, 2nd ed. (Zurich: Grethlein, 1923), 109.
2. *Wunder in Uns*, 2nd ed., with colorized redrawn illustrations signed "P.Fl." (Paul Flanerky).

3. Dust jackets, then a relatively new innovation, provided a conspicuous space for graphic design to occupy. So too did the covers of journals that before had often been used to list the table of contents. The term "pictorial turn" references the movement in cultural criticism to deprivilege the text as the object of analysis and to theorize image practice beyond the traditional close analysis methods of art history. See Mitchell, *Picture Theory*, chapter 1.
4. The chapter by Hermann Dekker was originally published, with the same title, in *Kosmos* 5, n.s. (1908): 28–31. Günther may have felt that it was out of date and so asked Hauri for something fresher. The three essays by Kahn were "The Mystery of the Thyroid," "On Shuddering," and "Degenerate Organs and Atavisms."
5. The crew is even more skeletal in the first edition of *Wunder in Uns*: four in the head; one in the torso.
6. See Philipp Sarasin, *Reizbare Maschinen: Eine Geschichte des Körpers, 1765–1914* (Frankfurt am Main: Suhrkamp Taschenbuch, 2001), and "The Body as Medium: Nineteenth-Century European Hygiene Discourse," trans. Brian Hanrahan, *Grey Room* 29 (2007): 48–65. Sarasin deals extensively with nineteenth-century hygiene discourse, which was rife with textual metaphors of the body as a technology. He does not deal much with images, and *Reizbare Maschinen* does not have a single illustration.
7. Winfield Scott Hall (and Jeannette Winter Hall?), "The Bodies We Live In: What Strange Land Is This?" *Pictured Knowledge*, 2nd ed. (Chicago: Compton-Johnson, 1917), 246–71 (and many editions thereafter). In addition to the first Compton-Johnson 1916 edition, many editions were published thereafter under the Compton imprint and later the Marshall Hughes Company of Kansas City. Chicago was the preeminent publishing center for illustrated encyclopedias aimed at the American market. No artist is credited; an endnote credits the writing and concept to Hall, who was dean of the medical faculty and professor of physiology at Northwestern University in Chicago, as well as the author of many books for youth on physiology and sexual purity. A short article in *Popular Science Monthly* 92 (1918): 22–23, reprints the pictures and says they are based on "drawings by Prof. and Mrs. Winfield Scott Hall" (referring to Jeannette Winter Hall, who is credited as a contributor in the front matter of *Pictured Knowledge*). She frequently collaborated on books and articles with her husband, but none of their other works, authored jointly or separately, even remotely resembles "The Bodies We Live In." The plates in the first edition of *Wunder in Uns* are closely copied from *Pictured Knowledge*; the plates in the second edition are colorized and redrawn with minor changes.
8. Miriam Eilers, message to author, 7 November 2013, notes that Kahn claimed to have started work on the first two volumes of *Das Leben des Menschen* in 1912, years before their initial publication (1921 and 1923) and well before the appearance of *Wunder in Uns* (1921). The influence of *Wunder in Uns* on Kahn becomes markedly apparent in 1926 with

the third volume of *Das Leben des Menschen* and "Der Mensch als Industriepalast," possibly because he had already prepared material for the first two volumes before the 1921 publication of *Wunder in Uns.*

9. "Bodies We Live In," 246.
10. Ibid., 247–48.
11. Ibid., 252.
12. Ibid., 260, 270.
13. Ibid., 256.
14. Ibid., 271.
15. Ibid.
16. Martin Kemp, "'The Mark of Truth': Looking and Learning in Some Anatomical Illustrations from the Renaissance and Eighteenth Century," in *Medicine and the Five Senses,* ed. W. F. Bynum and Roy Porter (Cambridge: Cambridge University Press, 1993), 94.
17. See Paul Saettler, *The Evolution of American Educational Technology* (Sacramento: Information Age, 2004), chapter 5. See also Henning Smidgen, "Cinematography without Film: Architectures and Technologies of Visual Instruction in Biology around 1900," in *The Educated Eye: Visual Culture and Pedagogy in the Life Sciences,* ed. Nancy A. Anderson and Michael R. Dietrich (Lebanon, N.H.: Dartmouth College Press, 2012), chapter 4; James R. Ryan, "On Visual Instruction," in *The Nineteenth-Century Visual Culture Reader,* ed. Vanessa R. Schwartz and Jeannene M. Przyblyski (London: Routledge, 2004), 146–51; and Daniel Joseph Walther, *Creating Germans Abroad: Cultural Policies and Settler Identities in Namibia* (Athens: Ohio University Press, 2002), 76.
18. Neil Harris, "Iconography and American History: The Halftone Effect," in *New Directions in American Intellectual History,* ed. John Higham and Paul Conkin (Baltimore, Md.: Johns Hopkins University Press, 1979), 196–211; Estelle Jussim, *Visual Communication and the Graphic Arts: Photographic Technologies in the Nineteenth Century* (New York: Bowker, 1974); Michael Winship, "The Rise of a National Book Trade System in the United States," and Megan Benton, "Unruly Servants: Machines, Modernity and the Printed Page," in *A History of the Book in America 4: Print in Motion: The Expansion of Publishing and Reading in the United States, 1880–1940,* ed. Carl F. Kaestle and Janice Radway (Chapel Hill: University of North Carolina Press, 2009). Gerry Beegan, *The Mass Image: A Social History of Photomechanical Reproduction in Victorian London* (New York: Palgrave Macmillan, 2008), gives a nuanced account of this transformation.
19. Harris, "Iconography," 199.
20. See Ellen Lupton and J. Abbott Miller, "A Time Line of American Graphic Design, 1829–1989," in *Graphic Design in America: A Visual Language History,* ed. Mildred Friedman (New York: Harry Abrams, 1989), 24–74; Michael Schudson, *Discovering the News: A Social History of American Newspapers* (1967; repr., New York: Basic Books, 1973); Allen Hutt, *The Changing Newspaper: Typographic Trends in Britain*

and America, 1662–1972 (London: Gordon Fraser, 1973); John E. Allen, *Newspaper Makeup* (New York: Harper Bros., 1936).

21. Frank Luther Mott, *A History of American Magazines, Vol. 4, 1885–1905,* and *Vol. 5, 1905–1930* (Cambridge, Mass.: Harvard University Press, 1957); Michele Bogart, *Artists, Advertising, and the Borders of Art* (Chicago: University of Chicago Press, 1995). For the early twentieth-century development of dust jacket design, see G. Thomas Tanselle, *Book-Jackets: Their History, Forms, and Use* (Charlottesville: Bibliographical Society of the University of Virginia, 2011).

22. The early twentieth century was a period in which marketing and economics took a psychological turn. In economic theory, Jeffrey Sklansky argues that there was a shift from a neoclassical "sovereign self" with preset wants and desires to a "psychological self" that could be formed and influenced by education, advertising, and technological invention to feel new feelings, new desires, and new wants; Jeffrey Sklansky, *The Soul's Economy: Market Society and Selfhood in American Thought, 1820–1920* (Chapel Hill: University of North Carolina Press, 2002). See also Thomas A. Stapleford's discussion of Hazel Kyrk, *A Theory of Consumption* (Boston: Houghton Mifflin, 1923), in "Market Visions: Expenditure Surveys, Market Research, and Economic Planning in the New Deal," *Journal of American History* 92 (2007): 421–22.

23. There were strong links between German and American applied psychology. One of the first influential figures in the field, Walter Dill Scott, author of *The Psychology of Advertising in Theory and Practice* (Boston: Small, Maynard, 1903) and many other works, studied with Wilhelm Wundt, the "father" of experimental psychology, in Leipzig in the 1890s. Hugo Münsterberg, author of *Psychology and Industrial Efficiency* (Boston: Houghton Mifflin, 1913), *Grundzüge der Psychotechnik* (Munich: J. A. Barth, 1914), and many other works, was another of Wundt's students. He was professor of psychology at Harvard, wrote in English and German, and was widely influential in both America and Germany. Scott, Münsterberg, and their successors, described themselves as practitioners of a scientific psychology and used the terms "suggestion," "instincts," "drives," "scientific," and "efficiency" (and variants) throughout their texts. See Matthew Hale Jr., *Human Science and Social Order: Hugo Münsterberg and the Origins of Applied Psychology* (Philadelphia, Pa.: Temple University Press, 1980); Andreas Killen, "Weimar Psychotechnics between Americanism and Fascism," *Osiris* 22 (2007): 48–71; and David P. Kuna, "The Concept of Suggestion in the Early History of Advertising Psychology," *Journal of the History of Behavioral Sciences* 12 (1976): 347–53.

24. See Corey Ross, "Visions of Prosperity: The Americanization of Advertising in Interwar Germany," in Swett, Zatlin, and Wiesen, *Selling Modernity,* 52–77.

25. See Ryan K. Anderson, "Merry's Flock: Making Something out of Educational Reform in the Early Twentieth Century," in *Education and the Culture of Print in Modern America,* ed. Adam R. Nelson

and John L. Rudolph (Madison: University of Wisconsin Press, 2010), 59–80, especially 60; William J. Reese, *America's Public Schools: From the Common School to "No Child Left Behind"* (Baltimore, Md.: Johns Hopkins University Press, 2005); John R. Thelin, *A History of American Higher Education* (Baltimore, Md.: Johns Hopkins University Press, 2004).

26. See Saettler, *Evolution of American Educational Technology*; Lynn K. Nyhart, *Modern Nature: The Rise of the Biological Perspective in Germany* (Chicago: University of Chicago Press, 2009), chapter 5.
27. J. Clayton Youker, ed., *Organized Sunday School Work in America, 1911–1914: Official Report of the 14th International Sunday School Convention* (Chicago: Executive Committee of the International Sunday School Association, 1914), 173.
28. Saettler, *Evolution of American Educational Technology.*
29. *Pictured Knowledge* (1917), xvi.
30. *Pictured Knowledge* (1917). The topical arrangement was designed to make the encyclopedia feel more like a mass-circulation magazine than a reference work in the alphabetic tradition of Diderot and d'Alembert's *Encyclopédie.*
31. For New York State and visual instruction, see Alfred Willis Abrams, "Visual Instruction in New York State," *Annals of the American Academy of Political and Social Science* 67 (1916): 270–72.
32. *Pictured Knowledge* (1917), xii.
33. R.E.B., "Personality: Switch Led to the Right Track," *New York Times,* 5 June 1960, is a human interest story on Bailey Howard, who started his career as a salesman in 1933 selling *Pictured Knowledge* door to door. The documentary film *Salesman* (dir. Albert and David Maysles, Charlotte Zwerin, 1968) shows the foot-in-the-door and sales-pitch work of door-to-door bookselling (in their case, illustrated Bibles and religious encyclopedias), a now-defunct occupational and marketing culture that persisted into the age of television but not the Internet.
34. *Pictured Knowledge* (1917), xiii.
35. Ibid., xiv.
36. G. Stanley Hall, "The Graphic Method," *Nation* 29 (1879): 238–39.
37. *Pictured Knowledge*, xiii, xiv.
38. Saettler, *Evolution of American Educational Technology*; Nyhart, *Modern Nature.*
39. *Pictured Knowledge* (1917), xiv.
40. *Ekphrasis,* a rhetorical term, usually designates the richly descriptive language that helps the reader or listener to imagine a painting or sculpture. It can also be generalized to mean language that creates a mental picture or scene.
41. "Kindergarten," *Pictured Knowledge* (1920), 897.
42. The evolutionary/historical argument is explicitly argued in F. Dean McClusky, C. F. Hoban, Daniel Knowlton, and Elda Merton, *Syllabus of a Proposed Textbook on Visual Instruction for Use in Teachers Training Schools* (n.p., 1930; mimeograph with printed cover), 15:

"The roots of modern visual instruction may be traced back through the history of the race. . . . In primitive times, boys were taught to hunt and fish; and girls, to cook, through imitation, observation and participation plus the necessary language explanations. . . . Early records were picture records; cavemen drew pictures to warn and inform. . . . Egyptian hieroglyphs are the transition from picture writing to early alphabets." A later edition of *Pictured Knowledge* lauds the ancient Greeks for their special contributions to this image-based civilizing process: "They were, more than any other nation in history, what we may call 'pictured knowledge' people." "Wonderful Pictured Knowledge of Long Ago," *Pictured Knowledge* (Kansas City: Marshall Hughes, 1926, 1937), 2083. In other words, the Greeks, in their use of pictures, as in their architecture, philosophy, and literature, gave birth to modernity.

43. Garry Cleveland Myers, foreword to *Pictured Knowledge,* new edition (Kansas City: Marshall Hughes, 1926, 1937), n.p. After Calvin Kendall's death, *Pictured Knowledge* was sold to the Marshall Hughes Company, which issued an expanded edition with additional material. In his foreword, Myers, the new editor in chief, added his own rationale for the encyclopedia and summarizes Kendall's aims and approach.
44. *Pictured Knowledge* (1917), 490, xiv.
45. Wilbur S. Jackman, "The Correlation of Mathematics," *Educational Review* 25 (1903): 249–64: "All learning . . . is simply image growth. . . . The thought-stuff that furnishes the basis for the image is derived from energy manifesting itself through the phenomena of nature of which man is a part. . . . The subjects through which the manifestations of energy are presented are natural science and history; and the subjects through which the mental reactions are directed include those by means of which the individual seeks to define his image. . . . Thought studies are those that have a knowledge content that is . . . independent of the process by which it was gained. Form studies include all the various modes of expression: making, modeling, painting, drawing, written language, associated with which are reading and spelling, oral language, gesture, form, and number. It is through the use of one or more or of all of these means of expression that the image emerges from vagueness towards clearness, and finally becomes a definite product in consciousness. Each mode of expression has a special function in the process of image growth which is unique and which cannot be performed wholly by any other mode" (250).
46. *Pictured Knowledge* (1917), xiv.
47. Ibid., 1863; *Pictured Knowledge* (1937), 1881.
48. *Pictured Knowledge* (1917), 689–707.
49. Ibid., 489. I have been unable to identify the nineteenth- or twentieth-century source that provides the basis of the 80 percent figure, but the physiological superiority of vision, linked to some factor of 80, seems to have been an article of faith for some educators and reformers. A review of Sylvanus Stall's *Five Minute Object Sermons for Children* (1908), for

example, praises the Reverend for "captur[ing] the city of the child-soul by assaulting not only ear-gate, but eye-gate as well. He appreciates the fact that the nerve from the eye to the brain is eighty times as powerful as that from the ear—that the eye-gate opens eighty times as wide as ear-gate, and is, therefore, an avenue of pre-eminent importance in reaching the child mind and child heart." *New York Observer and Chronicle*, 4 June 1908, 741. (Thanks to Jennifer Burek Pierce for providing this quote.) See also *American Home Missionary* 20 (1914): 28. The terms "eye-gate" and "ear-gate" come from John Bunyan, *The Holy War* (1682; repr., London, 1887), 270, probably via W. T. Stead, "The Mission of the Magic Lantern," *Review of Reviews* 2, no. 12 (1890): 561, and "The Mission of the Cinematograph," *The Americanisation of the World* (London: Review of Reviews Annual, 1902), 174, which make similar arguments about the need for a visual approach in religious and moral instruction.

50. *Pictured Knowledge* (1917), 489.
51. Ibid., 490.
52. Ibid., 490–92.
53. Ibid., 492.
54. Ibid., 492–93.
55. Ibid., 494.
56. Ibid.
57. Ibid., 495.
58. Ibid., 496.
59. Ibid., 496–97.
60. This pedagogical critique potentially could be elevated into a political critique: the valuation of word over image is exclusionary and undemocratic. For one thing, it was widely believed that image knowledge did not require a high degree of literacy. Posters and illustrated texts are thus necessary to educate the populace in hygiene, safety, and political ideology. The image (subordinated in the cultural hierarchy) and the masses (subordinated in the social hierarchy) become equivalent. However, on the ground, pedagogical reformers such as advertisers, commercial publishers, and political propagandists primarily valued image work for its instrumental uses in manipulating and engineering human subjectivity, not in setting it free.
61. *Pictured Knowledge* (1917), 509.

3. OCULARCENTRIC!

1. *Pictured Knowledge* (1917), 496.
2. "Scopic regime," a concept introduced in a seminal article by Christian Metz, has been adapted and critiqued by Martin Jay, Jonathan Crary, and other scholars. Crary argues that in the nineteenth century, emergent technologies (e.g., microscopes, X-rays, stereopticons, halftone reproduction, wood engraving, the automobile, lithography, and motion pictures), institutions (e.g., public school, museums, shop

windows, billboards, movie houses), and social practices of viewing combined to create a "scopic regime," a pervasive cognitive structure of seeing, reading, and interpreting. See Jonathan Crary, *Suspensions of Perception: Attention, Spectacle, and Modern Culture* (Cambridge, Mass.: MIT Press, 1993); Martin Jay, *Downcast Eyes: The Denigration of Vision in Twentieth-Century French Thought* (Berkeley: University of California Press, 1993), and "Photo-Unrealism: The Contribution of the Camera to the Crisis of Ocularcentrism," in *Vision and Textuality,* ed. Stephen Melville and Bill Readings (London: Macmillan, 1995), 344–60.

3. This description refers to the legion of illustrators who never gained any celebrity. America also had star illustrators—James Montgomery Flagg and Charles Dana come to mind—who had more latitude.

4. Other kinds of commercial artists had more leeway (which resulted in often brilliant work) in certain genres. Cartoonists were given license to parody, fool around with, or entirely disregard the conventions of perspectival realism and formal speech, and in that way became an independent source for a vernacular modernist aesthetics. Fashion and cosmetics illustrators, addressing women readers, were given the license to create pen-and-ink illustrations with sensual line and elongated distortion that signified au courant feminine elegance, a chic elegant minimalism that drew on Art Nouveau and other amenable art movements. Children's book illustrators could be fanciful in a variety of registers. There were also a few elite magazines that adopted a stylish look that relied on the wit and charm of the artwork.

5. T. Jackson Lears, *Fables of Abundance: A Cultural History of Advertising in America* (New York: Basic Books, 1994), 272, 298. Roland Marchand, *Advertising the American Dream: Making Way for Modernity, 1920–1940* (Berkeley: University of California Press, 1986), calls it "dramatic" (56) or "sentimental" (148) realism.

6. See Jeremy Aynsley, *Graphic Design in Germany, 1890–1945* (Berkeley: University of California Press, 2000); Ross, "Visions of Prosperity," 54–60; de Grazia, "Arts of Purchase," 221–57. The account given here and in succeeding paragraphs is largely derived from de Grazia's and Ross's discussion of the qualities, role, and practice of German poster and advertising artists, in comparison to American artists, in the interwar period. Eventually European and American practice converged as European artists and designers fled the Nazis, or, in the aftermath of World War II, sought new economic opportunities in America.

7. Ross, "Visions of Prosperity," 58, quoting Hanns Kropff, "Gotterdammerung der deutschen Reklamekunst," *Die Reklame* 17 (1924): 674–75 (Kropff's emphasis).

8. Going further, avant-garde modernists took commercial jobs and published in commercial venues, while lesser-light commercial artists took up modernist aesthetics, with varying degrees of commitment. Michael Cowan, *Walter Ruttmann and the Cinema of Multiplicity: Avant-Garde—Advertising—Modernity* (Amsterdam: Amsterdam University Press, 2014), shows that avant-gardists took commissioned

work with advertisers and argues that their gallery art and commercial work should not be cordoned off from each other.

9. Rudi Feld, "Schrauben Sie eine andere Birn ein . . . und Sie warden volle Kassen haben!" *Licht-Bild-Bühne,* 5 August 1930, 2, quoted in Ward, *Weimar Surfaces,* 170–71.

10. H. K. Frenzel, "Sensational News—Sensational Advertising," trans. E. T. Scheffauer, *Gebrauchsgraphik* 8 (1931): 9–16, quotation at 9.

11. Ross, "Visions of Prosperity," 52–77.

12. J. Murray Allison, "Continental Advertising," *Advertising World,* April 1927, 18, quoted in de Grazia, "Arts of Purchase," 235.

13. The publisher's flyer for *Der Mensch Gesund und Krank* (1939), and the sun girl illustration on which it is based, are stylistically identical to the Werbekraft GmbH advertisements of the design team of Plucer and Elfer. Elfer is almost certainly the Hungarian graphic designer Arpad Elfer (1910–99), who was then in Berlin; after fleeing Nazi Germany, Elfer went on to a long and distinguished career in England. The identity of Plucer is uncertain.

14. Hans Schleger, "Interview of the Month," trans. E. T. Scheffauer, *Gebrauchsgraphik* 8 (1931): 74–78.

15. Ross, "Visions of Prosperity," 60n28, quoting *Berliner Illustrierte Zeitung* 41 (1921) from Walter Schmitt, *Das Filmwesen und sein Wechselbeziehungen zur Gesellschaft. Versuch einer Soziologie des Filmwesens* (Freudenstadt: Oskar Kaupert, 1932), 113.

16. De Grazia, "Arts of Purchase," 225.

17. Fritz Kahn and Georg Piek-Patrik, *Rhythmische Musterkunst der Natur: Die Grundzüge eines neuen Stils* (Berlin, 1924), influenced by Ernst Haeckel, *Kunstformen der Natur* (1904).

18. Martin Kemp, "Style and Non-style in Anatomical Illustration: From Renaissance Humanism to Henry Gray," *Journal of Anatomy* 216 (2010): 192–208.

19. Ibid.

20. For the theatricality of early modern anatomical illustration, see Michael Sappol, *Dream Anatomy* (Washington, D.C.: GPO, 2006), "Fun with Science."

21. For the puzzling itinerary of the white-on-black nervous system diagram, see Ron Ladouceur, "I Speak to You through Electrical Language: Traveling into the Nineteenth Century with the 'Nervous Icon,'" *Textbook History,* 3 June 2012, http://www.textbookhistory.com/i-speak-to-you-through-electrical-language-traveling-into-the-nineteenth-century-with-the-nervous-icon-2/.

22. *Der Mensch Gesund und Krank* (1939), 1:1.

23. Hermann Dekker, *Lebensrätsel: Der Mensch biologisch dargestellt,* 2nd ed. (Stuttgart: Moritz, 1913).

24. J.-B. M. Bourgery, *Traité complet de l'anatomie de l'homme . . . ,* 8 vols. (Paris, 1831–54); illus. Nicolas-Henri Jacob. Kahn sometimes credited the sources he borrowed from. In *Das Leben des Menschen,* Kahn mostly credited Bourgery, but he stopped doing so in later works.

Kahn seems to have loved Bourgery's anatomy, perhaps because of its provocative showmanship and artistic mastery.

25. The first edition of *Das Leben des Menschen* was issued in these years: volume 1, 1922; volume 2, 1924; volume 3, 1926; volume 4, 1929; volume 5, 1931. Later editions often bear the same publication information as earlier ones, which makes it hard to know when specific revisions were introduced.
26. Even the bindings of *Das Leben des Menschen* become more modernist: later editions have spare, minimalist blue-gray cloth-bound covers with all capital letters in a Futura sans-serif typeface.
27. Fritz Kahn, *Das Buch der Natur* [The book of nature], 2 vols. (Zurich: Albert Müller Verlag, 1952); published in English as *Design of the Universe* (New York: Crown, 1954), this book on physics, chemistry, and earth science goes even further. Its illustrations, almost entirely commissioned for the book and not taken from other sources, have an uncharacteristically coherent modernist look.
28. For a more detailed account of Kahn's relationships with his artists, see Eilers, "Lebensbild(er) von Fritz Kahn," which is based on research in the *Kosmos*/Franckh archive. We will, I hope, gain a greater understanding as more material is digitized and comes online.
29. Alwin Freund-Beliani seems to have done well. A photograph of Senta Freund-Beliani, *Revue des Monats* 1 (1927): 1227, is captioned, "Wife of the well-known graphic artist" (the photo is by the writer Béla Bálasz) and shows a prosperous-looking young woman. The equally talented Roman Rechn, in his application to study at Bauhaus, 26 August 1924 (probably made just before he came to work for Kahn), complained that his "financial situation is pretty bad." Walter Gropius rejected Rechn's application to study at Bauhaus. I want to thank Miriam Eilers for generously directing my attention to the Bauhaus application documents.
30. Letter, Roman Rechn, Berlin, to editor, *Kosmos,* Stuttgart, 6 May 1936, in Franckh'sche Verlagshandlung Archives; the translation is mine. Rechn is referring to two vertical foldout color plates he made for volume 4, "Der Aufbau des Menschenhaares" (pl. 23) and "Das menschliche Rückenmark" (pl. 17). Schüler did the artwork for "Der Mensch als Industriepalast" and a full-color wall chart of "Blutkreislauf"in the style of "Der Mensch als Industriepalast" that was published in 1931 along with a pamphlet. The full text of the letter reads: "Zur Frage der Urheberschaft „meiner" Bilder im „Leben des Menschen" kann ich nur Folgendes sagen: dass diese, vom ersten bis zum letzten in der Idee und auch im Aufbau von *mir* entworfen worden sind. Seien es die Tafeln „das Rückenmark," „das Haar" oder die vielen anderen aus dem IV. und V. Bande. So hat auch Dipl.Ing. Fritz Schüler „seine" Tafeln und Bilder ideenmäßig vollkommen selbstständig freischöpferisch geschaffen; vom „Menschen als Industriepalast" bis zu dem „Blutkreislauf." Dr. Kahn gab lediglich das Thema und stand

uns als scharfer Kritiker zu *Seite* oder *gegen* uns. Er war ja gar nicht in der Lage, auch nur die primitivste Skizze zu Papier zu bringen; als spitzer Analytiker in seiner Kritik zwang er uns gerade dadurch zu nie rastender Tätigkeit des Geistes. Nicht die Ausführung einer Zeichnung dauerte oft wochenlang, sondern die Durcharbeit immer neuer Ideen zu *einem* Problem. *Wie* gross ist allein der Ideen-Aufwand, welcher in „meinen" Umschlägen zum „Leben des Menschen" steckt.—Nein!—Dr. Kahn gab uns das Thema—die Ideen aber hatten wir." I want to thank Miriam Eilers, who found this letter and generously shared it with me.

31. *Life Magazine,* 19 April 1943, 65.
32. New York *Sunday Mirror,* 7 April 1943, 6–8.
33. Frank Hartmann, "Visualizing Social Facts: Otto Neurath's ISOTYPE Project," in *European Modernism and the Information Society*, ed. W. Boyd Rayward (Burlington, Vt.: Ashgate, 2014); Jordi Cat, "Otto Neurath" and "Supplement: Otto Neurath—Visual Education," *Stanford Encyclopedia of Philosophy* (2009), http://plato.stanford.edu/entries/neurath/index.html and http://plato.stanford.edu/entries/neurath/visual-education.html.
34. Whitney Davis, "Visuality and Pictoriality," *RES: Anthropology and Aesthetics* 46 (2004): 9–31, quotation at 19. Invoking Panofsky, Ernst Cassirer, and Martin Heidegger, Davis shows both the necessity and difficulties of such a distinction. I would add that historical conditions—the proliferation of images in newspapers, magazines, posters, and film—must have in some way influenced Panofsky and his fellow theorists of *Weltbildung* and *Weltanschauung* to make such concepts foundational elements of their philosophy and disciplinary methods.
35. The concept of "period eye" comes from Michael Baxandall, *Painting and Experience in Fifteenth-Century Italy* (1972; repr., Oxford: Oxford University Press, 1988), which argues that in a given time and place, people see the world around them in historically specific ways that influence and are influenced by visual practice in painting, sculpture, architecture, writing, and other cultural productions. He defines the "period eye" as "a stock of patterns, categories and methods of inference; training in a range of representational conventions; and experience, drawn from the environment" (31).
36. There was also much variation in Kahn's written words, due in part to his own inconsistency but also his reliance on sources who may not be in consistent agreement.
37. This is a potential characteristic of every illustrated text. In this respect, Kahn is a bit like Vesalius, whose Latin text was only selectively read (and never translated in full in many European languages until the twentieth and twenty-first centuries), while his illustrations traveled far and wide, and exerted enormous influence.

4. VARIETY SHOW

1. Other signatures that appear in *Das Leben des Menschen* and articles published in the 1920s and early 30s, and in *Kosmos* and other periodicals: Donalies, H. Orlowski, M. Landsberg, Ewald Streletzki, Bernhardt (sometimes B), wB, Churfurst, G. Edelstein, Paul Flanerky (usually as "P.Fl."), Frohse (an anatomical wall-chart company), H. Grundner, Grü, TNvH (sometimes TNH), Krugel, and A. B. Menninger.
2. Eugene S. Ferguson, "The Mind's Eye: Nonverbal Thought in Technology," *Science*, 26 August 1977, 827–36.
3. The phrase is taken from Gertrude Stein, *Picasso* (London: B. T. Batsford, 1938), 12.
4. Except for Plucer and Elfer, whom we can link to only one Kahn illustration. See *Gebrauchsgraphik* 8 (1931): 9–16; and this volume, Plate 8, Figure 3.4, and discussion in chapter 3.
5. For a discussion of Kahn's efforts to brand himself, see Cornelius Borck, "Communicating the Modern Body: Fritz Kahn's Popular Images of Human Physiology as an Industrialized World," *Canadian Journal of Communication* 32 (2007): 495–520.
6. Le Corbusier, *Towards a New Architecture* (London: John Rodker, 1931), frontispiece.
7. Roth, "Skyscrapers," *Berliner Börsen-Courier*, 12 March 1922, 111–12, in Roth, *What I Saw*. The skyscraper was never built.
8. John Tagg, "The Discontinuous City," in *Visual Culture: Images and Interpretation*, ed. Norman Bryson, Michael Ann Holly, and Keith Moxey (Middleton, Conn.: Wesleyan University Press, 1994), 84–85. I have substituted the italicized *images* for Tagg's "texts."
9. Le Corbusier, *Vers une Architecture* (Paris: G. Crès et Cie, 1924): "La maison est une machine à inhabiter." Le Corbusier repeats the maxim three times in the book. The 1931 English edition translates it as, "The house is a machine for living in" (4, 95, 107).
10. Martin J. S. Rudwick, "The Emergence of a Visual Language for Geological Science, 1760–1840," *History of Science* 14 (1976): 149–95, is a seminal work that suggests how one might begin such a project.
11. Roland Barthes, "The Plates of the *Encyclopedia*," trans. Richard Howard, in *The Continental Aesthetics Reader*, ed. Clive Cazeaux (London: Routledge, 2000); John Bender and Michael Marrinan, *The Culture of Diagram* (Stanford, Calif.: Stanford University Press, 2010).
12. Hugo Gernsback's proto–science fiction popular science magazine, *Science and Invention*, also occasionally featured X-ray cutaway illustrations. See, e.g., Frank R. Paul, illus., "Science in New Apartment House De Luxe," *Science and Invention* 9 (1922): 809. Thanks to Miriam Eilers for pointing out this connection.
13. *A Four Part Discussion of the Dynamic Forces that Are Bringing About the Dawn of a New Era in World Commerce; Transportation, Communication, Money, Goods* (New York: Importers Guide, 1935).

14. Henri Cartier-Bresson, *The Decisive Moment* (New York: Simon and Schuster, 1952). Obviously Kahn and his artists were not directly influenced by Cartier-Bresson's writings or work. Rather, they were applying the same principles, which were in wide circulation well before Cartier-Bresson's 1952 publication.
15. Fritz Kahn, *First Aid Popular* (New York: Friedrich Krause, 1942), publisher's blurb, inside cover.
16. Kahn, *Man in Structure and Function* (1943), 1:vi.
17. Wendell J. S. Kreig, M.D., review of Kahn's *Man in Structure and Function*, in *Saturday Review*, 17 August 1943, 11. The illustration was reprinted in Kahn's 1943 *Man in Structure and Function*, a revised translation of *Der Mensch Gesund und Krank*.
18. Barbara Maria Stafford, *Body Criticism: Imaging the Unseen in Enlightenment Art and Medicine* (Cambridge, Mass.: MIT Press, 1991).
19. Scott McCloud, *Understanding Comics: The Invisible Art* (New York: William Morrow, 1994), is a model inventory and analysis of visual devices, as well as a classic in the history of visual explanation.
20. *Der Mensch Gesund und Krank* (1939), 1:10–17. Most of this material on radiant energy was edited out of the American translation, *Man in Structure and Function*.
21. *Der Mensch Gesund und Krank* (1939), 1:15: "Man lege, nachdem diesen Abschnitt zu Ende gelesen, das Buch beiseite und suche sich Wahrheit, die neu gefundene, vorzustellen; man durchträume in seiner Phantasie die Welt, in der man lebt—die strahlende!"
22. *Das Leben des Menschen* (1931), 5:61. The slit-lamp ophthalmoscope, now a fixture of the standard eye examination, seemed like a marvel of modern science when it was first devised; Gullstrand was awarded the Nobel Prize in Medicine or Physiology in 1911. Kahn shows a playful photograph of the oddly photogenic device on page 65, a few pages later.
23. This is an updated, more dynamic version of an older illustration, "The Doctor of the Future," *Das Leben des Menschen* (1924), 2:227 (with artwork by "Krugel").
24. *Der Mensch Gesund und Krank* (1939), 1:16: "Denn auch das Innere Körpers strahlt, das Hirn, das Blut, der Muskel und die Drüse, jede Zelle unseres Körpers ist eine mikroskopische Sendestation—der Mensch lebt nicht nur in einer strahlenden Welt, sondern ist selbst ein strahlenden Wesen! Das is die grosse, vielleicht die grösste Entdeckung der modernen Wissenschaft, und in 50 Jahren werden vielleicht all Bücher, die bis heute über die Natur geschrieben wurden, veraltet sein, weil die Menschen, die sie verfassten, nichts von den Strahlen wussten." For growth rays, see "Die Zellestrahlung des Menschen" [Human cell radiation], *Der Mensch Gesund und Krank* (1939), 1:31.
25. *Der Mensch Gesund und Krank* (1939), 1:15–17.
26. Iridology had only limited acceptance among physicians. Its diagnostic power could only be substantiated for a few diseases. Although Kahn

presented himself as an authoritative voice on the science of the human body, like many writers on popular medicine, he occasionally strayed from medical orthodoxy.

27. Michael Baxandall, *Painting and Experience in Fifteenth-Century Italy* (1972; 2nd ed., Oxford: Oxford University Press, 1988), uses "period eye" to denote ways of seeing that are hard for us to discern but central to people in other times and places, arising from their unique historical experience and habitual practice.
28. The quotation comes from Hitler's courtroom testimony after the Munich putsch; quoted in Steve Coll, "Hitler and the Muslims," *New York Review of Books*, 2 April 2015, 62.
29. W. Kaempffert, "Science, the Machine and Democracy," *Science and Man*, ed. Ruth Nanda Anshen (1942; repr., New York: Greenwood Press, 1968), 135–50, especially 150.
30. Martin Heidegger, "The Age of the World Picture" (1938), in *The Question Concerning Technology and Other Essays*, trans. William Lovitt (New York: Harper and Row, 1977), 130.
31. Lorraine Daston, "Objectivity and the Escape from Perspective," *Social Studies of Science* 22 (1992): 597–618; quotation at 600.
32. *Der Mensch Gesund und Krank* (1939), 2:137, trans. George Rosen, in Kahn, *Man in Structure and Function* (1943), 2:472.
33. Kahn, *Man in Structure and Function* (1943), 1:161–62.
34. Jay David Bolter and Richard Grusin, *Remediation: Understanding New Media* (Cambridge, Mass.: MIT Press, 2000), 48–59.
35. For "mechanical objectivity," see Lorraine Daston and Peter Galison, *Objectivity* (New York: Zone Books, 2007).
36. Roland Barthes, "The Reality Effect," in *The Rustle of Language*, trans. Richard Howard (Oxford: Blackwell, 1986), 141–48; *Camera Lucida: Reflections on Photography*, trans. R. Howard (New York: Farrar, Straus and Giroux, 1981), 76. Barthes applied the term "reality effect" to the literary naturalism of Flaubert and other writers, who achieved it through overelaborate written descriptions of scenes and characters, "full of futile detail." I am not arguing that photography is an unproblematic mirror of reality, only that, notwithstanding its critics, it was regarded as a technology of representation of the real. If critics complained that photographs lacked verisimilitude, that they deceived, or that through retouching or other means their producers conspired to deceive, then this complaint was predicated on the near-universal tendency to give credence to the representational accuracy of the photograph—a credence that was never accorded to even the most brilliant trompe l'oeil. For historical and recent critiques of photographic realism, see Martin Jay, "Photo-Unrealism: The Contribution of the Camera to the Crisis of Ocularcentrism," in *Vision and Textuality*, ed. Stephen Melville and Bill Readings (London: Macmillan, 1995), 344–60.
37. Martin Kemp, "Style and Non-style"; "'The Mark of Truth': Looking and Learning in Some Anatomical Illustrations from the Renaissance

and Eighteenth Century," in *Medicine and the Five Senses*, ed. W. Bynum and R. Porter (Cambridge: Cambridge University Press, 1993), 85–121; Kemp, "'A Perfect and Faithful Record': Mind and Body in Medical Photography before 1900," in *Beauty of Another Order*, ed. A. Thomas (New Haven, Conn.: Yale University Press, 1997), 120–49, especially 123.

38. "Die Photographie," in *Das Ornament der Masse* (Frankfurt am Main: Suhrkamp Verlag, 1963), 35, quoted in Jay, "Photo-Unrealism," 348.

39. See Mia Fineman, *Faking It: Manipulated Photography before Photoshop* (New York: Metropolitan Museum of Art, 2012).

40. See Jeremy Aynsley, *Graphic Design in Germany, 1890–1945* (Berkeley: University of California Press, 2000), especially 152–53. According to Aynsley, there was a "rich exchange between the applied arts [commercial and political graphics] and fine arts" (7–8). In the 1890s and early 1900s, photographic manipulations that were innocent of any modernist aesthetic flourished in illustrated posters, newspapers, magazines, advertisements, and books. These influenced modernist photomontagery and were in turn influenced by it. Key figures in development of the (now canonical) political/aesthetic photomontage include Gustav Klucis, Valentina Kulagina, El Lissitzky, Hannah Höch, Raoul Hausmann, John Heartfield, Alexander Rodchenko, and László Moholy-Nagy, as well as film directors Walter Ruttmann, Dziga Vertov, and Sergei Eisenstein.

41. Rudi Feld, "Schrauben Sie eine andere Birn ein . . . und Sie warden volle Kassen haben!" *Licht-Bild-Bühne* 186 (5 August 1930): 2, quoted in Ward, *Weimar Surfaces*, 170.

42. Hadwig Kraeutler, *Otto Neurath—Museum and Exhibition Work: Spaces (Designed) for Communication* (Frankfurt am Main: Peter Lang, 2008); Michael Cowan, *Walter Ruttmann and the Cinema of Multiplicity: Avant-Garde—Advertising—Modernity* (Amsterdam: Amsterdam University Press, 2014).

43. In 1911, the Prussian Ministry of Culture commissioned graphic artist Ludwig Sütterlin to develop a modern handwriting script. Trester sometimes used it as a flowing midlevel display script to add a free-flowing element to balance Futura-based type and his precision-ruled diagrams. Fraktur was the house type used in Kosmos/Franckh'sche Verlag publications. In the typographical politics of the day, it was associated with German nationalism, in opposition to the Roman fonts used in France. Kahn's Kosmos publications used Fraktur in the text, but never in the frame of the illustration. After Kahn went over to Albert Müller Verlag, a Swiss publisher, Fraktur disappears entirely from his publications.

44. Barthes, "Plates of the *Encyclopedia*," 399. See also Bender and Marrinan, *Culture of Diagram*, 28.

45. Frank Bunker Gilbreth and Lillian Moller Gilbreth, *Process Charts* (New York: American Society of Mechanical Engineers, 1921), http://www.archive.org/stream/processchartsoogilb#page/n1

/mode/2up. For a useful short introduction to the history of the flowchart, see http://en.wikipedia.org/wiki/Flowchart.

46. Willard C. Brinton, *Graphic Methods for Presenting Facts* (New York: Engineering Magazine Company, 1914), 16–19.
47. Evart Routzahn and Mary Routzahn, *The ABC of Exhibit Planning* (New York: Russell Sage Foundation, 1918); Walter Dill Scott, *The Theory of Advertising: A Simple Exposition of the Principles of Psychology in Their Relation to Successful Advertising* (Boston: Small, Maynard, 1903, and many editions thereafter).
48. For the now-classic manifesto of the diagrammatic mathematical approach, see Norbert Wiener, *Cybernetics* (New York: Technology Press, 1948).
49. Adolphe Quetelet, *Sur l'homme et le développement de ses facultés, ou Essai de physique sociale*, 2 vols. (Paris: Bachelier, 1835).
50. Routzahn and Routzahn, *ABC of Exhibit Planning*, 73–74.
51. John Dalton, *A New System of Chemical Philosophy* (Manchester, 1808); "History of Molecular Theory," http://en.wikipedia.org/wiki/History_of_molecular_theory. Visualization was more widespread in the subfield of crystallography. For the early history of systems of notation in chemistry, see Matthew Daniel Eddy, *The Patchwork Picture: Science, Education, and the Visual Foundations of Knowledge, 1760–1820* (Chicago: University of Chicago Press, forthcoming), and "The Graphic Culture of Experimentation: Tables and Figures as Diagrams in the Medical Lectures of Joseph Black," in *Three Hundred Years of Chemistry at the University of Edinburgh*, ed. Robert Anderson (Edinburgh: Birlinn, 2015).
52. Kahn introduced the molecular coding system in "The Digestive System," *Das Leben des Menschen* (1926), 3:81–83, 85. He may have adapted it from a contemporary source, perhaps emanating from nutritional hygiene discourse, but that's only a guess.
53. E-mail message to author, 19 February 2016.
54. See, e.g., Schmitson's work in Otto Bollinger, *Atlas und Grundriss der pathologischen Anatomie* (Munich: Lehmann, 1896); Georg Sultan, *Grundriss und Atlas der speziellen Chirurgie* (Munich: Lehmann, 1922–23); Eduard von Hoffmann, *Atlas der gerichtlichen Medizin* (Munich: Lehmann, 1898); Franz Mraček, *Atlas and Epitome of Diseases of the Skin . . .* (Philadelphia: W. B. Saunders, 1905).
55. Faith A. Wallis suggests that the first title might allude to Joseph von Eichendorff's 1826 *Aus dem Leben eines Taugenichts* [From the life (memoirs) of a ne'er-do-well], a picaresque satire that was popular in nineteenth-century Germany; Wallis to author, 18 October 2007. There are at least eight such illustrations in *Das Leben des Menschen* (1924), 2:290–93 and pl. 30–33. Another appears in volume 3 (1926), pl. 6. Several of the pictures in these series are reprinted, with Schmitson's signature erased and some modification, in *Der Mensch Gesund und Krank* and later works.

56. Kahn and Schmitson might have also been aware of Edgar Rice Burroughs's Pellucidar series (1914–29), which is also set in the interior of the earth, figured as a primordial junglelike environment where dinosaurs still hold sway.
57. See Heinrich von Staden, "Science as Text, Science as History: Galen on Metaphor," *Clio Medica* 28 (1995): 499–518; Marri Lynn, "Vesalius and the Body Metaphor," *Public Domain Review*, 18 April 2013, http://publicdomainreview.org/2013/04/18/vesalius-and-the-body-metaphor.
58. William A. Alcott, *The House I Live In; or, The Human Body . . .* (Boston: Light and Stearns, 1837), 13–21; see also Michael Sappol, *A Traffic of Dead Bodies: Anatomy and Embodied Social Identity in Nineteenth-Century America* (Princeton, N.J.: Princeton University Press, 2002), chapter 4.
59. There was one major exception: in Vesalius and other early modern anatomies, artists often showed the dissected or skeletal body posing in front of distant architectural ruins as a poetic parallel to the ruin of the dead and anatomized body.
60. "Ueber die innere Architectur der Knochen und ihre Bedeutung für die Frage vom Knochenwachsthum," *Archiv für pathologische Anatomie und Physiologie und für klinische Medicin* 50 (1870): 389–450; English translation, M. O. Heller, W. R. Taylor, N. Aslanidis, and Georg N. Duda, "The Classic: On the Inner Architecture of Bones and Its Importance for Bone Growth," *Clinical Orthopaedics and Related Research* 468 (2010): 1056–65; Richard A. Brand, "Biographical Sketch: Julius Wolff, 1836–1902," *Clinical Orthopædics and Related Research* 468 (2010): 1047–49. One source might be G. B. Gruber, *Kosmos: Handweiser für Naturfreunde* 4 (1907): 321–23, which reprints photographs of bone cross-sectional specimens and diagrams taken from Wolff. In other words, Kahn's publisher probably had these pictures on file.
61. Unless, of course, it is a photograph of something old or an old photograph; then it may simultaneously signify modern and not modern.
62. See Bernhard Schapiro, *Präjaculin. Kombiniertes Epiphysen-Präparat gegen Reizzustände am Genitale und Hypererotismus* (Hamburg: Chemische Fabrik Promonta, 1932). Courtesy of the Leo Baeck Institute, New York.
63. Erving Goffman, *Stigma: Notes on the Management of Spoiled Identity* (New York: Simon and Schuster, 1963).
64. Borck, "Communicating the Modern Body," 15. I'm indebted to Borck for much of the discussion here on Kahn's sexology, but Borck slightly overstates his case. Kahn used representations of attractive female figures in several places in the same way that advertisers did, as a modern way of recruiting the viewer's gaze, and in some places he situated female figures in close proximity to modern appliances or industrial products.

5. KAHN'S TAKE AWAY

1. I gathered this list from a combination of bibliographic sources, especially WorldCat. Apart from "Der Mensch als Industriepalast," Kahn's most widely translated works are *Unser Geschlechtsleben* [Our sex life], *Der Mensch Gesund und Krank* [Man in health and sickness], and *Das Atom* (which was on physics and nuclear energy).
2. According to *The Human Factory: Sketch of the Chemical Functions of the Body* (Stuttgart: Franckh'sche Verlag, 1938), an English-language pamphlet supplement in the Deutsche Nationalbibliothek, Frankfurt, the poster was "also available in Arabic, Chinese, French, Hebrew, Italian and Spanish at the same price." An insert in the supplementary pamphlet of an English-language version published in Capetown, South Africa, lists a Romanian version (Cartea Romanesca, Bucharest) and a Spanish-language version (Lutz Ferrando and Cia, Buenos Aires; and Pablo Ferrando, Montevideo).
3. See, e.g., "Chelovek: dvorets industrii" [Man [as] palace of industry], *Gigiena i zdorov'e rabochei i krest'ianskoi sem'i* 3 (February 1927): cover; and a redrawn, recolored unauthorized Farsi "Der Mensch," in *The Book of Health* (Tehran, 1967), via Planche Scolaire Iranienne and Piccsy.com, http://piccsy.com/2010/11/planche-scolaire-iranienne-le-livre-de-sante-v4-1967.
4. Panel, "Ren ti hao xiang gong chang" [The human body is like a factory], 53 × 78 cm (poster); Physiology and Hygiene Association, Shanghai, 1933, https://www.nlm.nih.gov/hmd/chineseposters/understanding.html.
5. See, e.g., Evart Routzahn and W. W. Peter, *The Health Show Comes to Town: The Story of a Campaign of Health Education which Included the Platform Use of Mechanical Devices and Dramatic Action* (New York: Russell Sage Foundation, 1920).
6. A. R. Brubacher, Educators Association, *The Volume Library: A Concise, Graded Repository of Practical and Cultural Knowledge Designed for Both Instruction and Reference* (New York: Educators Association, 1932), 979. The illustration is an outlier: the book's other images show no Kahnian influence whatsoever.
7. Harold F. B. Wheeler, ed., *The Miracle of Life* (London: Oldhams Press Limited, n.d., ca. 1938), 341, 354, quotation from 341. Many other illustrations in the volume show a Kahnian influence.
8. "Model Portrays Man as Factory," *Popular Science Monthly* 127, no. 6 (December 1935): 30.
9. C. E. Cummings, *The Human Factory: New Exhibits in Hall of Man* (New York, ca. 1939): "The actual model . . . was constructed in the Museum laboratories under the Direction of Dr. Cummings. Joseph Morgenroth and his son prepared the static portion, and the mechanical and moving parts were produced by Manley S. Blakeslee"; Bruno Gebhard Papers, Dittrick Medical History Center, Cleveland, Ohio. The Buffalo Museum of Science still has the "Human Factory." After refurbishment and rehousing, it went back on display in spring 2014,

not as a historical object but as a working science lesson. Thanks to Linda Lohr and Kathy Leacock for providing information on this.

10. "The Total Number of Blood Cells," *Das Leben des Menschen* (1924), 2:309; artist: Georg Helbig.
11. "The Blood Cell Total," *Gigiena i Zdorov'e* 7 (April 1936): 16.
12. Bruno Gebhard et al., *Wunder des Lebens* (Stuttgart: Union Deutsche Verlagsgesellschaft, 1936), 100. Artist: E. A. Mühler.
13. Gerhard Venzmer, Fritz Gross, and Ludolf Fischer, eds., *Hand- und Lehrbuch der Krankenpflege* (Stuttgart: Franckh'sche Verlagshandlung, 1938–39), 1:72. Artist: "Oelfinger." As discussed later in this chapter, Franckh was forced to take Kahn's works out of circulation and replace them with "Aryan"-authored, Nazi Party–approved books on health and the human body, which closely resembled Kahn's in structure and content. These books were heavily illustrated and even featured illustrations taken from Kahn, along with redrawn and revised versions of Kahn's pictures.
14. "Four Times Around the Earth!" *Der Mensch Gesund und Krank* (1939), 1:223, artist uncredited. In the 1924 version, the blood cells encircle the earth "three-and-a-half" times; in the 1939 version, Kahn rounds up to four (perhaps to make the headline catchier).
15. *Wonder Engine of the Body: The Human Heart* (New York: American Heart Association, 1951; sound; black and white; 12:00; 16mm). Animator: Bray Studios. A few other animated images in the film also appear to be based on pictures from *Das Leben des Menschen.*
16. "Wonders of the Human Body Reproduced in Museum by Machines and Electricity," *Popular Science,* August 1935, 36–37; Marian Niquette and William J. Buxton, "Sugar-Coating the Educational Pill," in *Patronizing the Public: American Philanthropy's Transformation of Culture, Communication, and the Humanities,* ed. W. J. Buxton (Lanham, Md.: Lexington Books, 2009), 153–94.
17. For GeSoLei, see Paul Weindling, *Health, Race and German Politics Between National Unification and Nazism . . .* (Cambridge: Cambridge University Press, 1993), 413–14, 423; see also chapter 1, note 13, of this book.
18. Bruno Gebhard, ed., *Wunder des Lebens* (Stuttgart: Union Deutsche Verlagsgesellschaft, 1936), 35, 47, 71, 100. Das Wunder des Lebens opened in Berlin in 1935.
19. This took place in many countries, including the United States, Britain, and France. See Julie K. Brown, *Health and Medicine on Display: International Expositions in the United States, 1876–1904* (Cambridge, Mass.: MIT Press, 2009); and "Connecting Health and Natural History: A Failed Initiative at the American Museum of Natural History, 1909–1922," *American Journal of Public Health* 104 (2014): 1877–88.
20. Niquette and Buxton, "Sugar-Coating," 177; Robert Rydell, Christina Cogdell and Mark Largent, "The Nazi Eugenic Exhibit in the United States," in *Popular Eugenics: National Efficiency and American Mass*

Culture in the 1930s, ed. Susan Currell and Christina Cogdell (Athens: Ohio University Press, 2006), 361–65.

21. Erin McCleary and Elizabeth Toon, "'Here Man Learns About Himself': Visual Education and the Rise and Fall of the American Health Museum," *American Journal of Public Health* 102 (2012): 27–36.
22. "Scientific Peep Show Tells How Our Bodies Work," *Popular Science Monthly,* February 1938, 54–55.
23. Margaret Manson, *Your Health a Guide to the Medicine and Public Health Building at the New York World's Fair, 1940 . . .* (New York: American Museum of Health, 1940), 40–41; *Illustrated London News,* 15 May 1948: cover; "Mac the Mechanical Man," New York World's Fair 1939–40 records, New York Public Library, Manuscripts and Archives Division, MSS coll. 2233; Alex Mold, "Picturing Public Health: Visual Culture and Public Health in Post-war Britain," paper presented at the American Association for the History of Medicine annual meeting, New Haven, Conn., 2 May 2015.
24. Peter E. Fäßler, "Eine symbiotische Beziehung? Zur Kooperation zwischen Deutschem Hygiene-Museum und NS-Regime," in *Gesundheit und Staat: Studien zur Geschichte der Gesundheitsämter in Deutschland, 1870–1950,* ed. Axel C. Hüntelmann, Johannes Vossen, Herwig Czech (Husum, Germany: Matthiesen, 2006), 63–74. Thanks to Thomas Steller (e-mail, 23 January 2014) for this reference and for his help in understanding what happened at the museum when the Nazis took control.
25. Eckart mockingly quoted Kahn's 1920 book, *Die Juden als Rasse und Kulturvolk* [The Jews as race and culture] (Berlin: Welt-Verlag, 1920), in an essay, "Der Bolschewismus von Moses bis Lenin: Zwiegespräch zwischen Hitler und mir" [Bolshevism from Moses to Lenin: Dialogues between Hitler and me], published in 1935 with an introduction by Alfred Rosenberg.
26. This was reported by Kahn in an interview in the *New York [Sunday] Mirror,* 11 July 1943, 6–10; clipping in Ser. II/2/A fol. 2/15, Fritz Kahn Papers, Leo Baeck Institute, New York. I have not been able to independently verify this claim, but it seems quite likely, given that the guidelines for book burning specified all books by Jewish authors.
27. To tie up loose ends, the publisher also attempted to get some of Kahn's artists to acknowledge the publisher's and author's rights. Roman Rechn did not agree, but Franckh used his work anyway. See letter, Roman Rechn, Berlin, to editor, *Kosmos,* Stuttgart, 6 May 1936, in Franckh'sche Verlagshandlung Archives. I am here again indebted to Miriam Eilers, who has generously shared some of her research with me. For a more detailed account of Kahn's experience with the Nazis, negotiations with Franckh and reaction to Venzmer, see Miriam Eilers's forthcoming work.
28. Gerhard Venzmer, *Der Mensch und sein Leben: Ein Volksbuch vom menschlichen Körper* (Stuttgart: Kosmos/Franckh'sche Verlagshandlung, 1939).

29. In the Weimar period, when other Franckh authors or newspapers used images from *Das Leben des Menschen,* the credit usually read, "Aus Kahn, Das Leben des Menschen." By my rough count, about 40 percent of the illustrations in *Der Mensch und sein Leben* directly derived from *Das Leben des Menschen* (including some photographs that Kahn took from other publications); some 5 to 10 percent are indirectly derived.
30. Gerhard Venzmer, Fritz Gross, and Ludolf Fischer, *Hand- und Lehrbuch der Krankenpflege,* 2 vols. (Stuttgart Franckh'sche Verlagshandlung, 1938, and many editions thereafter). Venzmer continued his career as a writer of popular medical books and articles into the postwar era, but he dropped the illustrations directly derived from Kahn.
31. "The Human Body: Pictures Explain Its Structure and Functions," *Life,* 19 April 1943, 64–71; *New York [Sunday] Mirror,* 7 November 1943, 6–10; clipping in Ser. II/2/A fol. 2/15, Fritz Kahn Papers, Leo Baeck Institute, New York.
32. Ibid., 65. Chapter 3 discusses Rechn's disagreement with Kahn in greater detail. Again, I am indebted to Miriam Eilers for sharing her research with me on this point.
33. Even before the full Nazi takeover, in the early 1930s Venzmer was authoring slender illustrated books for Franckh on medical topics. These featured original artworks that showed a Kahnian influence but were not plagiarized.
34. See, e.g., Jeffrey Herf, *Reactionary Modernism: Technology, Culture, and Politics in Weimar and the Third Reich* (Cambridge: Cambridge University Press, 1984). See also Zygmunt Bauman, *Modernity and the Holocaust* (Cambridge: Polity Press, 1990). I thank Miriam Eilers for helping me with this issue.
35. Fritz Nonnenbruch, *Die dynamische Wirtschaft* (Munich, 1936), 153, quoted in Herf, *Reactionary Modernism,* 209.
36. Joseph Goebbels, "Deutsche Technik," March 1939, 105–6, and *Reden* (Berlin, 5 June 1943), both quoted in Herf, *Reactionary Modernism,* 196.
37. The group most estranged from the consensus on modernity was ultramontanist Catholicism—the most consistently antimodernist of all twentieth-century political positions. Between 1920 and 1960, Catholic school textbooks on health and hygiene were mainly designed to indoctrinate students in Catholic sexual and reproductive morality. These tended to be devoid of illustration: anatomical diagrams could provoke sexual feelings and lead to subversive thoughts, and suggested that doctors and scientists could rival priests and bishops as authorities on the body. Even so, conservative Catholic parties and institutions selectively adopted a modern look and sound to better combat or co-opt the secular nation-state and secular parties—and perhaps also to better fight in factional contests for power within the faith.
38. The library catalog of the National Library of Medicine lists 142 publications by Adrián Vander in its collection, spanning the 1920s to

the 1970s. See, e.g., *Moderna medicina natural; nuevos tratamientos curativos,* 12th ed. (Barcelona: Sintes, 1950).

39. Adolf Nichtenhauser Papers, 4:260, National Library of Medicine. *Enemy Bacteria* (U.S. Navy, Walter Lantz Studios, 1945). Animators: Shamus Culhane, Myron "Grim" Natwick. Layout artist: Arthur Heinemann. Director: Dick Lundy. Voices: Mel Blanc (best known for voicing Warner Bros. cartoon characters such as Bugs Bunny and Daffy Duck).
40. *A Few Quick Facts about Fear* (1945; Army–Navy Screen Magazine; United Productions of America; dir. Zack Schwartz; black and white; sound; 3:34); *The Traitor Within* (1946; American Cancer Society; John Sutherland Productions; dir. George Gordon; color; sound; 10:00); *Rodney* (1950; National Tuberculosis Association; Film Graphics Inc.; dir. Lu Guarnier; color; sound; 10:00); *Man Alive* (1952; American Cancer Society; UPA; dir. William T. Hurtz; color; sound; 12:00); *Oral Hygiene: Swab Your Choppers* (1948; U.S. Navy; UPA; color; sound; 7:00); *Multiple Screening* (1950; Pennsylvania Department of Health; Communication Material Center, Columbia University Press; Tempo Productions; dir. David Hilberman; black and white; sound; 9:00).
41. James Burkhart Gilbert, *Redeeming Culture: American Religion in an Age of Science* (Chicago: University of Chicago Press, 1997), 234, estimates that the Bell Science series was watched by over 5,000,000 schoolchildren and 500,000 college students.
42. See David Cantor, "*Man Alive!* (1952): Cartoon Fun with Cancer, Cars and Companionate Marriage in Suburban America," *Medical Movies on the Web* (National Library of Medicine, 2014), https://www.nlm.nih.gov/hmd/collections/films/medicalmoviesontheweb/pdf/manalive.pdf. Some older animators also embraced the new style, notably Shamus (aka James) Culhane, a veteran of silent-era cartooning, who had also worked with Disney and the Fleischers.
43. György Kepes, *Language of Vision* (Chicago: Paul Theobald, 1944); John Hubley and Zachary Schwartz, "Animation Learns a New Language," *Hollywood Quarterly* 1, no. 4 (July 1946): 63–68.
44. *A Few Quick Facts about Fear.*
45. Kepes, *Language of Vision,* 200–09.
46. Ibid., 130.
47. Sloane also did commissioned work for the American Museum of Natural History and the Smithsonian Institution. See "Eric Sloane, N.A. (American, 1905–1985)," http://weatherhillfarm.com/, http://www.ericsloane-awareness.com/bio.htm, and http://en.wikipedia.org/wiki/Eric_Sloane.
48. H. A. Carpenter, G. A. Bailey, and P. E. Smith, *Adventures in Science with Ruth and Jim* (Boston: Allyn and Bacon, 1948); review, *Science Education* 33 (1949): 77–78.
49. See R. Roger Remington, *Lester Beall: Trailblazer of American Graphic Design* (New York: Norton, 1996); R. Roger Remington and Robert S. P. Fripp, *Design and Science: The Life and Work of Will Burtin* (Burlington, Vt.: Lund Humphries, 2007).

6. "TO PICTURE THE BODY"

1. For a discussion that foregrounds the normally almost invisible depth/surface binary, see Tony Blackshaw and Tim Crabbe, *New Perspectives on Sport and "Deviance": Consumption, Performativity and Social Control* (New York: Routledge, 2004), 195–97.
2. Colin McGinn criticizes "expositors of neuroscience" who "engage in misguided homunculus talk" and mistakenly insist that neurons and groups of neurons make decisions and have intentions; McGinn, "Storm Over the Brain," *New York Review of Books,* 24 April 2014, 62.
3. See Pierre Bourdieu, *The Logic of Practice,* trans. Richard Nice (Cambridge: Polity, 1990), who uses "habitus" as a term for a habitual, customary cultural praxis that may be "coherent without springing from an intention of coherence and a deliberate decision; adjusted to the future without being the product of a project or a plan" (51). Bourdieu: "The conditionings associated with a particular class of conditions of existence produce *habitus,* systems of durable, transposable dispositions, structured structures predisposed to function as structuring structures, that is, as principles which generate and organize practices and representations that can be objectively adapted to their outcomes without presupposing a conscious aiming at ends or an express mastery of the operations necessary in order to attain them" (53).
4. My argument here overlaps Barbara Maria Stafford's account of "distributed agency" in *Echo Objects: The Cognitive Work of Images* (Chicago: University of Chicago Press, 2007), 98–104, in its emphasis on mimesis and intersubjectivity in speech and representation, but focuses on social, technological, and political aspects of subject formation. Unlike Stafford, I deliberately black box biophysical mechanisms of mind. For a recent neuroscientific account of how the self becomes consolidated into a singularity, see Michael S. Gazzaniga, *Who's in Charge? Free Will and the Science of the Brain* (New York: Ecco, 2011), and Gazzaniga, Richard Ivry, and George R. Mangun, eds., *Cognitive Neuroscience: The Biology of Mind,* 4th ed. (New York: Norton, 2013). John Sutton and Kourken Michaelian, "Distributed Cognition and Memory Research: History and Current Directions," *Review of Philosophical Psychology* 4 (2013): 1–24, argue that recent critiques of neuroscience fail to take into account the last decade or two of work on "distributed cognition" (which makes arguments that largely align with the argument offered here). But it remains the case that the most widely publicized neuroscientific accounts of human behavior and subjectivity tend to focus on functional magnetic resonance imaging research and internalist explanations of mind. See also John Sutton and Nicholas Keene, "Cognitive History and Material Culture," in *The Ashgate Research Companion to Material Culture in Early Modern Europe,* ed. David Gaimster, Tara Hamling, and Catherine Richardson (Burlington, Vt.: Ashgate, 2015).
5. Douwe Draaisma, *Metaphors of Memory: A History of Ideas about the Mind* (1995), trans. Paul Vincent (Cambridge: Cambridge University

Press, 2000), 11–13. Usages that appear to be purged of metaphor almost always turn out to have originated as metaphor—origins that were originally apparent but that have become invisible (or that were brought over from Latin or Greek or some other language in order to be rendered invisible), a dead metaphor. For embodied metaphor, dead and alive, see George Lakoff and Mark Johnson, *Metaphors We Live By* (Berkeley: University of California Press, 1980).

6. Michel Foucault, *The Order of Things: An Archaeology of the Human Sciences* (1966; New York: Vintage, 1970), 319.
7. Ibid.
8. Ibid., 318.
9. Michel Foucault, *The Birth of the Clinic: An Archaeology of Medical Perception,* trans. Alan Sheridan Smith (1963; New York: Vintage, 1994), 163.
10. See Martin J. S. Rudwick, *Scenes from Deep Time: Early Pictorial Representations of the Prehistoric World* (Chicago: University of Chicago Press, 1995).
11. Wolfgang Schivelbusch, *Licht, Schein und Wohn* (Berlin: Ernst and Sohn, 1992), 133, quoted in Ward, *Weimar Surfaces,* 203.
12. Quoted in Draaisma, *Metaphors of Memory,* 54–55.
13. John William Draper, *History of the Conflict between Religion and Science,* 8th ed. (New York: Appleton, 1881), 301.
14. My argument is influenced by Nancy Leys Stepan, "Race and Gender: The Role of Analogy in Science," *Isis* 77 (1986): 261–77; Lakoff and Johnson, *Metaphors We Live By;* and Draaisma, *Metaphors of Memory.* See also Klaus Hentschel, ed., *Analogien in Naturwissenschaften, Medizin und Technik* (Stuttgart: Wissenschaftliche Verlagsgesellschaft, 2010; *Acta Historica Leopoldina* 56 [2010]).
15. Francis Bacon, *De argumentis scientiarum,* bk. 4, chapter 3, *The Works of Francis Bacon,* ed. R. L. Ellis, D. D. Heath, and J. Spedding (New York: Hurd and Houghton, 1862–79), 1:353–55, quoted in Daniel Heller-Roazen, *The Inner Touch: Archaeology of a Sensation* (New York: Zone Books, 2009), 177.
16. William A. Alcott and other nineteenth-century writers and lecturers spread the anatomical body = house analogy to a mass audience by conjoining anatomical illustration and pictures of buildings; see Sappol, *Traffic of Dead Bodies,* 168–211, and chapter 4 here. Both Galen and Vesalius used analogy liberally, a common practice in Greco-Roman and early modern medicine; see Marri Lynn, "Vesalius and the Body Metaphor," *Public Domain Review,* 18 April 2013, http://publicdomainreview.org/2013/04/18/vesalius-and-the-body-metaphor; and "Pipes, Reins, and the Cerebral Winepress: Mechanical Metaphor in Vesalius' *Fabrica,*" *Wonders and Marvels,* July 2007, http://www.wondersandmarvels.com/2012/07/metaphor-vesalius-fabrica.html.
17. Francis M. Walters, *Physiology and Hygiene for Secondary Schools* (Boston: D. C. Heath, 1909), iv. Walters argued that pupils should instead be given "laboratory work" and "make experiments

and observations." Kahn would have endorsed that call, but as a complement, not an alternative, to metaphorical explanation.

18. James Cabell Minor (author, illustrator), *The Plan o' the House o' Man, Sir! or, The Parts Water and Position Play in the Prevention and Treatment of Physical Disorders of the Body* (Little Rock, Ark.: Central Printing, 1912); Louis Chauvois (author, illustrator), *La machine humaine enseignée par la machine automobile* (Paris: G. Doin, 1926); Arthur Keith, *The Engines of the Human Body . . .* (London: Williams and Norgate, 1919); Wilfred T. Grenfell (author, illustrator), *Yourself and Your Body* (New York: Charles Scribner's Sons, 1924). These were matched by a parallel literary universe of mechanical men: L. Frank Baum's Tin Man and Tik-Tok, and Karěl Capek's robots. See also Hanns Günther, *Automaten* (Stuttgart: Kosmos/Franckh'sche Verlagshandlung, 1930).
19. That maxim had many variants: "an ounce of pictures is worth a pound of words" or "a ton of talk."
20. Quoted in Hale, *Human Science and Social Order,* 106.
21. Alfons Labisch, "Doctors, Workers and the Scientific Cosmology of the Industrial World: The Social Construction of 'Health' and the 'Homo Hygienicus,'" *Journal of Contemporary History* 20 (1985): 599–615; Andreas Daum, *Wissenschaftspopularisierung im 19. Jahrhundert bürgerliche Kultur, naturwissenschaftliche Bildung und die deutsche Öffentlichkeit, 1848–1914* (Munich: R. Oldenbourg Verlag, 1998); Paul Weindling, *Health, Race, and German Politics between National Unification and Nazism, 1870–1945* (Cambridge: Cambridge University Press, 1989).
22. Alfred Grotjahn, *Die hygienische Forderung* (Königstein: Langewiesche Brandt, 1921), 179, quoted in Labisch, "Doctors, Workers," 605.
23. Otto Neurath, "Bildliche Darstellung sozialer Tatbestände" [Pictorial representation of social facts], *Aufbau* 8 (1926): 170–74; *"Bildstatistik nach der Wiener Methode"* [Pictorial statistics according to the Viennese method], *Die Volksschule* 37 (1931): 12. Neurath called his system ISOTYPE, the International System of Typographic Picture Education.
24. Martin Heidegger, "The Age of the World Picture" (1938), in *The Question Concerning Technology and Other Essays,* trans. William Lovitt (New York: Harper and Row, 1977), 130.
25. Ludwig Wittgenstein, *Philosophical Investigations* (1953), trans. G. E. M. Anscombe (New York: Macmillan, 1968), §115. See S. W. Patterson, "A Picture Held Us Captive: The Later Wittgenstein on Visual Argumentation," *Cogency* 2 (2010): 105–34. Wittgenstein again: "The human body is the best picture of the human soul" (*Philosophical Investigations,* 4:25). (Consider this inversion: the best picture of the human body is the human soul.)
26. See Davis, "Visuality and Pictoriality."
27. Kaempffert, "The Vision of a Blind Man," *Popular Science Monthly* 88 (1916): iii–xi, quotation at xi.

28. Gustave Le Bon, *The Crowd: A Study of the Popular Mind* (1895; New York: Viking, 1960), 41–43, quoted in Jonathan Crary, *Suspensions of Perception: Attention, Spectacle, and Modern Culture* (Cambridge, Mass.: MIT Press, 1993), 245.
29. Louis Haugmard, "L'Esthétique du cinématographe" [The aesthetic of the cinematograph], *Le Correspondant*, 25 May 1913, 762–71, in *French Film Theory and Criticism: 1907–1929,* ed. and trans. Richard Abel (Princeton, N.J.: Princeton University Press, 1993), 77–85, quotations at 83–85. "The 'masses' are like grown-up children who demand a picture album to leaf through in order to forget their miseries" (80).
30. Le Bon, *Crowd*, 3.
31. Martin Jay, *Downcast Eyes: The Denigration of Vision in Twentieth-Century French Thought* (Berkeley: University of California Press, 1993).
32. Mitchell, *Picture Theory,* 91.
33. Logan Clendening, "Guide and Chart for the Human Interior," review of Kahn, *Man in Structure and Function,* in *New York Times Book Review,* 4 April 1943. At the time, *Fortune* was celebrated for its innovative modernist covers, interior illustrations, and page designs. Doré was then still celebrated as one of the great illustrators of the nineteenth century.
34. Mabel S. Ulrich, review of Kahn, *Man in Structure and Function* (1943), clipping from unidentified magazine, ser. II/2/A fol. 2/15, Fritz Kahn Papers, Leo Baeck Institute, New York. Ulrich was active in many civic and suffrage organizations. She was also a bookstore owner, sold art and antiques, and served as state director of the WPA Federal Writers' Project from 1935 to 1938. As a young woman in the 1910s, she delivered public lectures on sex hygiene, abortion, and other challenging topics.
35. Ernst Haeckel's *Kunstformen der Natur* [Art forms of nature] (1899–1904) was a big influence on early twentieth-century design and popular science publications. Kahn collaborated with designer Georg Piek-Patrik on *Rhythmische Musterkunst der Natur* (Berlin: Der Konfektionär, 1924) to demonstrate "the basic principles of a new style" based on images of anatomical sections, cells, and vein and nerve trees as pattern sources for wallpaper, carpet, fabric, furniture, lamps, blankets, and stockings. For Tchelitchew, see Meredith Mendelsohn, "Sex-Ray Vision," review of Pavel Tchelitchew: Landscape of the Body, Katonah Museum of Art, Katonah, N.Y., 14 June–6 September 1998, in *Artnet,* 27 August 1998, http://www.artnet.com/magazine_pre2000/features/mmendelsohn/mendelsohn8-27-98.asp.
36. *New York [Sunday] Mirror,* 11 July 1943, 6–10; clipping, ser. II/2/A fol. 2/15, Fritz Kahn Papers, Leo Baeck Institute, New York.
37. Wendell J. S. Krieg, M.D., review of Kahn, *Man in Structure and Function,* in *Saturday Review,* 17 August 1943, 11.
38. Ralph W. Gerard, review of Kahn, *Man in Structure and Function,* in *Scientific Monthly* 57 (1943): 177–79, quoted in Borck, "Communicating the Modern Body," 503.

39. *Quarterly Review of Biology* 18 (1943): 385–86, quoted in Borck, "Communicating the Modern Body," 507.
40. I'm alluding here to Lakoff and Johnson, *Metaphors We Live By*.
41. Quoted in Killen, *Weimar Psychotechnics*, 63, citing Robert W. Chestnut, "Psychotechnik: Industrial Psychology in the Weimar Republic, 1918–1924," in *Proceedings of the Annual Convention of the American Psychological Association* 7, pt. 2 (1972): 781–82, quotation at 782.
42. Barthes, "Plates of the *Encyclopedia*," 398; Catherine Gallagher and Stephen Greenblatt, *Practicing New Historicism* (Chicago: University of Chicago Press, 2000), 15.
43. Lewis Mumford, *Technics and Civilization* (New York: Harcourt Brace, 1934), 359.
44. Neurath, "Bildliche Darstellung sozialer Tatbestände."
45. W. J. T. Mitchell, *What Do Pictures Want? The Lives and Loves of Images* (Chicago: University of Chicago Press, 2005).
46. I leave open here the precise mechanisms of ocularcentric subjectivity and the history of theories of ocular subjectivity, which would require a discussion of Rudolf Arnheim's mid-twentieth-century work, and then the cognitive turn and its impact on art theory. See Arnheim, *Art and Visual Perception* (Berkeley: University of California Press, 2004); Stafford, *Echo Objects: The Cognitive Work of Images*, (Chicago: University of Chicago Press, 2007); Martin Jay, *Downcast Eyes: The Denigration of Vision in Twentieth-Century French Thought* (Berkeley: University of California Press, 1993).
47. Walter Benjamin, "Children's Literature," broadcast on Southwest German Radio, Frankfurt, 15 August 1929, in Lecia Rosenthal, ed., *Radio Benjamin*, trans. Lisa Haries Schumann, (London: Verso Books, 2014), 252.
48. See, e.g., John Medina, *Brain Rules* (Edmonds, Wash.: Pear Press, 2014, http://brainrules.net/vision), which cites recent neuroscientific research and evolutionary psychology to argue that "vision trumps all other senses and boring things are hard to remember": "We are incredible at remembering pictures. Hear a piece of information, and three days later you'll remember 10% of it. Add a picture and you'll remember 65%. Pictures beat text as well in part because reading is so inefficient for us. Our brain sees words as lots of tiny pictures, and we have to identify certain features in the letters to be able to read them. That takes time. Why is vision such a big deal to us? Perhaps because it's how we've always apprehended major threats, food supplies and reproductive opportunity." Cited in Alphachimp, *The Alphachimp Rockstar Resource Guide*, TEDMED ed. (Nashville, Tenn.: Alphachimp Learning Systems, 2012), 1:6.
49. Gilles Deleuze and Félix Guattari, *Anti-Oedipus* (1972), trans. Robert Hurley (New York: Penguin Classics, 2009).
50. For biopolitics, biopower, and the current moment in historical perspective, see Roger Cooter and Claudia Stein, *Writing History in the*

Age of Biomedicine (New Haven, Conn.: Yale University Press, 2013). For emancipation, see Bennett, *Enchantment of Modern Life.*

51. Roland Barthes, *Roland Barthes by Roland Barthes,* trans. Richard Howard (New York: Hill and Wang, 1977), 161. I have slightly altered Howard's translation.

EPILOGUE

1. Bruno Latour, "Visualisation and Cognition: Drawing Things Together," in *Knowledge and Society Studies in the Sociology of Culture Past and Present,* ed. H. Kuklick and E. Long (Greenwich, Conn.: Jai Press, 1986), 6:1–40, quotation at 6:19. All italics are Latour's.
2. Ibid., 6:19.
3. Ibid., 6:5–13.
4. Ibid., 6:7, 19.
5. Alfred Döblin, "Metapsychologie und Biologie," review of *Das Leben des Menschen,* vol. 1, *Neue Rundeschau,* December 1923, 1231.
6. Latour, "Visualisation and Cognition," 6:11, 20.
7. Ibid., 6:18–20. Latour lists two other important features: inscriptions and images "can . . . be made part of a written text" and "merge[d] with geometry" (6:20).
8. Ibid., 6:9.
9. For intersubjectivity and objects, see Arjun Appadurai, ed., *The Social Life of Things* (Cambridge: Cambridge University Press, 1986); Mitchell, *What Do Pictures Want?*; John Sutton, "Porous Memory and the Cognitive Life of Things," in *Prefiguring Cyberculture: An Intellectual History,* ed. D. Tofts, A. Jonson, and A. Cavallaro (Cambridge, Mass.: MIT Press, 2002), 130–41.
10. Philipp Sarasin, "Body as Medium," trans. Brian Hanrahan, *Grey Room,* October 2007, 49.
11. Susan Leigh Star and James Griesemer, "Institutional Ecology, 'Translations' and Boundary Objects: Amateurs and Professionals in Berkeley's Museum of Vertebrate Zoology, 1907–39," *Social Studies of Science* 19 (1989): 387–420, quotation at 393.
12. Ibid.
13. Michel Foucault, *Technologies of the Self,* ed. Luther H. Martin, Huck Gutman, and Patrick H. Hutton (Amherst, Mass.: University of Massachusetts Press, 1988).
14. For a Foucauldian/Marxian critique of the neuroscientific turn, see Roger Cooter with Claudia Stein, *Writing History in the Age of Biomedicine* (New Haven, Conn.: Yale University Press, 2013).
15. Sutton and Keene, "Cognitive History and Material Culture."
16. In other words, neuroscientific findings on neural plasticity and fixity, and multiple consciousness, if we can read them against the grain, make a rapprochement with cultural studies possible—even necessary. That is, they easily allow for, and from the standpoint of critical theory and cultural studies require, the constitutive contribution of external social

relations, cultural productions, and political structures. See T. Kircher and A. David, eds., *The Self in Neuroscience and Psychiatry* (Cambridge: Cambridge University Press, 2003).

17. See, e.g., Michael Gazzaniga, *Tales from Both Sides of the Brain: A Life in Neuroscience* (New York: Ecco Press, 2015).

18. Ian Hacking suggests some of these conceptual difficulties in *Rewriting the Soul: Multiple Personality and the Sciences of Memory*, rev. ed. (Princeton, N.J.: Princeton University Press, 1998).

19. For "habitus," see Bourdieu, *Logic of Practice.*

20. For a provocative short discussion of the historical vagaries of narration, see Steve Almond, "Riff: Once Upon a Time, There Was a Person Who Said 'Once Upon a Time,'" *New York Times Sunday Magazine*, 13 January 2013, 44–45.

INDEX

Page numbers in *italics* refer to illustrations.

Abrams, Alfred Willis, 39
Adventures in Science with Ruth and Jim (Figure 5.19), *159*
aesthetics: xii, 2, 4, 7, 12, 14–15, 21, 23–24, 50, 52–54, 56, 58–59, 62, 65–66, 69, 71–72, 104, 109, 118, 123, 153, 156, 159, 168, 175, 201, 203; bioaesthetics, 59; functionalist, 4, 24; modernist, 11, 15, 52, 65, 109, 114, 118, 142, 151, 157–58, 216, 223; movements, 6–7; new, 153, 156, 201
agency, 27, 54, 165, 168, 180–82, 207, 231
Åhrén, Uno, 8
"Air: Before Inhalation/After Exhalation" (Figure 4.27), *113*
"Air: Before Inhalation/After Exhalation" revised (Figure 4.28), *113*
Alcott, William A., 125, 232
Allison, J. Murray, 54
Americanism, 24, 28, 212
Amerikanismus. See Americanism
architecture, xiv, 24, 31, 94, 120, 135, 142, 144, 214, 219; and the body, 30, 71, 74, 125; Neues Bauen, 25, 146
Arnheim, Rudolf, 68
Arntz, Gerd, 67
assembly line, xiii, 7, 28, 107, 189, 190
Atkinson, Eleanor Stackhouse, 39

Bacon, Francis, 170, 172
Barthes, Roland, 107, 180, 184, *222*
Bauhaus, xi, 4, 25, 52, 72, 92, 107, 123, 144, 146, 155, 173, 181, 205, 218
Baxandall, Michael, 69
Beall, Lester, 158, 160
Beattie, John, 146
Benjamin, Walter, 11, 17, 183
Berlin, 8, 12–13, 27, 56, 60, 74, 138, 142, 206, 208–9, 217, 227
Berliner Illustrierte Zeitung (newspaper), 3, 15, 57, 206, 217
Berlin im Licht festival (1928), 90
Bernath, Erwin, 67
"Biological Effects of Ultraviolet Light, The" (Schüler) (Figure 4.14), 91, *93*
biology, 4, 7, 168, 178; microbiology, 124, 146, 168; modern, 1, 5, 84
"Biology of the Smell of a Roast, The" (Figure 4.5), 79, *81,* 83–84
"Blood Crystals" (Helbig) (Figure 6.5), *176*
Blut und Zelluloid (Jacob), 23

body: anatomical, 7, 232; corporeal, 18, 35, 98, 105, 122, 125; industrial, xiii, 5, 7, 18–19, 22, 30–32, 34, 50–51, 74, 78–79, 84–85, 90, 100, 134, 153, 165; as machine, 78–79; parts, xiv, 4, 23, 25, 34, 84, 95, 98, 100, 118, 125, 186; processes, xiv, 18, 63, 83–84, 98, 107–8, 111, 118, 120. *See also* body factory
body factory, xii, 71, 78–80, 84, 86, 92, 120, 131–32, 134–35, 144, 151, 153, 169
bodyscape, 71–72, 120, 123–24, 131, 151, 153
"Body We Live In, The." See *Pictured Knowledge*
Bolter, Jay David, 100
Borck, Cornelius, 132, 194, 200, 225
boundary objects, 188–89
Burtin, Will, 158, 160

cartoons, 37, 87–88, 97, 137, 154–55, 161; animation, xii, 88, 141, 152, 156–57; cartoonists, 151, 153, 158, 216
Cassirer, Ernst, 68
China, xii, 133–34
city, 4, 9, 74, 77–78, 145, 215
Clark, Andy, 190
Clendening, Logan, 175, 234
commercial art, 4, 15, 17, 26, 37–38, 40, 51, 62, 114, 117–18, 133, 154, 173, 181, 215, 217; artists, xi, 3, 52–54, 73, 95, 151, 216; enterprise, 42–43; and fine art, 11, 52, 54, 59, 114, 223; illustrated text, 12, 37, 187, 215, 219; illustration, 52, 59, 94, 118, 123, 151
communism, xv, 23, 150
conceptual illustration, 1–3, 32, 49, 132, 152, 157, 164, 181, 199; scientific, xi–xii, 7, 29, 165, 172, 182–83; and visual rhetoric, 7, 29–30, 150. *See also* metaphor
"Construction of the Nervous System, The" (Trester) (Figure 4.25), *110*, 111
cross-section, xi, 29, 30, 32, 36, 78–84, *81–83*, 107, 131, 182, 186, 225
Crowd, The. See *Psychologie des Foules*
cutaway. *See* cross-section

"Daily Growth of Hair, The" (Figure 4.20), *103*
Dalton, John, 118, 224
Daston, Lorraine, 98
Davis, Whitney, 68
Dekker, Hermann, 30, 32, 63, 65, 210
DHM (Deutsches Hygiene-Museum), 17, 57, 105, 141–42, 146; "Eugenics in the New Germany," 142
diagrams, 1, 7, 19, 26, 29, 67, 72, 81, 83–84, 89, 95, 97, 101, 103, 105–7, 109–12, 118, 120, 126, 131. *See also* conceptual illustration
diaspora, xv, 9–10, 12, 133
Diderot, Denis, 81, 180, 213
"Digestion in the Mouth" (Trester) (Figure 4.35), *121*
"digestive zones in this region are, The" (Figure 4.31), *116*
display models, 82, *136*
Döblin, Alfred, 17, 23, 27, 186, 207
"Doctor of the Future, The" (Figure 4.13), 91, *92*
Draaisma, Douwe, 167
Draper, John William, 170
Dynamische Wirtschaft, Die (Nonnenbruch), 149

Eckart, Dietrich, 146, 228
economic crisis, 9, 15, 60, 191
Eichbaum, Edgar, 14, 205
Einstein, Albert, 13, 198
electric, 7, 25, 37, 74, 91
embodied life, xii, 5, 88, 95, 120, 143, 180. *See also* self, the

encyclopedia, xi, 3, 5, 32, 37, 39–41, 46, 134, 138, 183, 210, 213–14
Encyclopédie (Diderot), 180
Enemy Bacteria (Figure 5.15), *152,* 153
ethnicity, xv, 9–10, 96, 98
eugenics, 17, 142, 146–47, 150
"Examination of the Back of the Eye II" (Figure 6.9), *184*
exhibitions, xii, xiv, 4, 17, 37, 40, 100, 105, 134–36, 141–46, 173, 181; display, 79, 81, 142; freak show, xiv; Hall of Man, 135; health, 17, 105, 138, 142, 161; interactive, 31, 142–45

factory, 7, 18–19, 26, 135, 145, 167
"Factory within the Human Body, The" (Figure 5.2), *135*
"Fairy-Tale Journey on the Bloodstream 2, A" (Schmitson) (Figure 4.37), 121, *123*
Feld, Rudi, 54–55, 105
Few Quick Facts about Fear, A (Hubley and Schwartz) (Figure 5.16), 153, 155, *156*
film. *See* motion pictures
First Aid Popular (Kahn), 84, 198
"Five Main Lines of the Spinal Cord, The" (Figure 4.8), *85*
flashlight, 90–91
flowchart, 26, 72, 79, 83–84, 107–11, 119, 144, 153, 157, 224
"Food Journey through the Intestines, The" (Figure 4.23), 107, *108*
Ford, Henry, 31, 209; Fordism, 26
Foucault, Michel, 167–68, 170, 189
"Four Times around the Earth!" (Figure 5.8), *140*
"Four Ways to Deliver Drugs" (Figure 4.33), 118, *119*
Franckh (Franckh'sche Verlagshandlung), 13, 15, 23, 30, 60, 63, 65, 139, 147, 201, 206, 218, 227–29
Frau als Hausärztin, Die (Dückelmann), 14, 62
Frenzel, H. K., 54–55
Freud, Sigmund, 56, 190, 208
Freund-Beliani, Alwin, 66, 72, 77–78, 90–91, 218
"From the Life of a Wandering Cell IV" (Schmitson) (Figure 4.36), 121, *122*
futuristic, 118, 125, 135, 142, 164, 176

Gallagher, Catherine, 180
gaze, 15, 45, 50, 56, 168, 179
Gazzaniga, Michael, 191–92, 231
Gebhard, Bruno, 139, 142
Gebrauchsgraphik (magazine), 52, 54, 73
Gerard, Ralph W., 178
Germany, xv, 3–4, 12–17, 25–26, 37, 41, 49, 51–52, 61, 65, 73, 90, 137, 142, 147, 172–73, 212, 217, 224; Weimar, 24, 27–28, 51, 133
Gesamtkunstwerk, 12, 67, 109
Giese, Fritz, 179
"Gigantism Due to Hyperactivity of the Pituitary Anterior Lobe" (Figure 4.40), *128*
Gilbreth, Frank, 107–9
Gilbreth, Lillian, 107–9
Gitelman, Lisa, 6
Goebbels, Joseph, 149–50
Gombrich, Ernst, 69
Great War (World War I), xv, 15, 29, 78, 102, 173, 197, 204
Greenblatt, Stephen, 180
Griesemer, James, 188–89
Gropius, Walter, 25, 155, 183, 205, 218
Grotjahn, Alfred, 173
Grusin, Richard, 100
Günther, Hanns, 29–30, 32, 210

Haas, Walter de. *See* Günther, Hanns
halftone, 7, 36–37, 39–40, 44, 45, 60, 64, 73–74, 100, 102, 127, 137–38, 157, 211

Hall, G. Stanley, 40–41
Halle, 12, 197
Hand- und Lehrbuch der Krankenpflege (Venzmer), 147, 227
Harris, Neil, 37
Haugmard, Louis, 174
Hauri, Hans, 30, 32, 210
health, xiv, 12, 14–15, 36, 51, 62, 111–12, 127, 138, 147, 150–53, 163, 173, 175, *229*; and film, 139, 152, 154; and *Lebensreform,* 17; public, 35, 134, 136, 142, 146, 157, 172
Heartfield, John, 50, 104–5
Hebart, G. F., 41
Heidegger, Martin, 98, 173; and the world picture, 174, 219
Helbig, Georg, 15, 72, 79–80, 137–38, 141, 176, 206
Herf, Jeffrey, 149
Hirschfeld, Magnus, 130–31
historicism, xii, 8, 149–50, 205
History of the Conflict between Science and Religion (Draper), 170
History of the Royal Society (Sprat), 170
Hitler, Adolf, 23, 61, 95, 146, 149, 222
Hofmann, August Wilhelm von, 118
Hofmann, Walter, 14
homunculus, 19, 21, 76, 79, 89, 144, 207, 231; and identity, 188, 189; and *Mensch*-kins, 18–19, 21; and proxy, 18, 41, 167–69; and self, xiii, 78, 165, 180, 187
House I Live In, The (Alcott), 125
"How we see" (Figure 6.3), *169*
Hubley, John, 154–56
"Human Cell Radiation" (Figure 4.21), *104,* 105
"Human Respiration" (Helbig) (Figure 4.4), 79, *80*
"Human Skeleton Is Like an Automobile Chassis, A" (Chauvois) (Figure 6.4), *171*
iconography, 50, 59, 87, 95, 97, 108, 143, 155–56, 180
iconophilia, xv, 38, 180
identity, xv, 6, 9, 19, 21, 22, 97, 172, 178, 180–81, 186, 188; and formation, xii, 181, 191; and modernity, 8, 161, 202
illustration(s): anatomical, xiii, xiv, 18, 62–63, 72, 123, 125, 132, 181, 189, 217, 232; commercial, 52, 59, 94, 118, 151; synoptic, 71, 95. *See also* conceptual illustration; metaphor
image growth, 5, 42, 214
industrial technology, 27, 149, 154, 178
industry, 40, 42–43, 45, 51, 82, 91, 127, 134, 142–43, 152
"Influence of Lacing on the Female Body, The" (Falguiére) (Figure 4.42), 129, *130*
"Inner Structure of Bones, The" (Figure 4.39), 125, *126*
"Interaction of Adrenaline and Insulin in Glucose Metabolism, The" (Figure 5.13), *148*
"Iris-Key, The" (Rechn) (Figure 4.16), 95, *96*
"It Smells of Roast!" (Figure 4.7), *83,* 84

Jackman, Wilbur S., 42
Jacob, Heinrich Eduard. See *Blut und Zelluloid*
Jones, Seymour, 40
Jugendstil, 4, 15, 72, 123–24
Junge, Friedrich, 41

Kaempffert, Waldemar, 38–39, 96, 127, 174
Kahn, Fritz: biography, xii; chronology, 197; and medicine, 88, 138, 142; and *mikrokosmos,* 120; portrait, *12*; readership, 5, 12–15, 38, 50, 65, 118, 174; and remediation, 100–101, 105; work with artists, xi, 3, 7,

18–19, 26, 52, 61, 63, 65–67, 69, 72–74, 79–80, 83–85, 92–93, 95, 98, 100–101, 104–5, 108, 112, 114, 117–19, 124–25, 145, 148, 151, 153–55, 160–61, 164, 183, 185, 200, 218, 221, 228; and zionism, xv, 12
Keene, Nicholas, 190, 231
Kemp, Martin, 36, 62, 102
Kendall, Calvin N., 39–40, 214
Kepes, György, 154–56, 160
Kern, Stephen, 8
Kinsey, Alfred, 130
"Knockout!" (Figure 4.10), 86, *87*
Kosmos (magazine), 2–3, 13, 15, 29, 63, 147, 197, 218, 220, 223, 228
Kracauer, Siegfried, 9, 26, 41, 102
Krieg, Wendell J. S., 176–77
Kropff, Hanns, 52–53

Lang, Fritz, 76, 105
Latin America, xii, 151
Latour, Bruno, 6, 185–88
Lears, T. Jackson, 52
Leben des Menschen, Das (Kahn), xiii, 2–3, 12–24, 59, 61, 63, 65, 71–73, 79, 85, 95, 106, 112–13, 117, 121–22, 125, 127, 132, 137–38, 142, 146–49, 175; dust jackets, *22*
Lebensrätsel (Dekker), 63
Le Bon, Gustave, 174
Le Corbusier, 74, 78, 183, 220
Life (magazine), 67, 148
Lisbon, 13

"Mac the Mechanical Man" (Figure 5.12), *145*, 228
Man in Structure and Function (Kahn), 1, 175, 178, 234. See also *Mensch Gesund und Krank, Der*
"Man in the Radiant World" (Figure 4.15), 93, *94*
"Man Walking Is a Pendulum Machine" (Figure 4.18), 98, *99*
Marinetti, Filippo, 8
market/marketing, 12–14, 15, 16, 37, 51, 56, 58, 65, 68, 72, 148, 178, 183–84, 203, 210, 212–13
Marxism, 25
masses, the, 9–10, 17, 23, 25, 42, 47, 57–58, 142, 172–74; mass society, 19, 21, 34, 38, 57, 141, 145, 204; salaried, 9, 26–27, 179
"Mekeo Negro with Drawstring Waist Belt" (Figure 4.41), *129*
Mensch als Industriepalast, Der (Kahn), xii, xiv, 1, 4–5, 18–19, 21, 23–26, 31, 61, 65, 73, 79, 84, 118–19, 132–36, 141, 145, 183, 201, 218; display model of, *136*; "Man [as] Palace of Industry," *134*
Mensch Gesund und Krank, Der (Kahn), 5, 50, 55, 57, 61, 64–65, 73, 75, 83, 84, 85–89, 92, 94, 97, 99, 101, 104, 108, 115, 117, 119, 124, 126–27, 132, 139–40, 147–48, 151, 163, 166, 169, 175, 177, 198, 205, 217, 221, 224
Mensch und sein Leben, Der (Venzmer), 147
metaphor, 50, 63, 65, 69, 72, 98, 102, 127, 139, 167, 170, 172, 179, 232; architectural, 125; body, 36, 84, 172, 210, 232; industrial, 30, 32, 67, 88–89, 153; visual, 15, 31–32, 61, 144, 151, 155–57, 172. *See also* body factory; bodyscape; *mikrokosmos*
Meyer, Eugene, 125
mikrokosmos, 120
Mitchell, W. T. J., 5, 174, 181, 201
modern, the, xii–xiii, xv, 5–7, 9, 11–12, 15, 17, 26, 59, 65, 91, 111, 127, 146, 149–50, 153, 158, 159, 163–64, 167, 170, 202; era, xii, 4, 41, 71, 98, 172–73; world, 51, 88, 91, 118, 160
modernism/modernist, xii, 4–5, 7, 11–12, 15, 50, 59, 60, 65–66, 72, 77, 78, 99, 102, 118, 123,

144, 149, 151, 155, 158–61, 183, 203–4
modernist styles: abstract expressionism, 12, 161; Art Deco, xi, 4, 57, 181; cubism, 155; Dada, 56–57, 79, 102, 104; figurative realism, 4, 7, 21, 52, 61, 71–72, 102, 123, 158, 161, 165, 175, 181, 216; functionalism, xi, 173; *Neue Sachlichkeit,* 4. *See also* Bauhaus; photomontage
modernity, xiv, xv, 1, 5–9, 11, 14, 17, 21–23, 25, 27, 29, 31, 36–37, 41, 43, 50–53, 54, 62, 67, 69, 71–72, 74, 77, 84, 86, 88, 92, 93, 95, 98–101, 105, 114–15, 118, 131, 142, 146, 149–51, 154, 157–58, 160–61, 164–65, 167–68, 172–73, 175, 178–80, 204, 208; and the body, 74, 115; scholarship on, 6–8; and visuality, xv, 30, 32, 40–41, 72, 77, 180
modernization, 6, 12, 36, 38, 41, 146, 150, 202
moderns, xii, 6, 8–9, 23, 68, 179, 202
motion pictures, xii–xiii, 7, 38, 71, 215; and animation, 152–57, 201
Multiple Screening (Hilberman) (Figure 5.17), 153, *157*
Mumford, Lewis, 181
"Muscle Fibers and Automobile Engine Show Substantial Similarity in Structure" (Figure 6.6), *177*
Mussolini, Benito, 149
Münsterberg, Hugo, 172, 212

Nacktkultur, 17, 55
National Socialism, 4, 25, 77, 146, 149, 150; and modernity, 146; and Nazis, xii, 4, 12, 23, 25, 65, 134, 139, 146–47, 149–50, 216–17, 227–29
Naturfreunde, 13–15, 31, 205
Nazism. *See* National Socialism: and Nazis
"Nerve Cell, The" (Figure 4.38), *124*
"Nerve Cord, The" (Figure 4.38), *124*
"Nervous System Is a 'Relay System,' The" (Figure 4.11), *89*
Neue Sachlichkeit (New Objectivity), 4
Neurath, Otto, 67–68, 76–77, 105, 155, 173–74, 181, 233
neuroscience, 165, 180, 189–91, 207, 231
new media, 37, 100
newspapers, 2, 13, 16–17, 38, 40, 73, 86, 102, 127, 130, 142, 173, 219, 223, 229
New York, xv, 12, 13, 25, 33, 39, 56, 67, 74, 76, 78, 89, 90, 136, 140, 142–43, 146, 148, 175, 194, 197
New York Mirror (tabloid), 67, 148, 175
Nichtenhauser, Adolf, 152–53
Nietzsche, Friedrich, 93, 167, 202
Nipperdey, Thomas, 15
Nonnenbruch, Fritz, 149
Novum Organum (Bacon), 170

O Corpo Humano (Figure 5.14), *151*
Our Sex Life. See *Unser Geschlechtsleben*
"Oxygen Cycle, The" (Shüler and Trester) (Figure 1.3), 19, *20*

Palestine, 4, 14, 147, 198
Panofsky, Erwin, 68
Paris, 25, 57, 147, 198
Parker, Samuel, 170
pedagogy, xiii, 37–38, 41–42, 45, 49, 141, 163, 172
performance, xii–xv, 6–8, 11, 17, 31, 111, 149, 150, 163, 183, 192
period eye, 69, 95, 219
Pestalozzi, Johann, 41
phonograph, 19, 28, 35, 74, 118
photography, 22, 35, 43, 84, 95, 100–102, 127, *222*
photomontage, xi, 4, 50, 71, 84, 100, 102–5, 125, 186, 223

"Physiology Made Simple" (Figure 5.11), *144*
"Physiology of The Knockout, The" (Figure 4.9), 85, *86*
pictoriality, 17, 49, 69, 149, 161
Pictorial Representation of Social Facts (Neurath), 181
pictorial turn, xiii, 30, 210
Pictured Knowledge (encyclopedia), xi–xii, 5, 29, 32, 34–37, 39–42, 46, 49–50, 52, 183, 210, 213–14
picture pedagogy, 49
Piek-Patrik, Georg, 59, 124
politics, xii, 25, 51, 84, 95, 150, 173, 178, 190, 204, 223; biopolitics, 184, 235
popular anatomy, xiv, 187
Popular Science Monthly (magazine), 38, 53, 83, 127, 136, 143, 174, 210
postwar era, 108, 146, 150, 229; and America, 141
Pound, Ezra, 49
Progressivism (American), xi, 35, 40, 50, 112
Psychologie des Foules (Le Bon), 174
psychotechnik, 36, 38, 179, 183; and psychotechnicians, 38, 59, 179, 209
public, the, xiv, 51, 58, 69, 88, 105, 111–12, 134, 141, 151, 154, 173, 175
Purkinje, Jan, 91
"Purkinje Vein Figure, The" (Freund-Beliani) (Figure 4.12), *90*, 91

Quetelet, Adolphe, 111. *See also* statistics

radiation, 88, 91–92, 104–5; radiant lines, 87
radio, 8–9, 13, 19, 25, 74, 90–92, 100; activity, 88; and advertising, 151; and communication, 98; transmission, 74, 89–90; waves, 88, 91
rays, 55, 88, 92–93
Rechn, Roman, 3, 22, 60, 66–67, 71–73, 95–96, 119, 123, 148–49, 218, 228–29
remediation, 100–101
Riou, Édouard, 122
Röntgen, Wilhelm, 88
Rosen, George, 148, 163
Roth, Joseph, vii, 8–9, 19, 27, 74, 155, 209
"Routing Diagram for Materials and for Printed Forms in a Manufacturing Plant" (Brinton) (Figure 4.24), *109*
Routzahn, Evart, 109, 112, 114
Routzahn, Mary, 109, 112, 114
Royal Society. See *History of the Royal Society*
Ruttmann, Walter, 105, 223

Salaried Masses, The (Kracauer), 9, 26–27, 179
Samuel, Horace B., 6, 8
Sarasin, Philipp, 187, 210
Schleger, Hans, 56–57
Schlesinger, Georg, 27, 209
Schmitson, Arthur, 72–73, 121–24, 149, 153, 224–25
Schüler, Fritz, 2, 4, 20, 51, 66–67, 72–73, 93, 119, 131, 133, 136, 149, 164, 218
Schwartz, Zack, 154–56
science, xiv, xv, 14, 16–17, 35, 58, 60, 69, 97–98, 100, 104, 107, 133, 141, 146, 149–51, 154, 157, 160, 167–68, 170, 172–74, 176, 178, 186, 204–5, 207–8, 214, 221; and future, 8; modern, xiv, 5, 12, 23, 92, 96, 163, 221; popular, 12–13, 15, 23, 30, 51, 59, 62–63, 68, 127, 134, 143, 234; racial, 147
science fiction, 88, 124, 220
Scope (magazine), 160–61
Scott, Walter Dill, 109, 212
searchlight, 78, 91
self, the, xiii–xv, 4, 19, 50, 78, 90, 113, 122, 134, 149–50, 161,

165, 168, 178, 180–81, 185–89, 191–92, 201, 212, 231; and selfhood, 180–81, 183; and self-improvement, 12, 14, 40, 205
"sensational finding!!, A" (Schüler) (Figure 6.1), *164*
"Sensory Organs of the Head, The" (Figure 4.32), *117*, 118
"Sensory Pathway, The" (Figure 1.2), *16*
sexology, xii, 4, 129, 147; and genitals, 131; and Fritz Kahn, 130–32, 225; and sex education, 147; visual, 129
silhouette figures, 76–77, 86, 100
"Skull-House Has Four Floors, The" (Figure 6.2), *166*
skyscraper, 27–28, 74–78, 90, 155, 209, 220
"Slit-Lamp Microscope, The" (Figure 6.7), *179*
"Smoking through the Eyes" (Grü) (Figure 6.8), *182*
"Sodium Chloride Cycle, The" (Trester) (Figure 4.34), *120*
"Sound Perception" (Schüler) (Figure I.1), *2*
"Speed of Thought, The" (Figure 4.17), *97*, 98
spotlight, 78, 116–17
Sprat, Thomas, 170
Star, Susan Leigh, 188–89
statistics, 17, 35, 111–12, 146, 151, 153
Stein, Gertrude, 49
students, xv, 9, 38, 41, 44–45, 179
Sturtevant, B. F., 82
subject formation, xiii, 21, 188, 191, 231
subjectivity, xii, 21, 50, 54, 105, 122, 165, 167, 173, 180, 182, 187, 190–91, 215, 231, 235
sunlight, 55, 88, 92
Surén, Hans, 23, 208
Sutton, John, 190

Tagg, John, 77
Taylor, Frederick Winslow, 26, 27, 107, 209
"Technical Description of the Main Processes in Life, A" (Kahn) (Figure 1.5), *24*
Technics and Civilization (Mumford), 181
television, 2, 71, 91, 100, 154, 169, 183, 192, 213
"Therapy for Hypererotic Arousal Conditions" (Schüler) (Figure 4.43), *131*
"Three Most Important Illnesses of the Spine, The" (Figure 4.19), *101*
"Total Number of Blood Cells, The" (Helbig) (Figure 5.4), *137*, 141
"Total Number of Blood Cells, The" (Soviet) (Figure 5.5), *138*
"Travels of Protein in the Body Is a Dice Game with Hundreds of Thousands of Combinations, The" (Figure 4.30), *115*
Trester, Otmar, 20, 66, 72, 105–7, 109–11, 119–21, 123, 149, 223. *See also* diagrams
"Twelve of twelve hundred" (Rechn) (Figure I.2), *3*
"20,000 Kilograms [of Atmospheric Pressure] on Our Shoulders!" (Figure 4.26), *112*
"Two Divisions of the Nervous System, The" (Trester) (Figure 4.22), 105, *106*, 107

Uhu (magazine), 3, 15, 206
Ulrich, Mabel S., 175, 234
universal, xiv, 4, 12, 21, 24–25, 67, 76, 95–96, 98–100, 107, 122, 173, 183, 209, 222; and universalism, 24, 52, 97–98, 111, 147, 173, 178, 204, 208
Unser Geschlechtsleben (Kahn), 130, 132, 147, 200
urban, 8, 27, 57, 72, 89, 165, 172, 181

utopia/utopianism, 4, 6, 9, 22–23, 27, 33, 50, 100, 104, 159, 173, 184, 204

Vander, Adrián, 151
Ventilation and Heating (Sturtevant) (Figure 4.6), *82*
Venzmer, Gerhard, 139, 147–49, 227–29
Verne, Jules, 122
Vesalius, Andreas, 36, 125, 219, 225, 232
visual explanation, 1, 28, 32, 105, 107, 131, 160–61, 164, 221. *See also* conceptual illustration

Walter Lantz Studio, 152. See also *Enemy Bacteria*
Walters, Francis M., 170, 172, 232
Warburg, Aby, 68
Ward, Janet, 24–25
Weber, Max, 27, 92
Weimar (era), 9, 24, 27–28, 51, 133, 203–4, 229
"Where in the Brain the Most Important Functions Take Place" (Figure 5.10), *143*
Wiener, Norbert, 111
Wittgenstein, Ludwig, vii, 173, 233
Wolff, Julius, 125–26, 225
women, 14–15, 52, 55, 117–18, 129, 132, 178–79
working class, xi, xiii, 4, 9, 18, 21, 23, 34, 37–38, 79, 83, 172–73, 179, 209
World War I. *See* Great War
World War II, 97, 150–52, 157, 183, 218
Wunder in Uns, xi, 29–33, 35, 53, 63, 79–80, 210–11

x-rays, xiv, 19, 88, 92, 104, 128, 154, 165, 215

Your Body in Flight (Sloane) (Figure 5.18), *158*

MICHAEL SAPPOL lives in Stockholm and is a fellow at the Swedish Collegium for Advanced Study, Uppsala. For many years he was curator, historian, and scholar in residence in the History of Medicine Division of the National Library of Medicine. He is author of *A Traffic of Dead Bodies* and *Dream Anatomy*, and editor of *A Cultural History of the Human Body in the Age of Empire* and *Hidden Treasure.*